The Ecology of Trees in the Tropical Rain Forest

Current knowledge of the ecology of tropical rain-forest trees is limited, with detailed information available for perhaps only a few hundred of the many thousands of species that occur. Yet a good understanding of the trees is essential to unravelling the workings of the forest itself. This book aims to summarise contemporary understanding of the ecology of tropical rain-forest trees. The emphasis is on comparative ecology, an approach that can help to identify possible adaptive trends and evolutionary constraints and which may also lead to a workable ecological classification for tree species, conceptually simplifying the rain-forest community and making it more amenable to analysis.

The organisation of the book follows the life cycle of a tree, starting with the mature tree, moving on to reproduction and then considering seed germination and growth to maturity. Topics covered therefore include structure and physiology, population biology, reproductive biology and regeneration. The book concludes with a critical analysis of ecological classification systems for tree species in the tropical rain forest.

IAN TURNER has considerable first-hand experience of the tropical rain forests of South-East Asia, having lived and worked in the region for more than a decade. After graduating from Oxford University, he took up a lecturing post at the National University of Singapore and is currently Assistant Director of the Singapore Botanic Gardens. He has also spent time at Harvard University as Bullard Fellow, and at Kyoto University as Guest Professor in the Center for Ecological Research.

CAMBRIDGE TROPICAL BIOLOGY SERIES



*Also in the series*

Tomlinson, P. B. *The botany of mangroves*
Lowe-McConnell, R. H. *Ecological studies in tropical fish communities*
Roubik, D. W. *Ecology and natural history of tropical bees*
Benzing, D. H. *Vascular epiphytes*
Endress, P. K. *Diversity and evolutionary biology of tropical flowers*

# The Ecology of Trees in the Tropical Rain Forest

I.M. TURNER
*Singapore Botanic Gardens*

CAMBRIDGE UNIVERSITY PRESS
Cambridge, New York, Melbourne, Madrid, Cape Town, Singapore, São Paulo

Cambridge University Press
The Edinburgh Building, Cambridge CB2 8RU, UK

Published in the United States of America by Cambridge University Press, New York

www.cambridge.org
Information on this title: www.cambridge.org/9780521801836



First published 2001
This digitally printed version 2008

*A catalogue record for this publication is available from the British Library*

*Library of Congress Cataloguing in Publication data*

Turner, I. M. (Ian Mark), 1963–
The ecology of trees in the tropical rain forest / I. M. Turner.
p. cm. – (Cambridge tropical biology series)
Includes bibliographical references.
ISBN 0 521 80183 4
1. Rain forest plants – Ecophysiology. 2. Trees – Ecophysiology – Tropics.
3. Rain forest ecology. 4. Forests and forestry – Tropics. I. Title. II. Series.
QK938.R34T87 2001
577.34–dc21 00-045508

ISBN 978-0-521-80183-6 hardback
ISBN 978-0-521-06374-6 paperback

# CONTENTS

# PREFACE

> It is the detail of the . . . tropical forest, in its limitless diversity, that attracts.
>
> F. Kingdon Ward (1921) *In Farthest Burma*, Seeley, Service & Co. Ltd., London.

Trees make a forest: they are both the constructors and the construction. To understand the forest we must know about the trees. This book is about the trees of the tropical rain forest. It was written with the aim of summarising contemporary understanding of the ecology of tropical rain-forest trees, with particular reference to comparative ecology. The analysis of patterns of variation among species is a valuable technique for identifying possibly adaptive trends and evolutionary constraints. It may also provide a means of classifying species in ecological terms. A workable ecological classification might mean that the rain-forest community could be conceptually simplified and made more amenable to analysis.

The organisation of the book follows the life cycle of a tree. The living, growing mature tree is introduced with reference to form and process. Reproduction, including pollination and seed dispersal, follows. Then come consideration of seed germination, seedling establishment and growth, and the completion of the life cycle. At each stage a range of different characteristics and phenomena relevant to tree species growing wild in the tropical rain forest are considered. I have tried to give some idea of what is typical, and what is rare, the range and central tendency exhibited among species, and whether discrete groupings, or a continuous variation, are observed within the forest, and also whether one character tends to be correlated with another. Finally, I have tried to bring all the observations together in a critical analysis of ecological classification systems for tree species in the tropical rain forest. I have deliberately avoided the 'historical approach' to reviewing the scientific literature. There are points in favour of following the chronological development of ideas in a particular field, but in this case I felt it was not absolutely necessary. Firstly, there are several excellent texts that summarise much of the older work on tropical rain forests, notably P.W. Richards' *The tropical rain forest* (Richards 1952, 1996) and T.C. Whitmore's *Tropical rain forests of the Far East* (Whitmore 1975, 1984). Secondly, I wanted to avoid the problems of interpreting history with the benefit of hindsight. Glimmerings of ideas that later became important can often be found by careful sifting through earlier writing, but at the time such works were published those ideas had little if any impact. Thirdly, I believe many readers are more interested in the contemporary state of knowledge and understanding than how we arrived at that position.

I estimate that there are some 50 000–60 000 tree species occurring in the tropical rain forests of the world. We have a detailed knowledge of the ecology of perhaps a few hundred of these at best. This book is therefore written from a perspective of abject ignorance, which I hope readers will bear in mind when consulting these pages.

# ACKNOWLEDGEMENTS

I could not have written this book without the opportunity provided by consecutive visiting fellowships in the USA and Japan that allowed me uninterrupted time to complete the first draft. My Bullard Fellowship at Harvard University ran from July to December 1997. I am very grateful for the hospitality of Harvard Forest and Harvard University Herbaria. My special thanks go to Professor David Foster and Professor Peter Ashton. I was a Guest Professor at the Center for Ecological Research of Kyoto University for the period February–June 1998 and was made welcome by all the staff and students. The Director, Professor Eitaro Wada, provided assistance in many ways including funding for drawing of many of the figures in the book. Dr Noboru Okuda skilfully prepared these. My stay in Japan would not have been so enjoyable without the many services and opportunities provided by Professor Tohru Nakashizuka. I took up a post with Singapore Botanic Gardens after my stay in Japan. I am grateful to the Director, Dr Chin See Chung, for encouraging me to complete the book and for supporting me in many different endeavours. Dr Tan Wee Kiat, Chief Executive Officer of the National Parks Board, the parent organisation of the Botanic Gardens, is also thanked for his support.

The research for this book involved raking through a plethora of publications on various aspects of tropical forests. I bothered many different people with requests for copies of articles, leads on references or visits to libraries. Special thanks must go to the various librarians at Harvard University whom I confronted with long lists of things to find, particularly the team, led by Judy Warnement, in the Botany Libraries, who were ever willing to meet the next bibliographic challenge.

I have asked many people to read and comment on all or parts of the book. Nearly all have willingly done so and provided much valuable and constructive criticism. I thank Peter Becker, David Burslem, Richard Corlett, Stuart

Davies, Peter Grubb, Bernard Moyersoen, Tim Whitmore for their help and apologise for not always taking their advice.

Among the figures appearing in the book a large number are reproduced from other publications. I acknowledge the following for granting permission to reproduce material: Academic Press (Fig. 4.2), the editors of *American Journal of Botany* (Figs. 2.15, 4.1), the Association for Tropical Biology (Figs. 2.16, 4.3, 4.9), the editors of *BioScience* (Fig. 3.4), Blackwell Science Publishers (Figs. 2.23, 2.31, 3.9, 5.8, 5.9, 5.14), Botanical Society of Japan (Fig. 5.7), the editors of *Bulletin of the Torrey Botanical Club* (Fig. 2.12), Cambridge University Press (Figs. 2.3, 2.8, 2.13, 2.14, 3.7, 5.6, 6.2), Ecological Society of America (Figs. 3.6, 4.8, 5.4, 5.11, 6.3), Heron Publishing (Fig. 2.10), the editors of *Journal of Tropical Forest Science* (Fig. 3.3), Kluwer Academic Publishers (Figs. 2.24, 4.11), the editors of *The New Phytologist* (Fig. 5.16), NRC Research Press (Fig. 5.5), Oxford University Press (Fig. 2.9), the editors of *Revista Biología Tropical* (Figs. 2.20, 6.4), The Royal Society (Fig. 5.15), Society of American Foresters (Fig. 2.7), Springer Verlag (Figs. 2.2, 2.6, 2.11, 2.22, 2.25, 4.12, 5.13, 5.17, 6.1), the editors of *Tropics* (Fig. 4.10), UNESCO (Fig. 5.10), University of Chicago Press (Fig. 2.26), Urban & Fischer Verlag (Fig. 2.30), Yale University Press (Fig. 3.2). Through the assistance of Professor Tohru Nakashizuka and Dr Shoko Sakai, I was able to use some of the photographs taken by the late Professor Tamiji Inoue.

# 1

# Introduction

## The tropical rain forest

Tropical rain forest is one of the major vegetation types of the globe (Richards 1996; Whitmore 1998). It is an essentially equatorial and strongly hygrophilous biome as its name suggests and is found on all the continents that the tropics touch. Tropical rain forest is defined physiognomically with typical features being a closed, evergreen canopy of 25 m or more in height dominated by mesophyll-sized leaves, with an abundance of thick-stemmed woody climbers and both herbaceous and woody epiphytes. Altitude has a marked effect on forest physiognomy above about 1500 m, and montane facies have to be distinguished. The so-called tropical diurnal climate has a temperature regime in which the major periodicity is the daily march from night-time lows to afternoon highs. The fluctuation through the year in mean monthly temperatures is usually of smaller magnitude than the typical daily temperature range. Temperatures usually average at around 27 °C at lowland weather stations in tropical rain-forest regions, and minima rarely, if ever, enter the chilling range below 10 °C. Rainfall is generally at least 2000 mm per annum, and a month with less than 100 mm is considered dry. Rain forests can withstand dry periods though prolonged, or particularly severe, droughts on a regular basis usually lead to drought-deciduous forest replacing the true rain forest. Many rain forests do persist despite annual dry seasons, though only if the trees have access to ground water in areas experiencing long periods without rain.

Edaphic factors including soil physico-chemical properties and drainage regime influence the floristic, physiognomic and structural characteristics of the tropical rain-forest community strongly. Forest formations can be recognised for major soil groups and inundation classes across the geographic range of tropical rain forest. For instance, heath forest that occurs on acidic, highly leached sands is readily distinguishable whether one is in Asia, Africa or America.

## Tropical trees

I include in the category of tree, any free-standing plant that attains a diameter at breast height (dbh) of at least 1 cm, as this has become the lower limit of inclusion for a global network of tropical forest plots (Condit 1995). As well as the arborescent dicotyledonous species the term tropical tree brings to mind, I also include gymnosperms, woody monocotyledons, tree-ferns and bananas and their kin (Table 1.1). The latter are herbaceous, but their large size means that they can be considered trees, at least in terms of the structure of the forest. All the other groups have many fewer species than the dicotyledons, particularly the non-angiosperm classes. However, palms are an important component of most tropical forests, and dominate some forest types. As with most ecological classifications, there are fuzzy edges to any definition of tree. Many large woody climbers have juvenile stages indistinguishable from tree saplings. Woody hemi-epiphytes, mostly figs (*Ficus* spp.), but also species of *Clusia*, some Araliaceae and a few others (Putz & Holbrook 1986), begin life as epiphytes, but grow roots down to the ground and become terrestrial. Their host tree may eventually die and the hemi-epiphyte is left mechanically independent: it has become a true tree.

Figure 1.1 gives a phylogeny of the seed plants to ordinal level with the pteridophytes indicated as a sister group. It is important to note that evolutionary relationships of some lines remain unresolved; for instance, it is not clear which of the gymnosperm clades gave rise to, or at least is closest to, the angiosperms. In addition, quite a number of angiosperm families are not yet placed within the orders, although these are mostly small. Many different evolutionary lines are represented among extant tropical trees (Fig. 1.1), but perhaps there is a concentration of species among the rosids and the more basal offshoots, orders Laurales and Magnoliales.

## Species

The species is most often the unit of comparative ecology, although intraspecific comparisons of different populations can also be rewarding. If we are comparing species we have to consider what we mean by species. The definition has been a thorny problem in biology for many years, with considerable debate over the relative merits of biological and evolutionary definitions of species. From a practical point of view, it has to be realised that ecologists researching in tropical rain forests are nearly always working with a taxonomic species definition. They are forced to employ the species concepts of the taxonomist who wrote the flora or monograph that they are using. In most cases these are based on circumscription of morphological

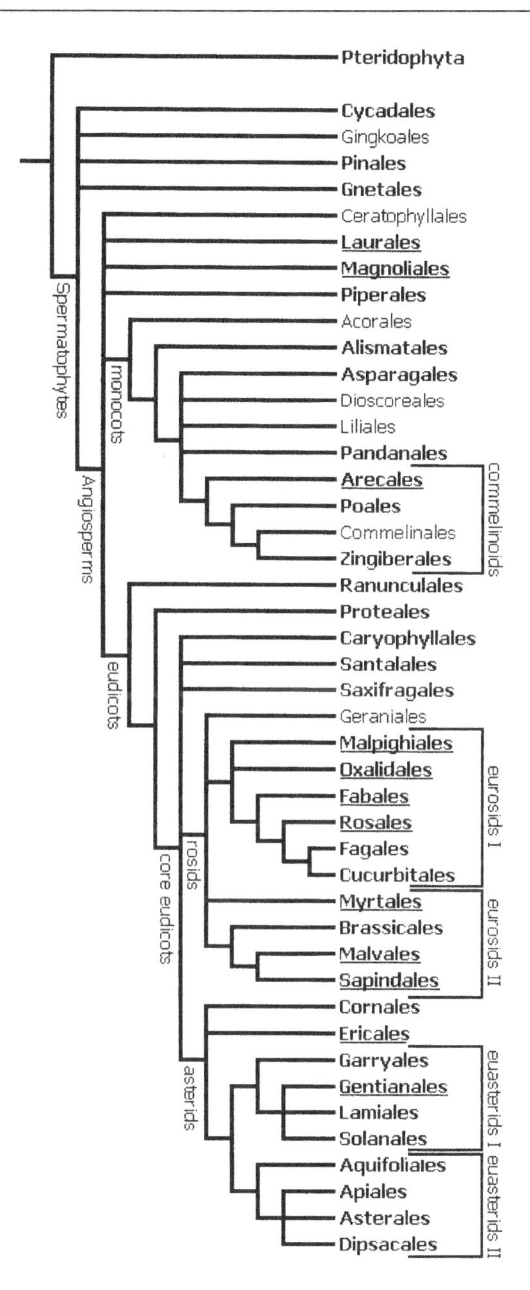


Figure 1.1 A phylogeny of the vascular plants to ordinal level (in part after Nyffeler (1999) based on Angiosperm Phylogeny Group (1998)). Names of orders given in bold indicate the presence of extant tropical tree species. Names underlined indicate that a majority of extant member species are tropical trees.

Table 1.1. *Taxonomic summary of extant tropical forest trees*

Inclusion of a genus implies at least one species considered to be a tree in tropical rain forest. Monocot families are presented in order of roughly decreasing degree of arborescence. Classification follows Brummitt (1992).

| Pteridophytes | Gymnosperms | | Monocots | | Dicots (important families) | |
|---|---|---|---|---|---|---|
| FILICOPSIDA | PINOPSIDA | GNETOPSIDA | Woody: | Herbaceous: | Anacardiaceae | Malvaceae |
| | | | | | Annonaceae | Melastomataceae |
| Cyatheaceae | Araucariaceae | Gnetaceae | Palmae | Musaceae | Apocynaceae | Meliaceae |
| *Cyathea** | *Agathis* | *Gnetum* | (Arecaceae) | *Ensete* | Aquifoliaceae | Monimiaceae |
| | *Araucaria* | (*G. gnemon* and *G.* | many genera | *Musa* | Araliaceae | Moraceae |
| Dicksoniaceae | | *costatum* seem to | | | Bignoniaceae | Myristicaceae |
| *Cibotium* | Cupressaceae | be the only species | Pandanaceae | Strelitziaceae | Bombacaceae | Myrsinaceae |
| *Culcita* | *Papuacedrus* (mtn) | that are trees, the | *Pandanus* | *Phenakospermum* | Boraginaceae | Myrtaceae |
| *Dicksonia* | *Widdringtonia* | rest are woody | *Sararanga* | *Ravenala* | Burseraceae | Ochnaceae |
| | (mtn) | climbers) | | | Capparidaceae | Olacaceae |
| Osmundaceae | | | Gramineae | Heliconiaceae | Celastraceae | Oleaceae |
| *Leptopteris* | Phyllocladaceae | | (Poaceae) | *Heliconia* | Chrysobalanaceae | Piperaceae |
| | *Phyllocladus* (mtn) | CYCADOPSIDA | bamboos | | Combretaceae | Pittosporaceae |
| also, but less convincingly | | | | Zingiberaceae | Cunoniaceae | Rhamnaceae |
| dendroid: | Pinaceae | Cycadaceae | Dracaenaceae | *Alpinia* | Dichapetalaceae | Rhizophoraceae |
| | *Pinus* (dry) | *Cycas* | *Cordyline* | | Dilleniaceae | Rosaceae |
| Blechnaceae | | | *Dracaena* | Araceae | Dipterocarpaceae | Rubiaceae |
| *Blechnum* | Podocarpaceae | Zamiaceae | | *Alocasia* | Ebenaceae | Rutaceae |
| *Brainea* | *Acmopyle* (mtn) | *Ceratozamia* | Velloziaceae | *Montrichardia* | Elaeocarpaceae | Sapindaceae |
| | *Afrocarpus* | *Chigua* | *Vellozia* | | Ericaceae | Sapotaceae |
| | *Dacrycarpus* | *Dioon* | *Xerophyta* | N.B. not all | Erythroxylaceae | Simaroubaceae |
| *in the broad sense, the | *Dacrydium* | *Lepidozamia* | | members of these | Euphorbiaceae | Sterculiaceae |
| genus is split into two or | *Falcatifolium* | *Zamia* | Cyperaceae | genera attain tree | Fagaceae | Styracaceae |
| more by some pteridologists | *Nageia* | | *Microdracoides* | size | Flacourtiaceae | Theaceae |
| | *Podocarpus* | | | | Guttiferae | Thymelaeaceae |
| | *Prumnopitys* (mtn) | | Xanthorrhoeaceae | | (Clusiaceae) | Tiliaceae |
| | *Sundacarpus* | | *Xanthorrhoea* | | Icacinaceae | Ulmaceae |

Taxaceae
*Taxus*

mtn = genera only represented in montane regions

dry = genera only represented in seasonally dry regions

Lauraceae
Lecythidaceae
Leguminosae (Fabaceae)
Loganiaceae
Magnoliaceae

Urticaceae
Verbenaceae
Vochysiaceae

variation to the degree that the taxonomist considered typical of a species. Taxonomists can vary in their opinion of where to draw these lines (see, for example, Wong (1996) for discussion of *Fagraea* in Borneo), leaving ecologists in the difficult position of choosing whom to follow.

Additional problems are faced with the identification of plants in the field. The high diversity makes identification difficult. It may be annoying and confusing to find that you have been using the wrong name for a species. For example, many papers were published concerning *Virola surinamensis* on Barro Colorado Island, Panama, but this was a mis-identification, and is correctly *Virola nobilis*. What is worse is when your species contains individuals of other species. This happened for a study of buttressed trees in Malaysia (Crook *et al.* 1997) where one pinnate-leaved species turned out to be several.

## Tropical rain-forest diversity

Tropical rain forests are the most diverse of terrestrial ecosystems. Many lowland forests contain more than 100 species among the trees of 10 cm dbh or over on 1 ha (Fig. 1.2), and in some more than 200 species may

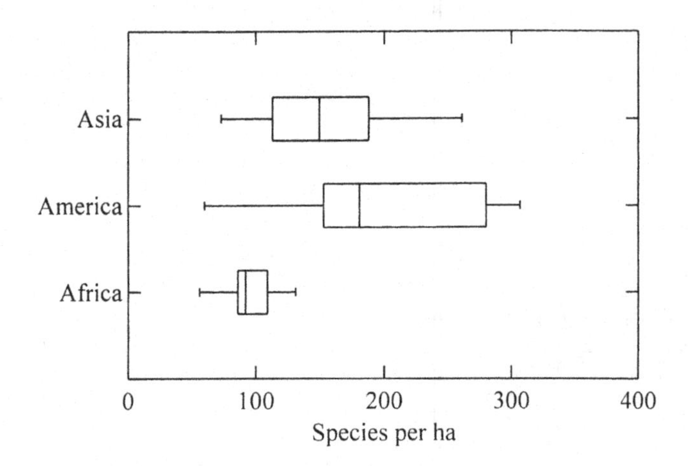


Figure 1.2 Box-and-whisker plot of species richness for trees greater than 10 cm dbh of rain-forest sites for the major tropical regions (Asia–Pacific, America and Africa). The line inside the box represents the mean value of the average number of species per hectare. The box extends for the range of 50% of the values above and below the mean. The whisker covers the complete range of the recorded values. Data from compilations by Phillips *et al.* (1994) and Turner (2001).

be found. Species richness rises very rapidly with area or number of individuals sampled in a forest (Fig. 1.3) and plots of 5 ha or more may be required to sample local diversity adequately. Topographic and edaphic variations will often lead to landscapes of patches of different forest communities that further add to the high diversity of a lowland tropical region. The high diversity of species within a particular forest frequently involves the co-existence of species in the same genus. For instance, of the 814 species recorded in 50 ha of forest at Pasoh, Malaysia, by Manokaran *et al.* (1992), 82% of the species had a congeneric present in the plot, with 70% having a congeneric in the same broad height category.

A commonly asked question, for which we are still seeking the answer, is: why are the tropical rain forests so diverse? Various responses have been put forward. One solution is to turn the question around. Why are extra-tropical regions so poor in species? There is clearly a general inverse relation between environmental harshness and diversity. The tropics have many species because it is easier to survive there than in less favourable environments.

However, this does not explain how all the tree species in the rain forest manage to co-exist. Classic ecological theory states that species can only co-exist if the levels of interspecific competition remain low enough to prevent competitive exclusion of some members of the community. One way in which this can be brought about is for all the species to occupy separate niches. Where do all the niches come from for the co-existence of hundreds of tree species on small areas of lowland tropical rain forest? Various mechanisms have been put forward by which tree species could be partitioning the environment and would thus exist in an equilibrium community. The light gradient from the shaded forest floor to the sunny canopy-top may allow specialisation in different heights at maturity (Kohyama 1993). The horizontal variation in light availability due to irregularity in canopy structure (presence of gaps, etc.) and in soil physical and chemical properties could also result in the subtle environmental variations needed for providing the many niches required.

However, the lowland tropical rain forest generally contains such a large number of species of very similar ecology that it is difficult to believe that strict niche partitioning is occurring. Another possibility is that factors other than niche partitioning may prevent competitive exclusion of species from the community. Compensatory mortality that places a ceiling on the local abundance of a species has been invoked as a means of species co-existence (Janzen 1970; Connell 1971). Species-specific causes of mortality, such as specialist seed and seedling predators and pathogenic diseases, may control the distribution of adult population density within the forest. Recent studies from Barro Colorado Island in Panama showed evidence for significant intraspecific density-dependent effects on recruitment for 67 out of 84 of the commonest tree species in the community (Wills *et al.* 1997). Givnish (1999) has argued that the fairly strong correlation between tropical rain-forest tree species diversity and total rainfall at a site is an indication of the increasing importance of compensatory mortality as dry spells become more infrequent in the local climate. The increasing rarity of periods of low rainfall allows invertebrates and plant-pathogenic organisms to maintain high population numbers and to provide the mechanism of compensatory mortality continuously.

Alternatively, the species in the forest may not have come to competitive equilibrium. Chance is seen as the major influence on community structure in this non-equilibrium view. There is growing evidence that recruitment limitation is common among species in the rain forest (Hubbell *et al.* 1999). That is, most species do not establish recruits successfully in all the sites in the forest that they are capable of occupying, either through failure in dispersal or through high mortality of juvenile stages. In a very species-rich community most species are so infrequent that no given pair of species will meet often enough for one to dislodge the other from the community through competi-

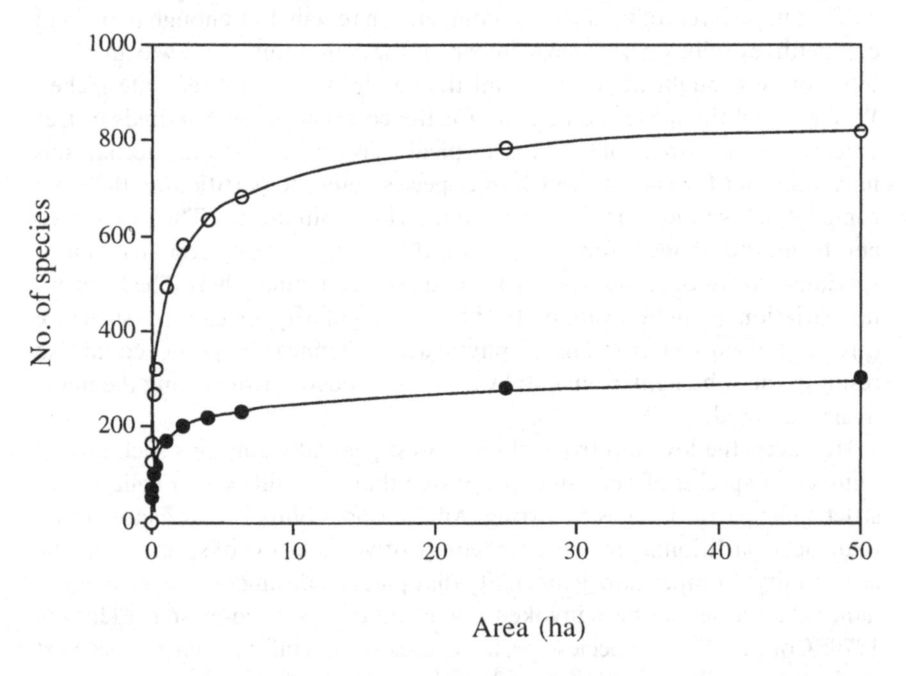


Figure 1.3 Species–area curves for 50 ha plots at Barro Colorado Island, Panama (closed circles), and Pasoh Forest Reserve, Malaysia (open circles). Trees ≥ 1 cm dbh. Data from Condit *et al.* (1996b).

tion. Recruitment limitation further increases the likelihood of an individual from a competitively inferior species reaching maturity by default.

In reality it seems likely that, to some extent, both equilibrium and non-equilibrium forces operate in the community simultaneously. The bulk of the diversity comes in the form of rare species that are probably not occupying separate niches. However, there seems to be more predictability in forest community structure than might be expected from a strongly stochastically driven system. Long-term studies of primary forest community composition show only minor variations, not random fluctuations, in species make-up and relative abundance over time (see, for example, Manokaran & Swaine 1994). Succession appears to follow relatively predictable trajectories in similar sites (Terborgh *et al.* 1996). We remain uncertain of what controls or constrains these processes. Hubbell (1997) has proposed that the main influences of chance are on the composition of the regional species pool and the relative abundance of species in the pool. These factors, together with environmental tolerances, are then responsible for community composition at any place in the region. Hubbell has produced some remarkably accurate simulations of community composition (dominance–diversity curves) from his model. However, the mechanics of the interface between regional and local community composition have yet to be explained in detail.

## Adaptation

The fit between organisms and their environment is one of the main fields of study of ecology. The working hypothesis is that natural selection favours the inheritance of features that suit an individual to its typical environment and way of life. These features are called adaptations. Comparative ecologists use multi-species studies to identify strong trends for the possession of a certain character, or suite of characters, among species in a particular environment. This provides circumstantial evidence that an adaptation may be involved. Proving that a feature is an adaptation is very difficult. It requires a demonstration that individuals with the feature are fitter than individuals that differ solely in not having the putative adaptation. Comparative ecologists tend to rely on the weight of circumstantial evidence for convergent evolution rather than attempt the full burden of proof. This has led to criticism of the so-called adaptationist programme: what Gould & Lewontin (1979) termed the Panglossian paradigm. They argued that the assumption of adaptation is too readily taken up, when in fact adaptation is only one of several possible reasons for the presence of a character. Extant species are not perfect; many features may be selectively neutral or arise through complex interactions among different genes and the environment.

For much of the content of the remaining chapters in this book, it is difficult to refute the charge of promoting the Panglossian paradigm because insufficient studies have been conducted to confirm adaptationist speculation. It would be very repetitive to include warnings concerning the assumption of adaptation at each mention of a supposedly advantageous feature, so the reader is reminded to maintain a degree of scepticism throughout. I strongly believe that the correlations evident from comparative studies are invaluable as a base from which to formulate hypotheses and tests of the evolutionary biology and ecology of tropical trees.

## The importance of phylogeny

There can be no doubt that a species is strongly influenced in many of its characters by its antecedents. We must therefore question whether the patterns ecologists see when they compare species, such as trees in the tropical rain forest, are not mostly reflections of phylogenetic relationships rather than recent ecological adaptations.

Can we estimate the degree of 'phylogenetic constraint' on the ecology of species? Or can we control for the influence of phylogeny when we design ecological experiments or observations? The answer to the first question is a qualified 'yes'. Techniques exist that attempt to partition interspecific variation into 'ecological' and 'phylogenetic' components, or at least filter out the phylogenetic effects as error in the statistical model. Two main approaches are available (Gittleman & Luh 1992). Autoregression techniques can be used that partition variance to different phylogenetic distances within the data set, generally by using either nested analysis of variance (ANOVA) or spatial statistics (Moran's *I*). The alternative is phylogenetically independent contrasts (PIC), which overcome the statistical dependence of species by restricting comparisons to adjacent branch pairs on the phylogenetic tree. The major problem with applying these techniques is that they require a phylogeny from which to work, although statistical ways round this have been suggested (Martins 1996). Only recently have detailed phylogenetic analyses of the higher plants become available, and these generally stop at the family level, although more are becoming available within orders and families. These phylogenies are mostly based on DNA sequences and rarely include more than one gene, often with no more than the barest minimum of sampling per taxon. Phylogenies drawn from data of this sort suffer from many interpretive problems (outlined by Donoghue & Ackerly 1996) and at present cannot be taken as definitive. It is not surprising therefore that some ecologists have argued that phylogenetic knowledge is still too fragmentary to consider any meaningful attempt at partitioning 'ecological' and 'phylogenetic' factors in comparative analyses of tropical trees; see, for example, Hammond & Brown (1995). Most others attempting phylogenetic analyses of comparative data for tropical trees have resorted to the use of classical taxonomies for the base phylogeny (Kelly & Purvis 1993; Kelly 1995; Metcalfe & Grubb 1995; Grubb & Metcalfe 1996; Osunkoya 1996). Most of these studies claim to use PICs in their comparative analysis, but this is not strictly true. PIC is correctly a comparison of sister, ultimate branches on the phylogeny. Instead these tropical studies contrast species within the same taxon, usually species in the same genus, but extending as far as species in the same order in one case (Metcalfe & Grubb 1995). This approach is equivalent to assuming a phylogeny for the taxon concerned of instantaneous radiation from the root species to all those currently extant, hardly a plausible evolutionary pathway. Another approach has been to assume that higher taxonomic ranks should possess a greater statistical independence from phylogeny than the species level, so means within genera (see, for example, Grubb & Metcalfe 1996) or families (Baker *et al.* 1998) are used as the basis for comparison.

What does the 'ecological' component of a character trait represent? If assessed by means of an accurate phylogeny, then it will represent the proportion of the character trait acquired subsequent to the last speciation event, which we can assume has a major adaptive element, particularly if repeatedly observed over many PICs. This does not mean that the 'phylogenetic' component is not adaptive. If a particular character state evolved in an ancestral species as an adaptation to its environment then it may still fulfil its adaptive role in the daughter species. However, phylogeny may constrain the range of variation that a descendant can exhibit, and what was once adaptive may no longer be, but merely represent the inertia of evolutionary change. We have returned to the thorny problem of the question of proof in the study of adaptation.

Does phylogenetic analysis get us any further in our understanding? To date the only area of research in tropical tree ecology where phylogenetic analysis has been applied more than once is the question of the relation between seed size and shade tolerance (Kelly & Purvis 1993; Kelly 1995; Metcalfe & Grubb 1995; Grubb & Metcalfe 1996). Studies of a range of tropical tree species at any particular site generally show that those believed to be light-demanding for regeneration have significantly smaller seeds on average than those that are shade-tolerant (Foster & Janson 1985; Metcalfe & Grubb 1995; Hammond & Brown 1995). But phylogenetically controlled analyses have failed to find significance, or even found larger seed size in light-demanders (see, for example, Fig. 1.4). This implies that there is a strong phylogenetic component to seed size that includes both inertia and inherited adaptation. The absence of the predicted trend in the ecological component may indicate that seed size is less adaptive than previously thought.

One of the problems with comparative analyses conducted thus far on tropical trees is that the data were usually collected with the aim of being as thorough as possible: as many species as were available were included. However, when phylogenetic effects are partitioned many of the data have little weight in the analysis because the species are too isolated phylogenetically from other species of contrasting ecology to provide much information. It would be more efficient to predetermine which species were to be contrasted. There has been an increasing frequency of published studies of congeneric species, which may be one way of efficiently tackling the conundrum of phylogenetic influence on ecology. The detailed investigations of the regeneration of *Shorea* section *Doona* in Sri Lanka (Ashton 1995; Ashton *et*

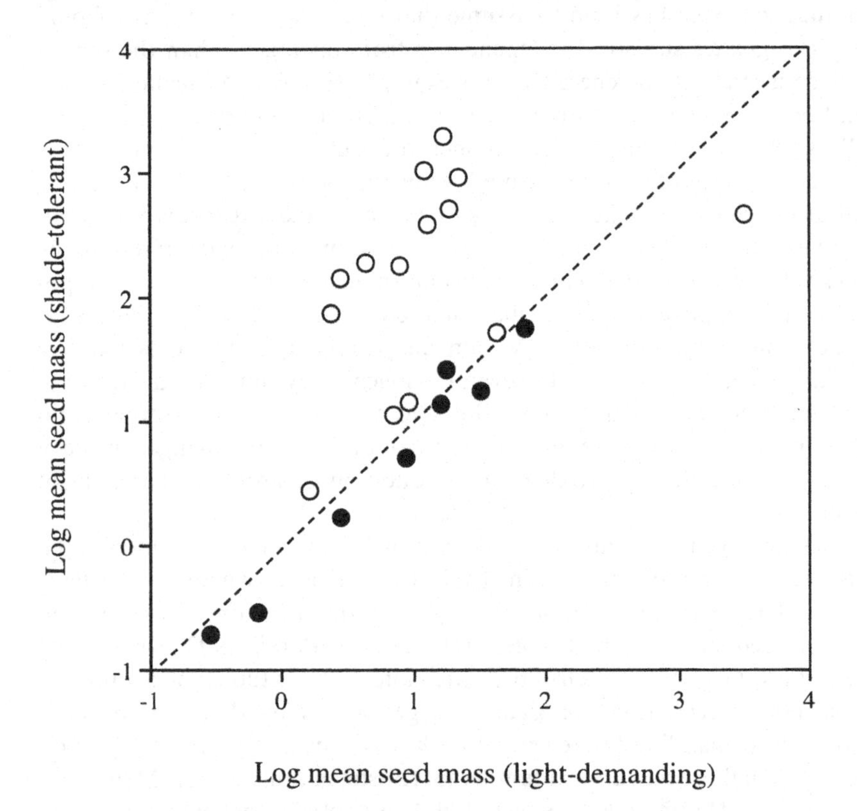


Figure 1.4 Scattergram of the mean seed dry mass values (mg) of shade-tolerant taxa relative to those of light-demanding taxa in phylogenetically controlled comparisons. Open and closed circles indicate genera within families and species within genera, respectively. After Grubb & Metcalfe (1996).

*al.* 1995; Gunatilleke *et al.* 1996a,b, 1997) and those of eleven sympatric and closely related species of *Macaranga* in Sarawak (Davies *et al.* 1998; Davies 1998; Davies & Ashton 1999) stand out as good examples.

## The ecological classification of tropical rain-forest trees

An ecological classificatory system for tropical trees is needed in order to organise our information about the species concerned. The systematisation of knowledge is one of the most important activities of science. A logical organisation efficiently condenses information into a series of generalisations, and allows predictions to be made about members of groups recognised by the system. Without such a system, ecology is reduced to natural history employing the scientific method. This is not meant to denigrate natural history. Natural history provides many of the observations that inspire the formulation of hypotheses, and data by which these hypotheses are tested, but it is science that organises the corpus of natural-history knowledge.

Diversity is the main hurdle faced in designing an ecological classification of tropical trees. The sheer number of tropical tree species means we have only just begun to have any information at all about a very large number of species. Ghana, in West Africa, is probably the only substantial tropical area where all the tree species have been considered ecologically (Hawthorne 1995). Therefore there is an inherent danger that we try to force new data into our pre-existing framework, when maybe we should re-examine the system of organisation we are using.

There are a variety of approaches that can be used to erect an ecological classification system. In classifying the classifications, as it were, one dichotomy might be between systems emphasising groups, and those that emphasise the means of differentiating the groups. Any system of classification has to include both elements, but it may be that the groups are defined by arbitrary points along the axes of differentiation, making the groups less important than the axes. This is the commonest state of affairs in plant ecological classification systems, because discrete groupings of species rarely occur. Axes of differentiation are often defined in terms of the characteristics of species that are thought to lie near the ends of each axis; for instance shade tolerance is usually defined by comparing its end points.

Another distinction might be between those systems derived from *a priori* data analysis and those proposed from general observations without recourse to quantitative data. Most of the classificatory systems proposed for tropical trees are of the second type. A good example is Swaine & Whitmore's (1988) pioneer versus non-pioneer classification of tropical trees. This was based on a wide range of observations made by many different foresters and ecologists, but did not involve any statistical tests of either the character correlations described or the group differentiation proposed. The alternative method of gathering data and performing multi-variate analysis to investigate trends and discontinuities has less commonly been done, although examples are available (see, for example, Condit *et al.* 1996a; Gitay *et al.* 1999). More studies along these lines will be of value to improving our current classification systems.

We will return to the ecological classification of tropical trees in Chapter 6.

# 2

# The growing tree

## Trees: form, mechanics and hydraulics

### Tree stature

Individual trees of a large range of size are to be found in the tropical rain forest (see the profile diagram in Fig. 2.1). Each tree species also has a characteristic size at maturity and species are often referred to various stature classes, such as understorey trees, canopy trees and emergents, but, as Figure 2.2 shows, there is no discrete clustering of species in size classes. Maximum diameter for species on 50 ha in Pasoh Forest, Peninsular Malaysia (Fig. 2.2), was approximately a truncated log-normal distribution with a modal class in the 10–20 cm maximum dbh range. A breakdown of species into height classes (Table 2.1) shows about half of the tree species to have maximal heights of 20 m or less. The tallest trees at Pasoh probably reach to about 60 m in height, showing that the community of understorey specialists in the forest is as rich in tree species as that of the canopy, and is definitely richer on a unit depth basis. A similar pattern is seen in the forest at La Selva, Costa Rica (Hartshorn 1980), and on Barro Colorado Island, Panama (Hubbell & Foster 1992).

The vertical distribution of tree species diversity in the forest may reflect the relative illumination at different heights above the forest floor. At Pasoh, relative illumination increases logarithmically with height (Fig. 2.3). The understorey species in the bottom 20 m of the forest rarely receive more than 5% of the radiation arriving at the top of the canopy. However, the rate of change of relative illumination with height is greatest near the forest floor. In other words, for a given height increase the relative, rather than absolute, increase in illumination is greater nearest the forest floor. It may be that this steep gradient facilitates partitioning of the light resource among species, and hence more species per unit depth as specialists than higher in the forest.

There has been a growing tendency in the tropical forest literature to use the terms shrub and treelet to categorise species in terms of maximum height, with 'shrub' taken to represent the smallest trees and 'treelet' the next size class up, with an arbitrarily set limit between the two. More traditionally, shrubs are distinguished as small woody plants either with multiple stems, or with branching very close to the ground. It has been claimed that 'true' shrubs are rare in tropical rain forests. Givnish (1984) pointed out that only 6 out of 95 non-climbing woody species of less than 10 m in height at maturity native to Ghana were true shrubs whereas temperate forests frequently have many shrubby species in the understorey. The multi-stemmed crown construction is probably a more efficient way of supporting a low, broad canopy than a single stem with wide-reaching branches. As yet there is no strongly supported hypothesis as to why shrubs are infrequent members of the rain-forest community.

The treelet is a life form perhaps commoner in the tropical rain forest than any other vegetation type: a single-stemmed, unbranched or sparsely branched tree of diminutive stature at maturity. D'Arcy (1973) proposed two main evolutionary origins of the treelets. They arose either through precocious flowering in woody groups, or through the development of woody stems in otherwise herbaceous evolutionary lines. The latter is well exhibited in Hawaii, where species of *Cyanea* and *Plantago* occur among the forest treelets.

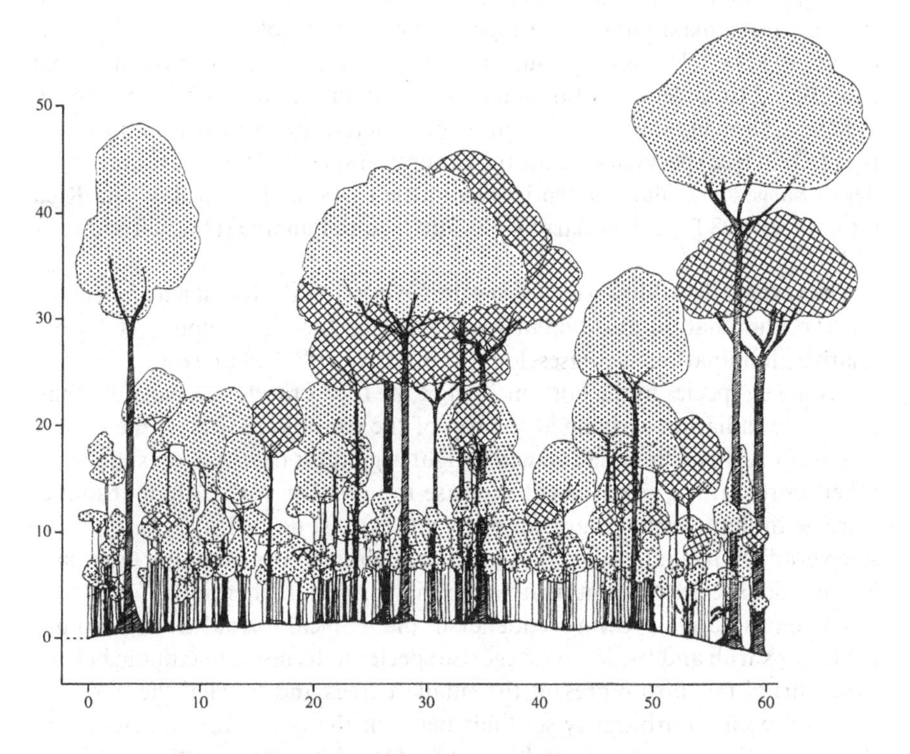


Figure 2.1 Profile diagram of the forest at Ulu Dapoi, Tinjar, Borneo. A plot 60 m × 8 m (200 × 25 feet) is shown. Courtesy of Professor P.S. Ashton.

Table 2.1. *Species diversity and tree frequency by stature class for Pasoh Forest Reserve, Peninsular Malaysia (stems ≥ 1 cm dbh for 50 ha of forest)*

| Stature | Definition | No. of species | Proportion of all species (%) | Trees per ha | Proportion of all stems (%) |
|---|---|---|---|---|---|
| 'shrub' | to 2 m tall | 54 | 6.6 | 303 | 4.6 |
| 'treelet' | 2–10 m tall | 111 | 13.5 | 1513 | 22.9 |
| understorey | 10–20 m tall | 283 | 34.5 | 2114 | 32.1 |
| canopy | 20–30 m tall | 317 | 38.7 | 1901 | 28.9 |
| emergent | > 30 m tall | 55 | 6.7 | 753 | 11.4 |

Data from Kochummen *et al.* (1990).

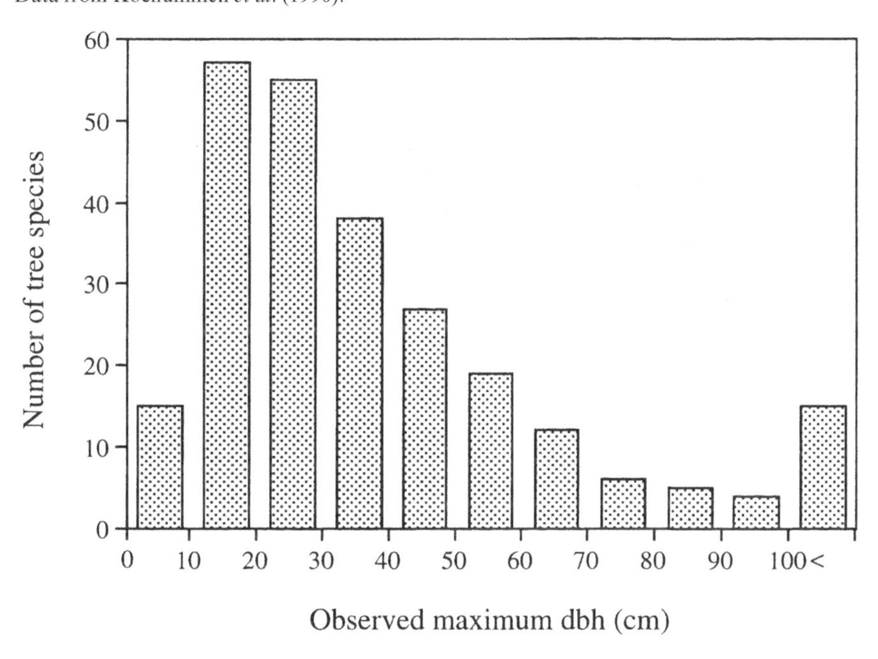


Figure 2.2 Distribution of the maximum trunk diameter for tree species with more than 300 individuals in a 50 ha plot at Pasoh, Malaysia. After Kohyama (1996).

I will use the terms shrub and treelet in their more exact morphological meaning in the text, and refer to pygmy trees when discussing the smallest of the arborescent taxa irrespective of architecture.

At Barro Colorado Island, the commonest species in the 50 ha plot was the treelet *Hybanthus prunifolius*, with nearly 40 000 individuals out of a total of 238 000 enumerated stems. One might argue that it is easier for pygmy trees to be common because they are smaller and hence more individuals can be packed into an area. This ignores the fact that there are small juveniles of large trees, which makes it theoretically possible for the largest tree species to have the greatest population densities. However, small-statured species must have the greatest scope for high densities of reproductively mature individuals. Burger (1980) noted that pygmy trees of the rain forest in Costa Rica tend to be from large genera such as *Piper* and *Psychotria*, with notably more species per genus in the flora than for larger trees. On a pantropical scale, these two mammoth shrubby genera are joined by others such as *Ardisia* and *Rinorea*, although a more detailed analysis is required to confirm that rain-forest pygmy trees generally have a greater species to genus ratio than larger life forms.

## Wood

Wood is derived by secondary thickening of root and stem axes. Secondary thickening involves the periclinal division of cells. This occurs at a peripheral cambium in typical trees, but at scattered sites through the axis in many dendroid monocots. Functionally, wood contains three main elements: xylem conduits (tracheids and/or vessels) that conduct the xylem sap, xylem fibres that provide most of the mechanical support (in gymnosperm wood the conduction and support roles are less distinctly demarcated as tracheids are

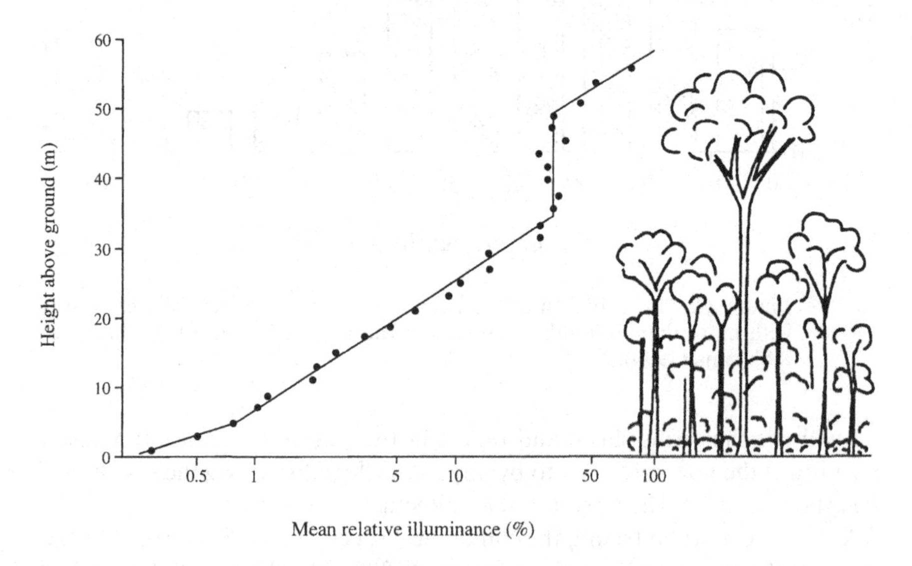


Figure 2.3 Vertical profile of mean relative illuminance (expressed as a percentage of light received above the canopy) at a rain-forest site at Pasoh, Malaysia. After Richards (1996).

solely responsible for both) and parenchyma cells that are probably important in the transverse conduction of water and storage, and possibly assist conduits to recover from cavitation. The first two cell types are very markedly elongated in the direction of the long axis of the tissue, the grain of the wood, with the conduits providing a pipework for the movement of water along the stem or root.

Wood is an excellent structural material (Jeronimidis 1980) because it combines strength and stiffness with a high resistance to the propagation of cracks in a relatively light composite that shows relatively little temperature dependence in its mechanical properties. It is, however, a highly anisotropic material. Its compressive strength is generally half the tensile strength, the strength across the grain is 10 to 100 times less than along it, and work of fracture (toughness) is only high across the grain.

The long columns of cells, particularly the fibres and tracheids, provide the strength of the timber along the grain, and the resistance of wood to crack propagation across it. The rays hold the axial fibres together, resisting torsional and shear stresses, and determine the transverse strength of the tree. Wood cells have cellulose microfibrils running helically along the long axis of the cell in the $S_2$ wall (Fig. 2.4). The orientation of the microfibrils at about 25° to the perpendicular from the long axis of the cell optimises their contribution to both tensile and compressive strength of the timber. Lignin plays an important role in cross-linking and cementing the cellulose components of the cell walls thus improving the stiffness and strength of the timber. The $S_2$ microfibrils are also important contributors to the resistance to cracking of timber across the grain. The avoidance of crack propagation is important because structural failure is usually brought about by crack formation at localised sites of high stress intensity rather than by forces exceeding the strength of the material. Lucas *et al.* (1997) proposed that the stress concentration at the tip of a crack propagating across the wood grain tends to cause the inward collapse of the $S_2$-microfibril layer of the fibre cells. The microfibrils pull the walls into the lumen along almost the complete length of the cell. This collapse of the fibre layer represents considerable use of energy, and plastic deformation may be responsible for up to 90% of the work done during fracture of the timber.

The mechanical properties of wood are strongly positively correlated with its density. The actual density of the solid component of wood varies little among tree species, being around 1.5 g cm$^{-3}$ for air-dry material (Williamson 1984; Detienne & Chanson 1996). It is the presence of spaces within and between cells that causes variation in wood density. Low-density timbers, such as balsa (*Ochroma pyramidale*), have thin-walled wood cells with large lumina and abundant spaces between the cells. The densest heartwood timbers approach the space-free density maximum. Strength properties are directly related to the amount of solid material present, but the cellular nature of wood does contribute to crack resistance (Lucas *et al.* 1997). Very dense timbers have little, if any, lumen to the fibres. Therefore the amount of plastic deformation possible during cracking is reduced and the work of fracture declines. Thus very dense timbers, such as lignum vitae (*Guaiacum* spp.), are very hard and strong, but apt to be brittle. Very low-density timbers have lower toughness than expected because their thin cell walls fracture before they can collapse.

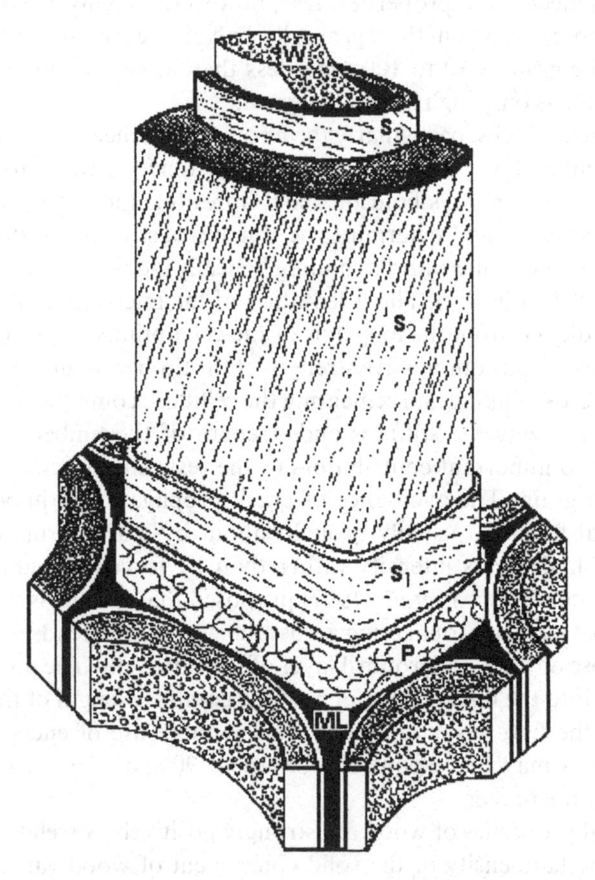


Figure 2.4 Schematic representation of the ultrastructure of the wall components of wood fibres. ML, middle lamella, P, primary wall, S, secondary wall, which is present as three layers, $S_1$, $S_2$ and $S_3$. The $S_2$ wall contains helically wound cellulose microfibrils. W represents the warty inner luminal surface. In some species the surface is smooth.

**Tropical timber**

The distinctive feature of the wood of tropical trees is the presence of wide (large internal diameter) vessel elements, generally with simple perforation plates (Carlquist 1988). The major advantage of wide vessels is increased rates of water conduction because fluid flows through pipes at rates proportional to the fourth power of their diameter. However, not all the vessels in tropical wood are wide. There are usually many small-diameter vessels also. These probably provide safety through redundancy to the conductive system of the tree, and may be important in the radial conduction of water (Tyree *et al.* 1994). Very wide vessels, of more than 100 μm diameter, are rare in trees, being common only in lianas that require high sap-flow rates in stems of limited cross-sectional area. The possession of more wide vessels would allow tropical trees even greater rates of sap flow through trunks and branches but there generally comes a point when delivery rates of water to the leaves would not be improved by increased conductivity of the stem and main branches because of the higher resistance to flow in the outer branches and twigs. Wide vessels have often been considered as more susceptible to cavitation than narrow ones. Tyree & Ewers (1996) found a weak negative correlation between applied pressure required to cause a 50% reduction in hydraulic conductivity and mean vessel diameter for a large selection of tree species. However, the mechanism of greater susceptibility to cavitation in wider xylem conduits is only clear in the case of freezing-induced embolism. Large air bubbles take longer to dissolve than small ones, increasing the likelihood of serious cavitation after the freezing of xylem sap (Tyree *et al.* 1994). Freezing temperatures are not experienced in the lowland tropics, but they are in the temperate zone, where trees tend to have vessels of smaller diameter (van der Graaff & Baas 1974). This is probably why temperate timbers of the same density as tropical ones consistently have 20% lower cross-grain work of fracture (Lucas *et al.* 1997). The greater vessel number in the temperate woods reduces their mechanical performance. It is questionable, however, whether there has been selection for a certain vessel size in tropical trees to improve the mechanical properties of the wood. Roots tend to have wider vessels than stems (Ewers *et al.* 1997), probably because roots are less liable to cavitation and are better able to recover from it.

Tropical forests are remarkable for the wide range of timber densities present, frequently including species of both very low (less than 0.3 $g\,cm^{-3}$) and very high (more than 0.7 $g\,cm^{-3}$) density (Williamson 1984). Wood densities are often reported as weight per volume at 12% moisture (Reyes *et al.* 1992; Detienne & Chanson 1996), though for ecological purposes the true density, oven-dry mass per green volume, is more useful. Reyes *et al.* (1992) provide a regression equation to make the conversion between the two types of density estimate. Multi-species samples from tropical forests show normal distributions of wood densities (Fig. 2.5) with means in the range 0.56–0.62 $g\,cm^{-3}$ (Reyes *et al.* 1992; Detienne & Chanson 1996). These studies provide support to the findings of Whitmore & Silva (1990) that South American, particularly Amazonian, timbers tend to be denser, or at least show a greater spread of densities with more very heavy woods present, than those of other tropical regions.

Wood density has some affinities with taxonomy. The Bombacaceae are notable for very light timber, particularly in genera such as *Cavanillesia, Ceiba* and *Ochroma.* Other low-density woods are found in genera such as *Cecropia, Musanga, Ficus* and *Sterculia.* Very heavy woods are often from the caesalpinoid subfamily of the legumes; *Swartzia* is a good example. The ebonies (*Diospyros*, Ebenaceae) are also renowned for their very heavy timber.

There is some evidence that the range of wood density in tropical forests narrows with elevation and widens with increased severity of drought (Williamson 1984). Tree species found in more exposed sites in the montane forests of Costa Rica mostly had higher wood densities than the species characteristic of more sheltered places (Lawton 1984). Presumably species with low-density, weak wood would be at a strong disadvantage under the greater wind loading of mountain ridges. The wood anatomy of species of the drought-deciduous forest at Chamela was compared with those from a wet lowland forest at Los Tuxtlas, also in Mexico (Barajas-Morales 1985). The evergreen-forest trees had softer, lighter-coloured and less dense wood with fewer crystalline and resinous inclusions than the deciduous-forest species. Average vessel diameter was lower in the trees from Chamela, as is typical for more seasonally dry forests (Tyree *et al.* 1994). However, in comparing 30 species that grew in both rain-forest and savannah sites in Ivory Coast, den Outer & van Veenendaal (1976) found a majority of species to have wider vessels in the drier savannah area. Overall the difference in average vessel diameters between the sites was not statistically significant.

Not all trees form heartwood: for instance palms, and a number of fast-growing soft-wooded dicot species e.g. *Dyera costulata,* do not. These sapwood trees can actively defend their wood through the agency of the living cells present throughout the trunk and are rarely, if ever, found hollow because any pest or disease that becomes established is likely to destroy the whole tree. The dead heartwood that constitutes the core of the trunk of most trees can only be protected from decay by the passive action of its constituents and by being covered by the overlying living sapwood.

Density may also be a measure of the degree to which woody organs are protected from attack by pests and pathogens. The mechanical properties of

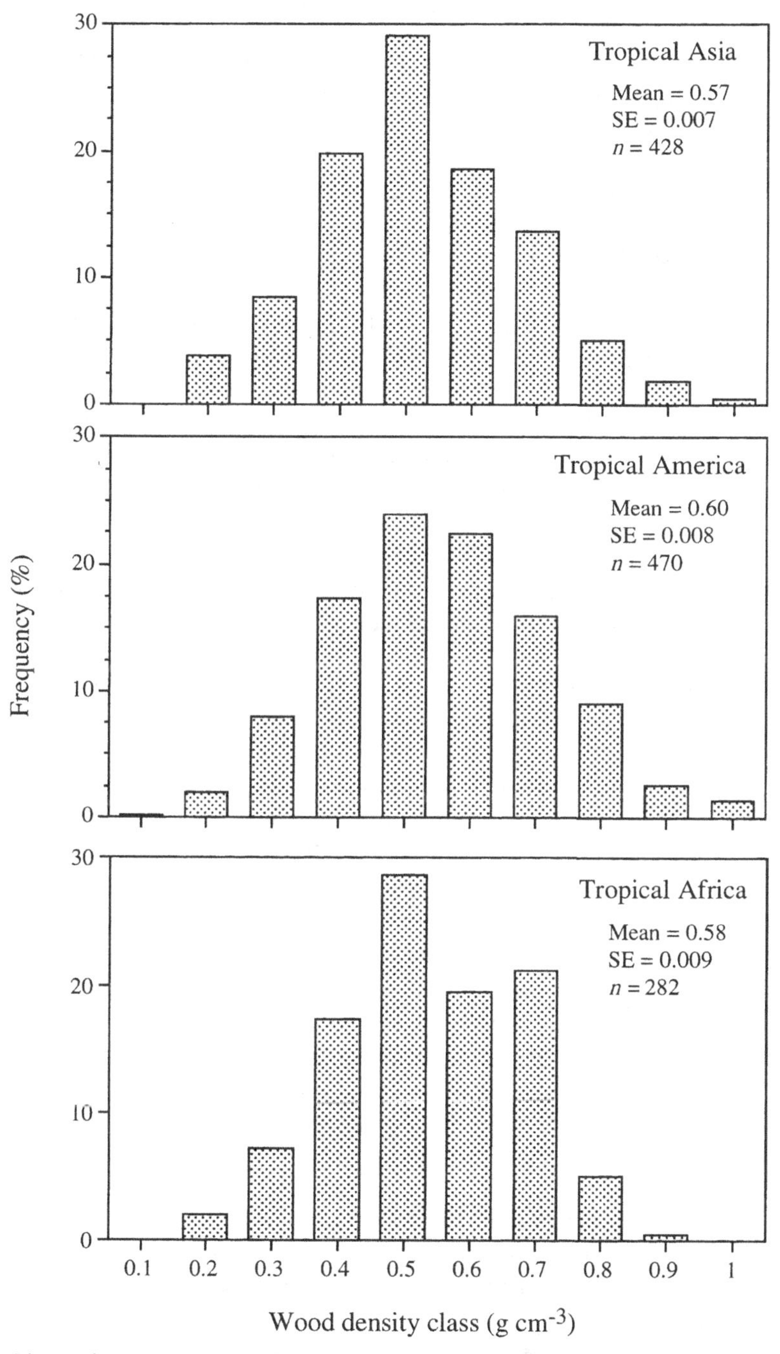


Figure 2.5 Frequency distribution of tropical forest species by wood-density class for three tropical regions. After Reyes *et al.* (1992).

wood at a smaller scale may help protect it from decay. Animals will find very dense, hard wood difficult to comminute prior to ingestion, and fungal hyphae will not be able to penetrate and break down the woody tissue. However, heartrot does occur and hollow trees are not uncommon in tropical forests.

Wood density has often been used by foresters and ecologists as a measure of a species' maximal growth rate and of its relative shade tolerance. Fast-growing, shade-intolerant species have low wood densities, whereas the wood of slow-growing, shade-tolerant species is heavy. There is surprisingly little reliable quantitative evidence to substantiate this assertion. Wood density showed a significant, though weak, negative correlation with average diameter growth rate across 122 species from French Guiana (Favrichon 1994). In a rain forest in Borneo, trees with a stem specific gravity (including bark) of 0.2–0.49 exhibited significantly faster relative stem diameter growth than those with stem specific gravity of more than 0.5 (Suzuki 1999). Seedling persistence in deep shade exhibited a significant positive correlation with adult wood density among 18 wind-dispersed trees from Barro Colorado Island, Panama (Augspurger 1984). Wood density was the best (though still weak) predictor of light requirement index for regeneration among tree species from Queensland, Australia (Osunkoya 1996), but the index was a ranking of the impressions of ecologists, and not a quantifiable measure of shade-tolerance hypothesised *a priori* to be relevant.

Most of the cells in wood—fibres, tracheids and vessel elements—are dead, but the ray parenchyma and other sapwood tissues are living and hence respiring. The rates of respiration of woody tissues of two tree species at La Selva, Costa Rica, have been measured (Ryan *et al.* 1994). The fast-growing *Simarouba amara* (1.24 $\mu$mol m$^{-2}$ s$^{-1}$) was found to respire at considerably faster rates than the slow-growing *Minquartia guianensis* (0.83 $\mu$mol m$^{-2}$ s$^{-1}$). When allowance was made for wood volume growth rates it was found that the two species had comparable maintenance respiration rates. These rates were twice those of temperate conifers. The high maintenance respiration of tropical trees is probably due to the high temperatures experienced. Rapid respiration is the main reason why, despite superior conditions for growth, wood production rates in tropical forests are not much larger than in temperate or even boreal forests (Jordan 1982).

### The mechanical design of trees

One of the key constraints on the form of a tree is its necessary ability to support its own weight and the range of external forces, particularly those generated by the wind, it is likely to meet during its lifetime. Mattheck & Kubler (1995) provide an excellent overview of how tree design helps solve these mechanical problems. Failure usually involves localised regions of excessive stress. The mechanical design of trees can be summarised as growth and allocation of material to maintain a uniformity of stress within the tree and avoidance of critical stresses at particular sites. This is achieved by a variety of mechanisms including:

1. *The passive bending of flexible parts.* Not resisting the external forces minimises the loading, but large heavy stems would buckle under their own weight if they were very flexible. Therefore, the bending strategy is confined to small trees, the ultimate branches of large trees and notably to bamboos (Mosbrugger 1990). Rheophytic trees and shrubs that survive periods of inundation in fast-flowing rivers use their flexibility, and tenacious root systems, to survive. Tree crowns require some flexibility to reduce their drag against the flow of the moving air on windy days. Under heavy wind loadings trees may produce 'flexure' wood (Ennos 1997), which has more markedly spiral grain. It is denser than normal wood, with shorter, thicker-walled cells. The cellulose microfibrils are also wound at a larger angle than usual. Theoretically, this should result in wood of lower stiffness with the same breaking strain. As yet, only the former has been demonstrated experimentally (Ennos 1997). Branches may have denser wood than tree trunks so that they can be relatively thinner and thus bend and 're-configure' more easily. Putz *et al.* (1983) noted that is was low-density-timber trees that tended to lose limbs in windstorms in Panama.

2. *The design of the wood as a cellular composite material.* The various components of the wood are arranged to meet different mechanical challenges. While the axial arrangement of fibres and tracheids provides most of the strength of the stem, ray fibres resist critical shear and torsional stresses. Monocotyledonous wood, with its diffuse sclerenchymatous material, lacks efficient cross bracing and is not good at resisting forces oriented in directions other than parallel to the main axis. This is probably why the dendroid monocotyledons are often unbranched and have limited crown development (Mosbrugger 1990). Branch insertion and spreading crowns inevitably generate lateral stresses. Palms overcome this limitation to some extent by having enormous leaves, their equivalent of whole branches. These are supported by highly fibrous leaf bases that clasp the main trunk tightly.

3. *The outer shape of the tree and the internal quality of the wood are optimised to meet the external forces.* Tree trunks are generally roughly circular in cross section. Angled stems would be disadvantageous because they would lead to high stress intensities at the angle. Buttresses are another example of using growth to overcome mechanical challenges (see below). Significant positive gradients in wood density from pith to bark were found in

16 out of 20 species of tree tested in Costa Rica (Wiemann & Williamson 1989). Palms also tend to have denser wood around the periphery of the base of the trunk (Rich 1987) (Fig. 2.6). Wiemann & Williamson (1989) found that these gradients were most marked in low-density timbers such as *Ceiba pentandra* (Fig. 2.7). Studies of such trees have shown that the wood density gradient is brought about by changes in the density of the wood produced as the tree ages, rather than by localised growth of denser wood (Rueda & Williamson 1992; de Castro *et al.* 1993). The sapling wood of the fast-growing *Laetia procera* had half the density of the adult tree, but the sapling and adult woods of the shade-tolerant *Dipteryx panamensis* were of the same density (King 1996). It is possible that the intense competition for height growth in

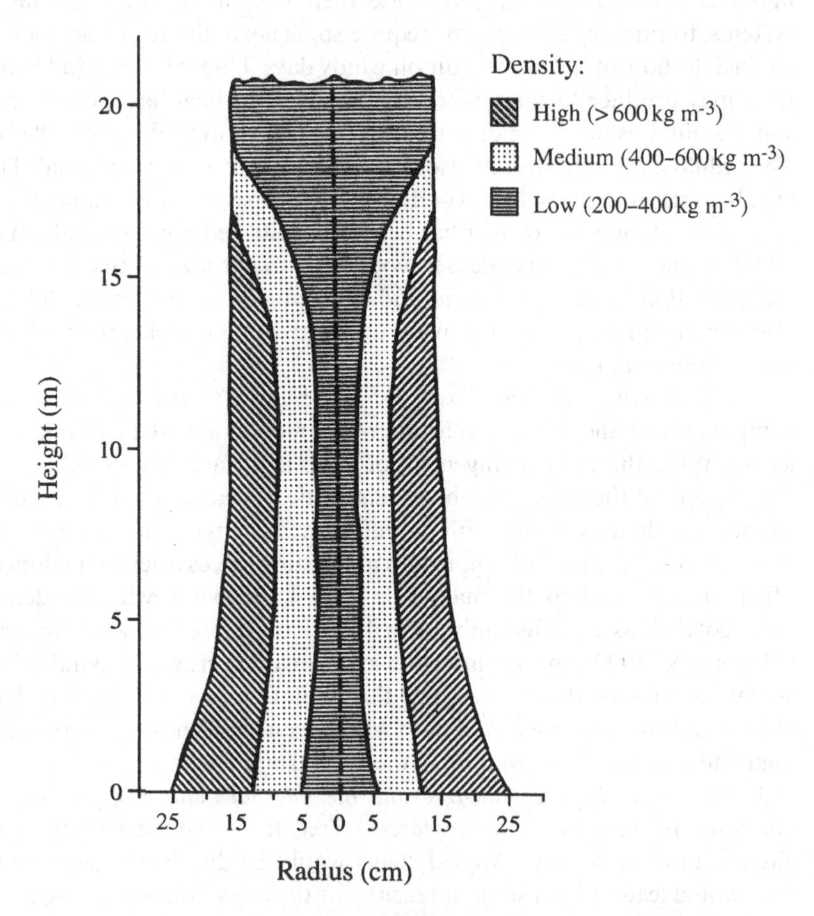


Figure 2.6 Schematic density distribution in the stem of the mature palm tree (*Cocos nucifera*). After Killmann (1983).

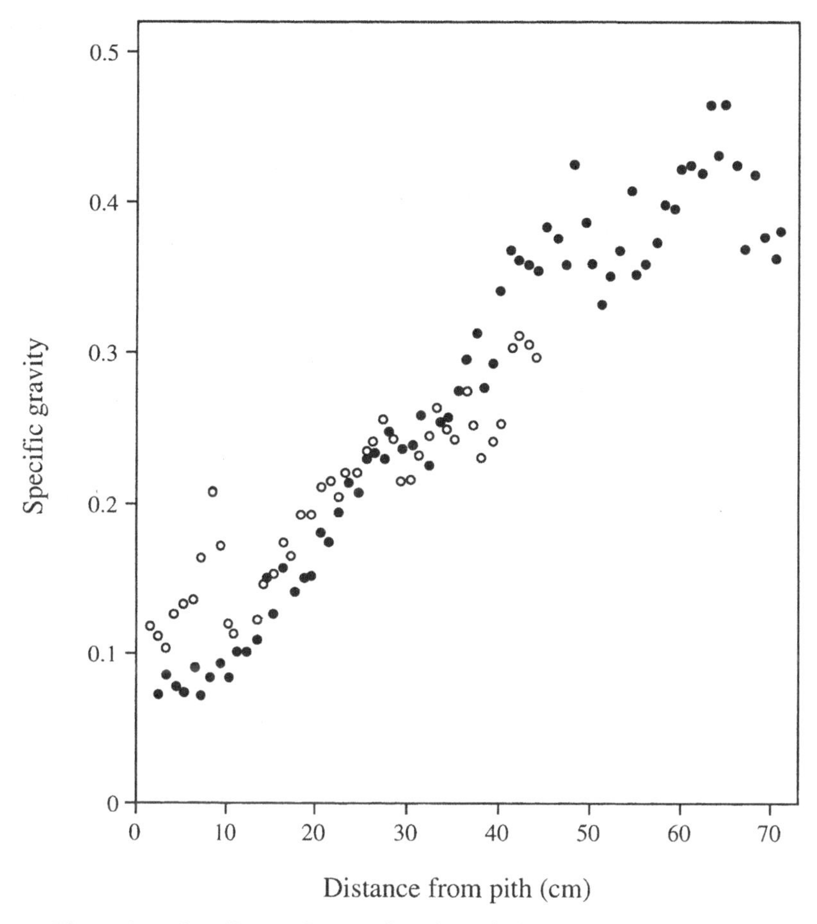


Figure 2.7 Specific gravity as a function of distance from the pith for two trees (different symbols) of *Ceiba pentandra*. After Wiemann & Williamson (1989).

juveniles of light-wooded tropical trees favours the 'low cost – high risk' use of very soft wood for stems and branches. The major stresses in tree trunks are produced through bending and occur in the outermost portions of the stem cross section. The central portion mainly supports the weight of material above. Removing the central portion, i.e. making a hollow tree, alters the mechanical properties of the trunk relatively little as long as the ratio of the thickness of the outer solid wall ($t$) to the diameter of the hollow ($D$) remains fairly large. The weakening of the stem by removing the central cylinder is made up for by the mass lost by its removal (Mosbrugger 1990). Hollow stems make mechanical sense and are found in some fast-growing species including members of the genera *Cecropia* and *Musanga*, where $t/D$ usually falls in the range 1/3 to 1/5. However, the engineering difficulty with hollow stems is branch insertion. A hollow branch meeting a hollow stem tends to lead to high stresses at the point of insertion. *Cecropia* generally has short, stiff branches to keep these stresses low, and may rely on pith turgor to reduce the risk of buckling in smaller branches. Large, old trees often have hollow trunks. They appear to be mechanically effective until $t/D$ approaches a critical value of 0.3 (Mattheck & Kubler 1995).

4. *Production of reaction wood at points of high stress intensity.* Reaction wood differs between conifers and dicotyledonous trees. Conifers produce compression wood that is rich in lignin and has a high microfibril angle in the $S_2$ walls of the wood cells. The conifer reaction wood has a high strength in compression and is produced at sites of high compression such as the underside of main limbs. Dicotyledons produce tension wood that has little lignin and a low microfibrillar angle. It is grown in sites of high tension stresses such as the upper sides of large branches.

5. *Internal pre-stressing of wood to counterbalance critical loads.* There is evidence that tree trunks are pre-stressed, with the surface of the trunk under tension and its core under compression. This improves the tree's performance against its most likely mode of general failure: the compressive stresses generated through bending on the side of the tree in the direction of the bend (the leeward side if it is the wind that is causing bending). If the wood in this region starts under tension it will take a greater force to reach critical compressive loads than for a trunk without pre-stressing. The pre-stressing will mean higher tension stresses on the windward side of the trunk, but wood has greater tensile than compressive strength.

### Buttresses

A prominent feature of the lowland tropical forest is the relatively high frequency of buttress roots growing from the base of large trees. In a survey of trees in a forest in the Malay Peninsula, 41% of individuals of 45 cm dbh or greater had buttresses to 1.3 m or higher (Setten 1953). A figure of 37% was found for the same tree-size and buttress-height groups in Venezuela (Rollet 1969). Eighteen out of 78 species (23%) of more than 10 cm dbh had buttresses at Kibale in Uganda (Chapman *et al.* 1998), with larger size classes tending to have a higher frequency of buttressing. Buttresses are variable in size and shape among species, being very tall in some. For instance, Rollet (1969) reported *Sloanea guianensis* and *Ceiba pentandra* to have them extending to more than 7 m up the tree base. Some tropical tree species are nearly always found with buttresses (88% of *Intsia palembanica* trees surveyed (Setten 1953)), some rarely if ever have them (0.8% of *Calophyllum*), and for the rest buttressing varies among sites and genotypes. The common use of buttress form as an aid to species identification in manuals of forest botany indicates a strong genetic control on buttress design.

Engineering arguments have frequently been put forward to explain why trees need buttresses. Their name, derived from the architectural devices used to keep buildings standing, implies a mechanical function. However, only recently have the principles of physics been applied to trees in order to develop a functional explanation of buttressing in trees (Mattheck 1993). The basic premise is that buttresses are part of the tree's mechanism for the efficient transmission to the ground of the stresses caused by loading due to wind (Ennos 1993a). Permanent anchorage of the tree necessitates the transmission of the stresses. If not the tree will fall. A tree which has a deep and thick tap-root immediately beneath the trunk can transmit forces directly into the ground. However, many trees have their largest roots radiating from the base of the trunk in the upper layers of the soil, creating a shallow plate of woody roots. Such root plates are easy to observe on wind-thrown trees in any tropical rain forest. Sinker roots, descending from the plate deeper into the soil, transmit the stresses generated in the aboveground parts of the tree into the ground. Mattheck proposed that very large forces would be generated at the junction between the base of the trunk and the lateral roots when the tree is subjected to loading by the wind. Such forces could easily snap or split the roots at the junction with the trunk. Responding to these stress concentrations, the juvenile tree develops outgrowths between the trunk and the main lateral roots that eventually become buttresses. These brace the lateral to the trunk and allow more efficient transmission of stress with a reduced stress concentration at the junction between trunk base and root.

Buttresses can probably function as both tension and compression elements depending on the direction of the wind (Fig. 2.8). On the windward side of the trunk the bending of the tree tends to pull the roots out of the ground. The buttresses transmit this tension smoothly to the roots, acting like guy ropes. The root system is pushed into the ground by the compressive forces generated on the leeward side of the tree. Here the buttresses act as props spreading out the stress. Ennos (1995) used electronic strain gauges to demonstrate that the qualitative patterns of strain within buttressed trees were as Mattheck had predicted, with the highest strains along the tops of the developing buttresses.

The best experimental test of Mattheck's hypothesis of the functional significance of buttresses was conducted on young trees in the lowland rain forest in the Danum Valley, Sabah, Malaysia (Crook *et al.* 1997). Two groups of trees were studied. *Mallotus wrayi* represented a typically unbuttressed species. A pinnate-leaved, typically buttressed, morphospecies was found to include individuals of several species of *Aglaia* and *Nephelium ramboutan-ake*. The buttressed trees did not always have sinker roots, and they did sometimes have tap-roots. The trees were subjected to imitations of severe

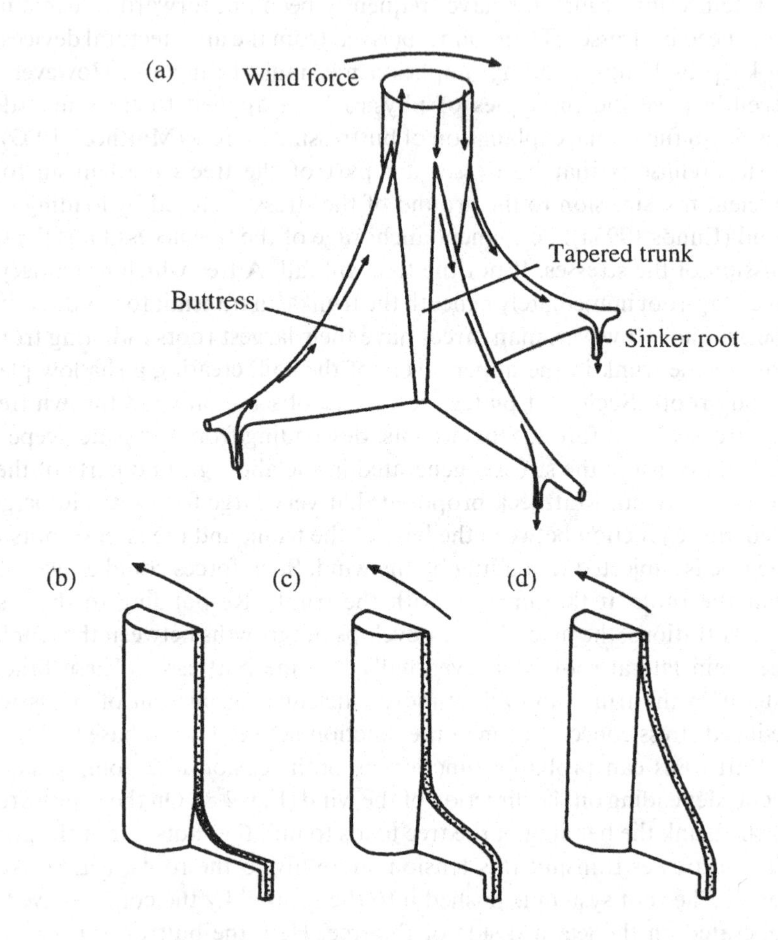


Figure 2.8 Mattheck's model for the function and development of buttresses. If a tree is pushed over by the wind (a) the bending force is transmitted smoothly to lateral sinker roots by the buttresses. Windward sinkers resist upward forces and the buttress is put in tension while leeward sinkers resist downward forces and the buttresses are put in compression. Parts (b), (c) and (d) show successive stages in Mattheck's simulation of buttress development. When the trunk is pulled over stress is concentrated at the top of the junction between the lateral root and the trunk (stippling). Growth in heavily stressed regions (b) results in the formation of buttresses (c, d) and a great reduction in stress concentrations. After Ennos (1995).

wind loadings by means of a winch. In buttressed trees without sinkers, failure was by breakage of leeward laterals and pulling out of windward ones (Fig. 2.9). With sinkers, windward laterals suffered delamination. The lateral roots were tall rectangles in cross-section, up to eight times tall as wide. This helps the roots resist bending up or down. Laterals were relatively ineffective at anchoring in non-buttressed trees. The laterals pulled out, or bent in compression, easily. The buttressed individuals were about twice as stable as the unbuttressed trees of the same size. The study confirmed Mattheck's predictions that buttresses could be effective both as tension and compression elements, stopping roots either being pulled out of the ground or pushed into it. The presence of sinkers on the buttresses improves their effectiveness in tension considerably, but even without sinkers they work well in compression. Possibly, as the trees grow larger more sinker roots develop. Reaction wood is not commonly found in the buttresses of tropical dicotyledonous trees (Fisher 1982). The increased tensile strength of reaction wood would possibly carry a disadvantage of reduced compressive strength.

Kaufman (1988) has argued that buttress formation is largely the result of a developmental crisis in the tree's life due to factors such as crown asymmetry or sudden exposure to high winds after gap formation or growth above the main canopy. The buttress may persist on the tree despite the reduction of the factor that led to its initial formation. In support of this theory, Chapman *et al.* (1998) found that understorey species at Kibale, Uganda, showed significantly less buttressing than canopy and emergent species, even after a correction for individual size had been made.

Crown asymmetry may be quite common in trees of the tropical rain forest. The average degree of crown asymmetry of 127 trees (greater than 20 cm dbh) observed at Barro Colorado Island was such that three quarters of the crown area of a typical tree was on the heavy side (Young & Hubbell 1991). The trees tended to develop crowns into gaps or away from larger neighbours. Emergents were typically more symmetrical than individuals of smaller stature. Over a period of 6.7 years the more asymmetric individuals had a higher chance of falling than those with more symmetrical crowns, and they tended to fall towards the heavy side (Young & Perkocha 1994). Buttress formation in the trees was more pronounced on the side of the trunk opposite to the heavy side of the crown. On large trees, buttresses often grew in size more quickly than total tree size (Young & Perkocha 1994). This would decrease the effective bole length of the trees concerned and might help reduce the risk of trunk buckling. The buttresses make the base of the trees stiffer and possibly allow greater heights and wider crowns to be achieved relatively safely.

Another possible advantage of buttresses is a reduced requirement for

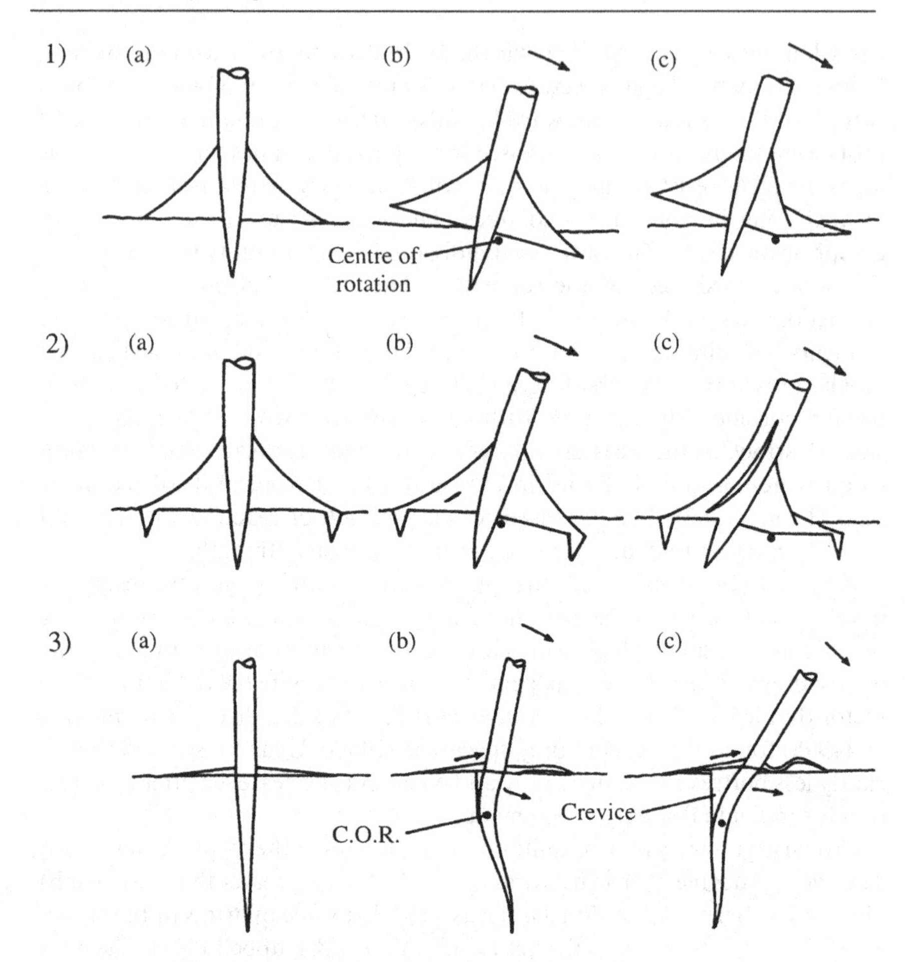


Figure 2.9 Trunk and root movements during anchorage failure of buttressed and non-buttressed trees. (1) Buttressed tree without sinker roots (*Aglaia affinis*), (2) buttressed tree with sinker roots (*Nephelium ramboutan-ake*). Note: sinker roots may be present or absent on both *Aglaia* and *Nephelium* species. (1a) Buttressed tree without sinker roots. The tree is anchored into the ground by the thick buttressed lateral roots and the tap-root. (1b) As the tree is pulled over, the trunk rotates about a point just on the leeward side. Initially, the roots firmly anchor the tree in the ground, the leeward lateral resisting bending being pushed into the ground and the tap-root resisting uprooting. The windward buttress, held in the ground by fine roots only, uproots easily. (1c) As the test proceeds the leeward buttress finally fails, breaking towards end of the buttressing. The centre of rotation (C.O.R.) changes so that the tree rotates about this leeward hinge and the tap-root is levered out of the ground or breaks. (2a) Buttressed tree with sinker roots. The tree is anchored into the ground by the thick buttressed laterals by their sinker roots and by the tap-root. As the tree is pulled over

material in the base of the trunk. In some species, the trunk has a narrower diameter beneath the top of the buttresses than above them. Presumably the transmission of stresses through the buttresses allows the tree to remain mechanically viable with less material in the lower butt. This can be seen at its most extreme in species with highly developed stilt roots. These, we must presume, act like flying buttresses, transmitting stresses widely to the extensive lateral root system. In some stilt-rooted species the primary stem is extremely thin beneath the stilt roots. It is much too narrow to support the weight of the tree above. The development of stilt roots allows palms to gain more height for a given stem diameter than those without stilt roots (Schatz *et al.* 1985). Fisher (1982) found reaction wood in the stilt roots of *Cecropia* species, possibly indicating the requirement for a greater tensile strength in buttresses of more limited cross-sectional area. Stilt roots are not a common feature of trees of well-drained lowland rain forest. For instance, only 5 out of 246 species surveyed in Gabon, West Africa, had stilt roots (Reitsma 1988). Species of swamp forest are much more frequently stilt-rooted.

Using an approach based on allometry, Ennos (1993b) has shown that the mechanical efficiency of a root-plate system increases with plant size because the anchorage produced by the root-soil plate (largely provided by its weight) increases with size faster than stem strength. To be effective in a large tree, a tap-root needs to be of similar dimensions to the trunk, or even larger. In

Caption for Fig. 2.9 (*cont.*)

(2b), the tree rotates about a point just on the leeward side. Initially the roots firmly anchor the tree in the ground, the leeward lateral resisting being pushed into the ground and the tap-root resisting uprooting. The windward buttress, held in the ground securely by the sinker roots, also withstands uprooting and instead begins to delaminate. As the test proceeds (2c) the leeward buttress finally fails, breaking towards the end of the buttressing. The centre of rotation changes so that the tree rotates about this leeward hinge and the windward root continues to delaminate. (3a) Non-buttressed tree (*Mallotus wrayi*). The tree is anchored into the ground by the tap-root and to a lesser extent the lateral roots. (3b) As the tree is pulled over, the tree rotates about a point just on the leeward side of the tap-root at a depth of *ca.* 0.5 m. The leeward laterals are pushed only slightly into the soil and then buckle whilst the laterals on the windward side resist being pulled up, acting in tension. The tap-root pushes into the soil on the leeward side both bending slightly and rotating above the centre of rotation and below this bends and moves slightly windward. A crevice is formed on the windward side as the tap-root rotates. As the test proceeds (3c) these root movements continue, the leeward laterals buckling, the windward laterals uprooting and the tap-root pushing into the soil, increasing the size of the crevice. After Crook *et al.* (1997).

tropical soils, it may be difficult for a tree to produce as large a tap-root to sufficient depth as is required for adequate anchorage, particularly in species with heavy timber. Crook & Ennos (1998) analysed the mechanical failure of tap-rooted *Mallotus wrayi* trees at Danum Valley in Sabah. In smaller trees the trunks snapped when heavily loaded by means of a winch, but in larger individuals failure was through the tap-roots being displaced. This indicates an inability by large trees to produce big enough tap-roots to anchor themselves effectively.

In conclusion, there is now evidence that buttresses are effective structures that play an important role in supporting and anchoring large trees. However, many questions remain to be answered. Do the many unbuttressed big trees in the forest all have large tap-roots? Are buttresses and root-plate systems cost-effective in terms of material required to build them? Do buttresses perform other roles such as occupying space on the forest floor to reduce the likelihood of establishment by competing trees?

**Leaning trees**

Most trees stand upright, even on sloping ground. This is because it is mechanically far more efficient to place the support directly under the crown and keep the main stresses within the line of the trunk and trunk base. In addition, leaning trees gain height at a greater construction cost per unit height, which will result in a slower rate of height growth as well. Loehle (1986) pointed out that most trees that do lean are small, or are fast-growing species with early reproduction, or lean into conditions of permanent high light such as out over rivers. The costs of leaning in small trees are less than in large ones because they can utilise the elastic properties of wood to support the lean with little extra investment in support. Large trees need to brace the tree with thicker trunks and more massive tree bases if they are to lean. These represent major investments that will only be worthwhile if the rewards of leaning are high. Fast-growing species may risk death through shading if they do not lean into nearby gaps. It is also possible that their low-density wood makes leaning less problematic. Ishii & Higashi (1997) came to similar conclusions in comparing small and large trees growing on slopes where the gradient of relative illumination is perpendicular to the slope. The quantitative model they derived has been criticised as unrealistic (Loehle 1997), even though its general predictions were found to be correct for a comparison of two species growing on slopes in warm temperate rain forest in Japan. None of these hypotheses have been explicitly tested for tropical rain forest trees. Riverbank trees have been noted as often leaning. For instance, *Dipterocarpus oblongifolius* is characteristic of the banks of some rivers in the Malay Peninsula and Borneo, and habitually leans, often out over the river (Whitmore 1975).

## Tree fall

Despite the ability of trees to grow in ways that reduce the likelihood of localised high stress intensities, mechanical failure is common in tropical rain forests. On Barro Colorado Island in Panama, a survey of 310 fallen trees (Putz *et al.* 1983) found that 70% of the trees had snapped trunks, 25% had uprooted and 5% had broken off at ground level. Uprooted trees tended to be larger than snapped ones, with trunks of greater diameter for a given height, and to have denser, stiffer and stronger wood. The degree of buttressing did not appear to be related to tendency to snap or uproot. Susceptibility to hurricane damage was found to be negatively correlated with wood density in tree species on Puerto Rico (J.K. Zimmerman *et al.* 1994). However, for five species from Kauai, Hawaii, stem elastic modulus for bending was a better correlate of the likelihood of snapping during a hurricane than wood density (Asner & Goldstein 1997). The stiff stems with low elastic modulus were more likely to survive the hurricane without snapping. The mechanical properties of the wood of a species influence its ability successfully to resist wind loading and other external stresses, but more research is needed to pinpoint which properties are the most influential.

It must not be assumed that fallen trees always die. Of 165 trees of 10 cm dbh and above that snapped in the period 1976–1980 on Barro Colorado Island, 88 (53%) re-sprouted from the broken base, of which 26 (16%) were still alive in 1987 (Putz & Brokaw 1989). In the Atlantic forests of Brazil, 82 out of 100 uprooted trees had new sprouts, mostly from the base of the trunk (Negrelle 1995). In both studies it was noted that pioneer species were poor at re-sprouting. Putz & Brokaw (1989) found that smaller trees were more likely to re-sprout successfully in the long-term. Negrelle (1995) noted that *Tapirira guianensis* was particularly good at re-growing after treefall, and was able to re-sprout from the whole length of its trunk.

## Sap ascent

One of the marvels of nature is the apparent ease with which big trees lift large volumes of water to great heights while they transpire. Large quantities of water are needed by a tree crown, not only to replace the water inevitably lost during the uptake of carbon dioxide from the atmosphere during photosynthesis, but also because the evaporative loss of water is a highly effective way of cooling leaves exposed to the intense tropical sun. The xylem sap ascends the tree through the tiny-diameter pipes made by columns of tracheids or vessel elements. The driving force for the ascent against gravity is the difference in free energy between water in the gaseous state in the atmosphere around the leaves and their transpiring surfaces and the liquid water in the roots. The widely accepted cohesion–tension theory of xylem sap ascent proposes that the evaporation of water from cell walls in the leaves creates microscopic menisci that produce very strong tensions in the continuous columns of water that stretch from these sites to the roots through the xylem conduits. The cohesive properties of water allow the columns under tension to be pulled up rather than break and hence provide the transpiration stream. The cohesion–tension theory predicts that high tensions should be found in the xylem. At least 1 bar (0.1 MPa) is needed to overcome gravity for each 10 m increment in height, and an equivalent amount is probably required to counter the internal resistance to flow of the conduits.

Measurements made with the Scholander pressure bomb were seen as a vindication of the cohesion–tension theory. The gas pressure applied in the bomb to express sap from the test shoot was assumed to equal the xylem tension, and the values measured generally fell within the range predicted by the cohesion–tension theory. However, the pressure bomb does not measure xylem tension directly. It is possible to construct a probe with a minute pressure sensor that can be inserted right into vessels to measure the internal pressure. Such devices have consistently failed to record large tensions in plant shoots (Zimmermann *et al.* 1994), including high up in a tropical rain-forest tree. Other arguments against the cohesion–tension theory include theoretical and experimental studies of water columns that make it seem unlikely that the predicted tensions could exist in xylem (Smith 1994). The proponents of the cohesion–tension theory have launched a series of counter-arguments and experimental demonstrations to back continued support of the theory. The pressure-probe method is susceptible to technical scepticism. The probe tip is little smaller than the bore of the conduits being measured, and the technique requires that the hole in the vessel wall created by the entry of the probe seals perfectly so that neither the pressure in the conduit alters, nor air enters and causes embolism. The fluid in the probe is liable to cavitation because of gas seeds on the internal surfaces (Wei *et al.* 1999). Milburn (1996) has also pointed out that the probe only accesses the surface vessels. These are liable to have lower tensions than average, either because they are still-living protoxylem or because water recycling from phloem transport, so-called Münch water, rehydrates the outermost vessels. Others have used centrifugation techniques to show that significantly negative water pressures exist and can be maintained in xylem and that they correlate well with pressure-bomb measurements (Holbrook *et al.* 1995; Pockman *et al.* 1995). They also argue that glass capillaries used in experimental studies of water columns and their cohesiveness may not be realistic analogues of xylem conduits.

Recently, Canny (1997a,b) has used a rapid freezing technique to visualise the activity of xylem conduits in sunflower petioles. The surprising result of this study was evidence of frequent cavitation of vessels at relatively low tensions and their rapid re-filling by the movement of water from parenchyma cells around the conduits. The proportion of vessels cavitated on a day course did not coincide with the change in xylem tension as measured by a pressure bomb: there were fewer vessels cavitated when the bomb balance pressure was greatest. This is at variance with classic cohesion–tension theory. Canny supports a compensating-pressure theory of xylem function (Canny 1995). This proposes that xylem operates at much lower tensions than cohesion–tension theory predicts because a compensatory positive pressure from surrounding cells in, for example, the stele, ray parenchyma or phloem, allows this to be possible. In addition, the marked ability to re-fill cavitated conduits means that water supplies can be maintained by the xylem when tensions exceed the normally low operating values. The compensating-pressure theory has in turn been met with scepticism. Its formulation does not comply with generally accepted models of biophysics (Comstock 1999). Killed stems do not show differences in vulnerability to embolism that would be expected if compensating pressure from nearby living tissues is needed to maintain the flow within the xylem (Stiller & Sperry 1999).

In conclusion, the recent doubt has spurred researchers to produce stronger validation for the cohesion–tension theory.

### Tree hydraulics

Techniques for accurately measuring the rate of sap flow in trees and the internal resistance of the plant to the flow are now available. Using the compensation heat-pulse velocity technique, Becker (1996) found that the daytime sap-flow rate of individual trees was strongly, linearly correlated to crown area, with different species, and different sites of different forest type, fitting the same regression (Fig. 2.10). The regression slope changed between wet and dry seasons. Similarly, Andrade *et al.* (1998) found that tree size was the main determinant of whole-tree water use in Panama, with a 35 m individual of *Anacardium excelsum* estimated to use 379 litres per day. Sap-flow rates correlate with incident solar radiation, which is the driving force for transpiration (Fig. 2.11).

Stems provide a water store that can be tapped for evapotranspiration when water demand is high. The size of the water store rises exponentially with tree size and the volume of water used from the stem store can represent 9–15% of the total daily water use (Goldstein *et al.* 1998). The stem is recharged with water during periods of low transpirational water demand.

Tropical tree species show a wide range in the values for the different measures of hydraulic conductivity that can be estimated (Tyree & Ewers 1996). In general, tropical trees have average to high values for stem conductivity in comparison with temperate trees and conifers. Fast-growing, soft-wooded species typically have the highest leaf-specific conductivity values, that is, stem conductivity per unit leaf area distal to the measured segment. Therefore they should be able to deliver large volumes of water quickly to each leaf and maintain high rates of evapotranspiration at relatively low water potential gradients. This is achieved not by having a high conductivity per unit sapwood cross section, but by having a high proportion of sapwood in the branch cross section.

The branch-specific leaf area to sapwood cross-sectional area ratio (LA/SA) appears to be an important determinant of branch transpiration rate. Individuals of various sizes of four different species growing in Panama followed essentially the same relationship between conductance and evaporative demand when scaled by LA/SA (Meinzer *et al.* 1997). In other words, transpiration rate per unit leaf area was identical when LA/SA was allowed for, and thus LA/SA was more influential on transpiration than stomatal conductance in many instances because of decoupling through low boundary

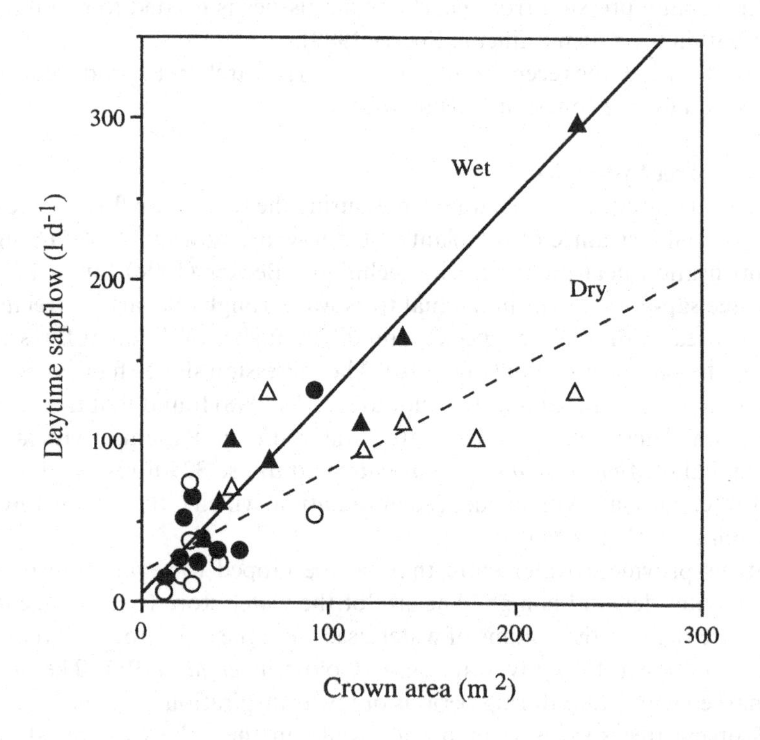


Figure 2.10 Relation between mean daily daytime sap flow and projected crown area for dipterocarp-forest trees (triangles) and heath-forest trees (circles) during wet (solid) and dry (open) periods in Brunei. The reduced major axis regression slopes of the wet and dry periods were significantly different. After Becker (1996).

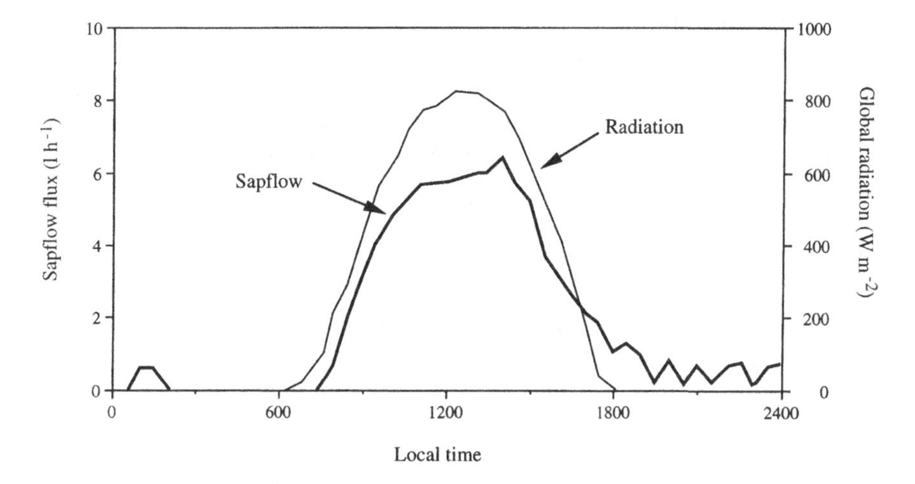


Figure 2.11 Daily course of short-wave radiation and sap flow in a *Neonauclea* tree for a sunny day during the dry season on Rakata (Krakatau), Indonesia. After Bruijnzeel & Proctor (1995).

layer conductance, particularly in still-air conditions.

Tropical rain-forest species appear to be highly vulnerable to xylem embolism, particularly in comparison with species from seasonally droughted forests (Tyree *et al.* 1998a). A comparison of species from mixed dipterocarp forest and heath forest from Brunei failed to show any significant difference in vulnerability to embolism between the sites (Tyree *et al.* 1998a). The freely draining sandy soils of the heath forest might have been expected to select for species with less vulnerable xylem than the dipterocarp forest.

The hydraulic conductance of leaves of tropical tree species has been found to span a similar range to that shown by temperate species (Tyree *et al.* 1999). The measured values are possibly low enough to indicate that, in some species, the leaf provides the largest resistance to water movement in the soil–plant continuum. Specific leaf area showed no obvious relationship to hydraulic conductance.

Using estimates of xylem sap-flow rates and nutrient concentrations, Barker & Becker (1995) were able to calculate the delivery rate of nutrients to the crowns of *Dryobalanops aromatica* trees in Brunei. The xylem sap concentrations of important elements were found to be such that N > K > Ca > P. The nutrient concentrations were not related to tree size, and often showed a positive correlation with sap-flow rate. Delivery rates per tree peaked at about 40 mmol $h^{-1}$ for N, 30 mmol $h^{-1}$ for K, 5 mmol $h^{-1}$ for Ca and 3 mmol $h^{-1}$ for P.

### Tree architecture

The form of construction of the above-ground parts of a tree – the general pattern in space of stems and branches – is governed by two factors: the genetic *bauplan* of shoot growth and branching and the fate of those shoots through their interaction with the environment. Hallé & Oldeman (1970, 1975) proposed a series of 23 architectural models for tree form, each named after a botanist who has contributed to the study of plant morphology. This system is a typology of the genetically controlled component of aerial axis organisation based on character states for number of main axes, branching, presence of resting periods in development, relative apical dominance and positions of inflorescences. Recently, Robinson (1996) has used symbolic logic to represent and analyse the Hallé & Oldeman system. He inferred the possible existence of several, as yet, unreported models.

Many trees in the forest do not appear to conform to one of the 23 basic models, even when allowance is made for loss of branches due to damage or disease. This failure to conform is often due to a process that Hallé & Oldeman called reiteration (Hallé *et al.* 1978). One or more meristems, instead of carrying on with the original model in the appropriate place, begins again, sometimes with a completely different model altogether. Reiteration appears to be an important process in the development of tree form. It allows the tree greater architectural flexibility.

Some trees never reiterate but conform to a single model throughout their lives. For large trees, Hallé (1986) referred to this as gigantism and pointed out three groups where this occurs quite commonly. These are conifers (e.g. *Araucaria* spp.), angiosperm families generally regarded as relatively primitive (e.g. Myristicaceae) and fast-growing, soft-wooded tropical trees such as *Cecropia* species. All the groups use simple, sparsely branched crowns to fill space with foliage. In the conifers, long-lived leaves persisting far back on the branches make them efficient light-harvesting units. In the fast-growing species, the huge leaves at the ends of the branches produce a crown with an outer covering of a layer or two of leaves that is supported with relatively low investment in wood. These tropical parasol trees can be considered neotenic: reproducing in their 'juvenile' architectural phase.

The architectural models are a useful tool in the description of tropical trees and have a clear phylogenetic component (Keller 1994), but as yet relatively little progress has been made in relating particular models with particular ecological roles (Bongers & Sterck 1998). This may be because the characters used to define the models are not precise enough to mirror any ecologically meaningful relationships of crown form. Two species of the same architectural model can look very different in the forest, and alternatively two species of different models can appear very similar.

This does not mean that tree morphology is not relevant to ecology. Branching patterns affect leaf display, efficiency of mechanical support and supply of water to foliage. All these factors are likely to act as selective pressures in the evolution of crown form. Comparison of two species of treelet in Peruvian lowland rain forest indicated the importance of crown form (Terborgh & Mathews 1999). *Neea chlorantha* was found in sites with more direct overhead illumination on average than *Rinorea viridifolia*, generally small gaps in the canopy. *Neea* was more highly branched with a shell of drooping leaves on the outside of the crown. *Rinorea* had whorled branches in tiers with planar foliage arrays. The latter design appeared more efficient at intercepting the larger amounts of lateral light available in forest understorey sites.

### Allometry

The change in relative dimensions of an organism as it grows, or across a range of organisms of different size, can reflect a range of constraints involved with growth. These can be physical where scale-dependent forces necessitate proportionally less, or more, investment in certain organs if the organism is to maintain a uniform likelihood of damage or destruction from the force concerned. Alternatively, the size-related investment in different parts may reflect scale-dependent evolutionary pressures to optimise performance at each growth stage. Allometric studies are also useful as empirical tools to gain insight into the confidence with which certain measures can be used as proxies for others. For example, many individual and stand characteristics are often estimated from tree bole diameter measurements.

The most frequently studied allometric relationship of trees is that between stem diameter ($D$) and height ($H$). A major goal has been to assess the magnitude of the allometric constant $a$ where $D \propto H^a$, and to ascertain its nearness to predicted values based on engineering or other principles. A linear relation ($a = 1$) between two allometric variables is referred to as geometric similarity. A free-standing column of uniform taper and material composition requires its diameter to increase faster than the height in order to maintain a constant mechanical stability. Engineering predicts $a = 1.5$ for this elastic similarity condition. Studies of the height–diameter relation for tropical trees over a range of heights, either for individual species or for multi-species samples, show that elastic similarity is probably a reasonable approximation for the height–diameter relation (Rich *et al.* 1986; King 1991a; O'Brien *et al.* 1995), as shown in Figure 2.12. This implies that young trees have relatively more slender trunks than large old ones. (This is a fact of which most people are consciously unaware yet use subconsciously in determining the relative height of trees when no scale object is available to view.)

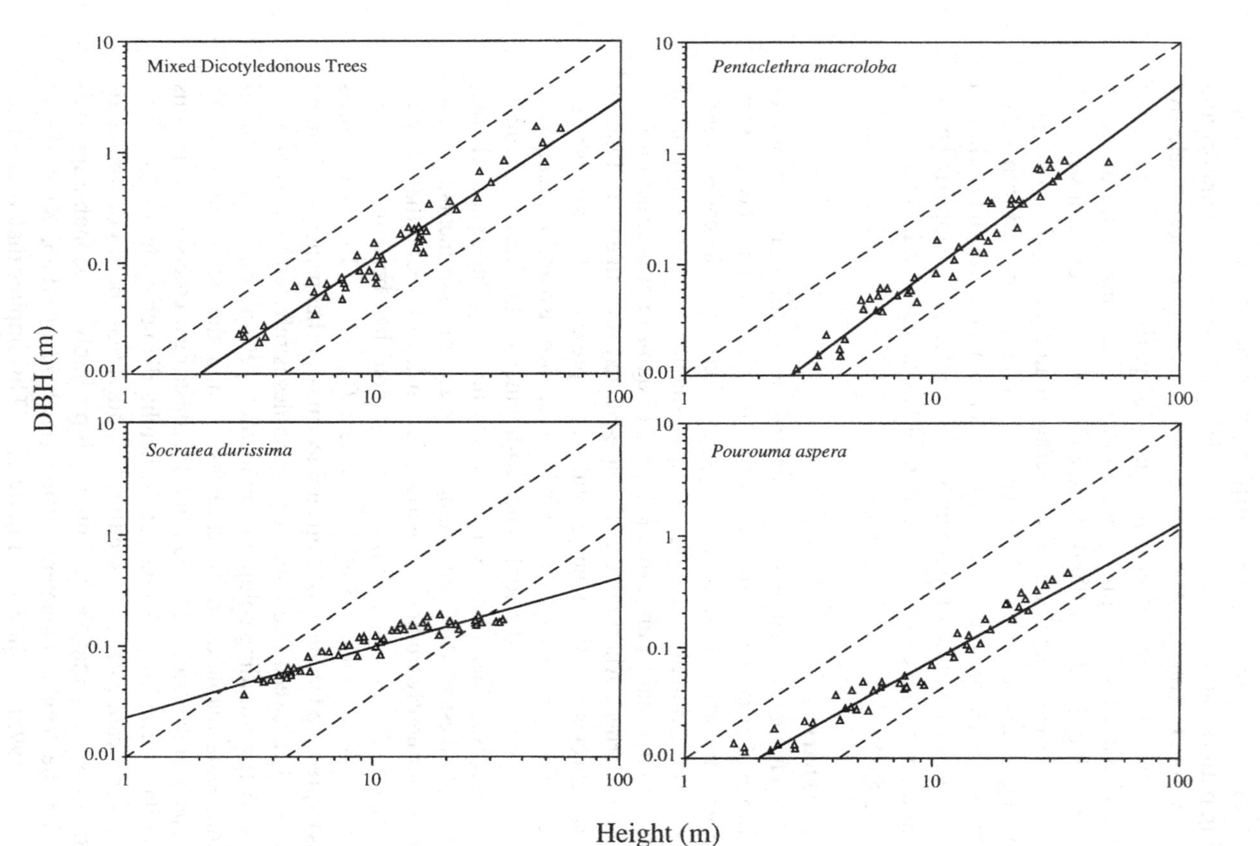


Figure 2.12 Allometry of stem diameter (dbh) and height for mixed dicotyledonous trees, the canopy dominant tree *Pentaclethra macroloba*, the gap-dependent tree *Pourouma aspera* and the arborescent palm *Socratea durissima* in tropical rain forest of Costa Rica. *P. macroloba* comprises 20% of the individuals. The solid line is a linear regression. The upper dashed line is the allometric curve for record-size North American trees and the lower dashed line a theoretical buckling limit for an 'average' tree, beyond which the tree will

One group of tropical trees that deviates considerably from the allometric relationship of the rest are palms. Rich *et al.* (1986) have shown that palm juveniles have much wider stems than dicot trees of the same height, but as mature trees the palm trunk diameters are similar, if not smaller, than those of dicots (Fig. 2.12). This pattern reflects anatomical constraints on the part of palms in thickening stems. The absence of a peripheral cambium means that palm stems have to be produced at almost their maximal diameter when they are formed beneath the apical meristem.

Even among dicot trees there is considerable variation in the height–diameter allometric relationship (O'Brien *et al.* 1995; King 1996; Bongers & Sterck 1998). Two species from French Guiana, *Vouacapoua americana* and *Dicorynia guianensis*, were found to have a near-linear relation between $H$ and $D$ for trees up to 25 m tall, but $D$ increased sharply thereafter (Bongers & Sterck 1998). Other deviations may be related to relative crown development. For instance, *Tetragastris panamensis* was consistently the shortest tree for a given stem diameter (despite having strong wood) among eight common species studied from Barro Colorado Island (O'Brien *et al.* 1995). This was probably because it had a proportionally larger crown than the rest. Adults of understorey species tended to be wider- and deeper-crowned than individuals of canopy species of the same height, significantly so at heights of 10 m (King 1996). The understorey species therefore tend to have thicker trunks at this size, although the denser wood of such species may mean that the diameter differences are less marked than might be expected (see below). Bongers & Sterck (1998) found that the crowns of *Dicorynia guianensis* and *Vouacapoua americana* tended to become relatively narrower and deeper with increasing tree height. Canopy openness (a measure of the light received by the tree) did not influence crown morphology of *Dicorynia,* but *Vouacapoua* showed crown narrowing and deepening in brighter conditions due to stronger apical dominance in development.

It is possible to predict the diameter at which a wooden column of standard mechanical properties and taper will buckle under its own weight. The stem diameter of a tree of a particular height can be compared with the buckling diameter as a measure of the safety margin in the tree's construction. Such analyses for tropical trees (Rich *et al.* 1986; Claussen & Maycock 1995; Thomas 1996a; King 1996) have found that safety factors are generally greatest for seedlings and saplings and decline thereafter (Fig. 2.13), although they may increase again for very big trees (King 1996; Bongers & Sterck 1998). The trunk diameters of eight species of canopy and emergent tree in the height range 6–24 m were 1.3–2.7 times the minimum critical diameter (King 1996). However, this approach to design safety can be criticised on several grounds. Firstly, such studies have generally assumed uniform wood properties with tree size but this may not be the case, particularly in species with low-density wood (see above). Indeed, Claussen & Maycock (1995) found that soft-wooded, fast-growing species followed a different allometric trajectory from more shade-tolerant hardwoods. The first group had larger diameters than the hardwoods as saplings of the same height, but generally narrower stems as taller trees. The narrower safety margin of the fast-growers may be a risk-taking strategy to increase investment in reproduction in large individuals. Secondly, the minimum diameter is derived for a crownless tree. King (1996) has shown that crown mass does affect the minimum critical diameter, although the allometric constant is 0.2, indicating that the influence is not large.

It is becoming clear that many species are not truly linear on the height–diameter log–log plot. There is a definite tendency to an asymptote in height, as can be seen in Fig. 2.14. As the trees reach very large size their height

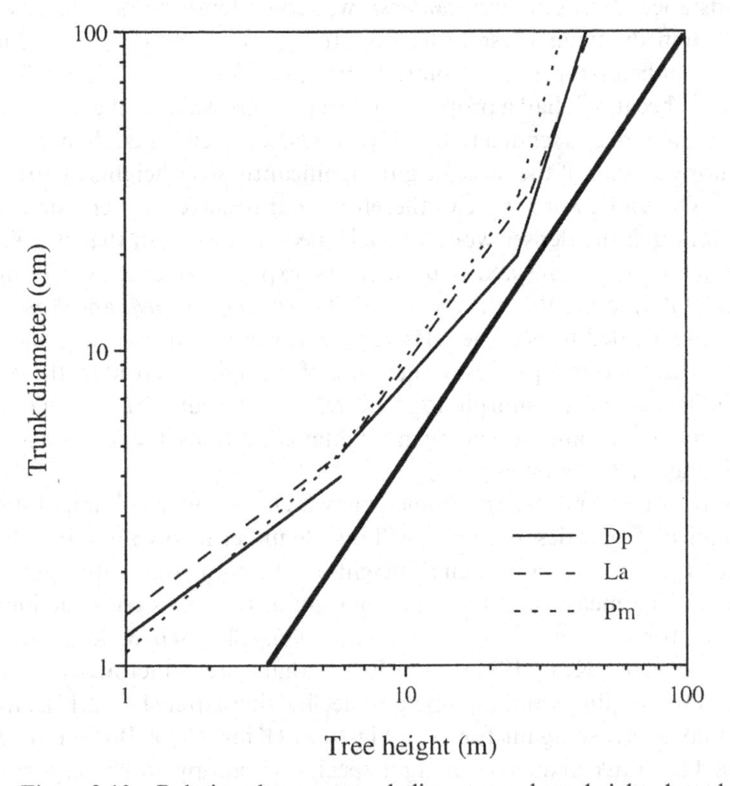


Figure 2.13 Relations between trunk diameter and tree height plotted on a log–log scale. The bold line indicates the minimum diameter required to prevent buckling in wooden columns. After King (1996). Dp, *Dipteryx panamensis*, La, *Lecythis ampla*, Pm, *Pentaclethra macroloba*.

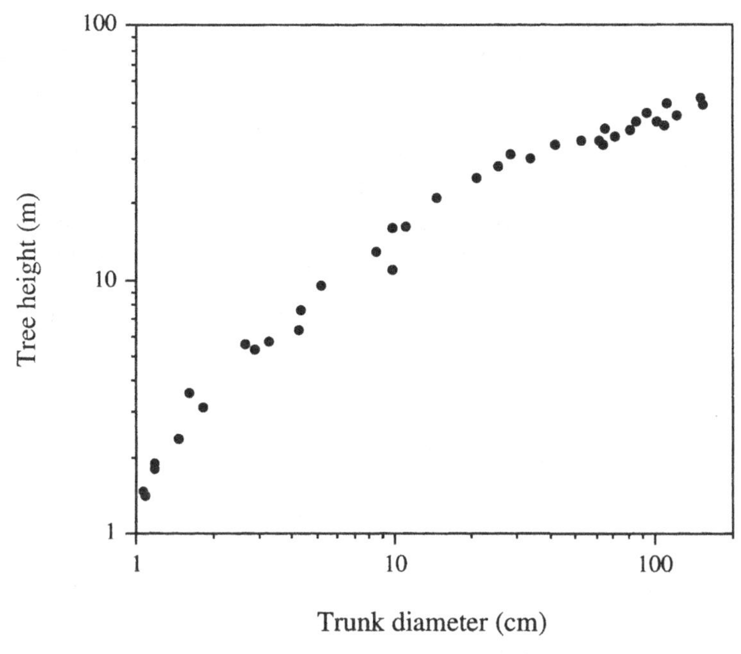


Figure 2.14 Trunk diameter against total tree height for the emergent species *Dipteryx panamensis*. After King (1996).

growth slows but diameter growth continues. There are several possible explanations for this (King 1996). Large trees are more exposed to strong winds and therefore require thicker stems to cope with the extra loading. Large trees also tend to develop wider crowns that would also need more support. Tree species may have a pre-determined maximal height and once this is attained height growth ceases. However, if the tree is to continue to live, it must carry on increasing in stem diameter to form new vascular tissues.

Several authors, e.g. Kohyama & Suzuki (1996) and Thomas (1996a), have pointed out the advantages of asymptotic height ($H'$) over maximal height ($H_{max}$) as a parameter for characterising tropical tree species. These include the probably greater statistical robustness of asymptotic height when sample size is small, and the way in which asymptotic height can disentangle physiological and demographic limitations to growth. Species with high adult mortality may be physiologically capable of reaching large size, but demographic factors mean that they rarely do so. Understorey species typically have lower asymptotic heights than canopy-top species (Kohyama & Suzuki 1996; Thomas 1996a).

Thomas (1996a,b) has studied the allometric relationships of nearly 40 species of tree from six genera in the lowland rain forest at Pasoh in Malaysia.

He found that species with large asymptotic heights showed high rates of diameter growth; in particular, those with *H'* greater than 30 m had considerably greater average diameter increments (Fig. 2.15). There was a negative relation between wood density and *H'* (Fig. 2.15): species of small stature had denser wood on average. On the height–diameter curves, the understorey species had steeper initial gradients than canopy species but their curves flattened sooner, crossing the canopy species lines. There was a strong tendency for a change in the slope of the *H–D* regression that coincided with the typical size at the onset of reproductive activity for the species. It is possible that allocation to reproduction is what reduces the rate of height growth.

### Bark

The successful use of bark characters to identify trees in the forest indicates both the considerable variation among species and the constancy of bark form within a species. However, there have been few studies of the functional significance of this variation in bark form. den Outer (1993) investigated the inner bark anatomy of 463 woody plant species from West Africa but came to no clear conclusions with regard to functionality. Bark thickness has been shown to increase with tree girth (Roth 1981; Hegde *et al.* 1998). Bark protects the sapwood from attack. Guariguata & Gilbert (1996) wounded (drilled holes 7 mm in diameter in) the trunks of seven tree species on Barro Colorado Island, Panama. Interestingly, the fast-growing, soft-wooded *Miconia argentea* showed the slowest wound closure rate.

## Roots

Our detailed knowledge of the tropical rain forest decreases both upwards and downwards from the forest floor. The relative difficulty of accessing the forest canopy means that we are only just beginning to learn of its biology. The subterranean portion of the forest is hidden from view and poorly understood. The excavation of the root systems of large trees is difficult and labour-intensive work.

Tree root systems can be divided into two portions, fine and coarse, based on diameter. Coarse roots are generally woody and provide a mechanical and conductive service to the tree. Water and mineral nutrients are taken up by the fine roots that ramify through the mineral soil and sometimes the litter layer as well. There are a complex set of interactions between tree fine roots and soil micro-organisms, particularly fungi, that can result in symbioses, including mycorrhizas. These play an extremely important role in mineral nutrition and possibly in other areas of tree physiology.

The morphology of tree root systems is complex. Jeník (1978 and

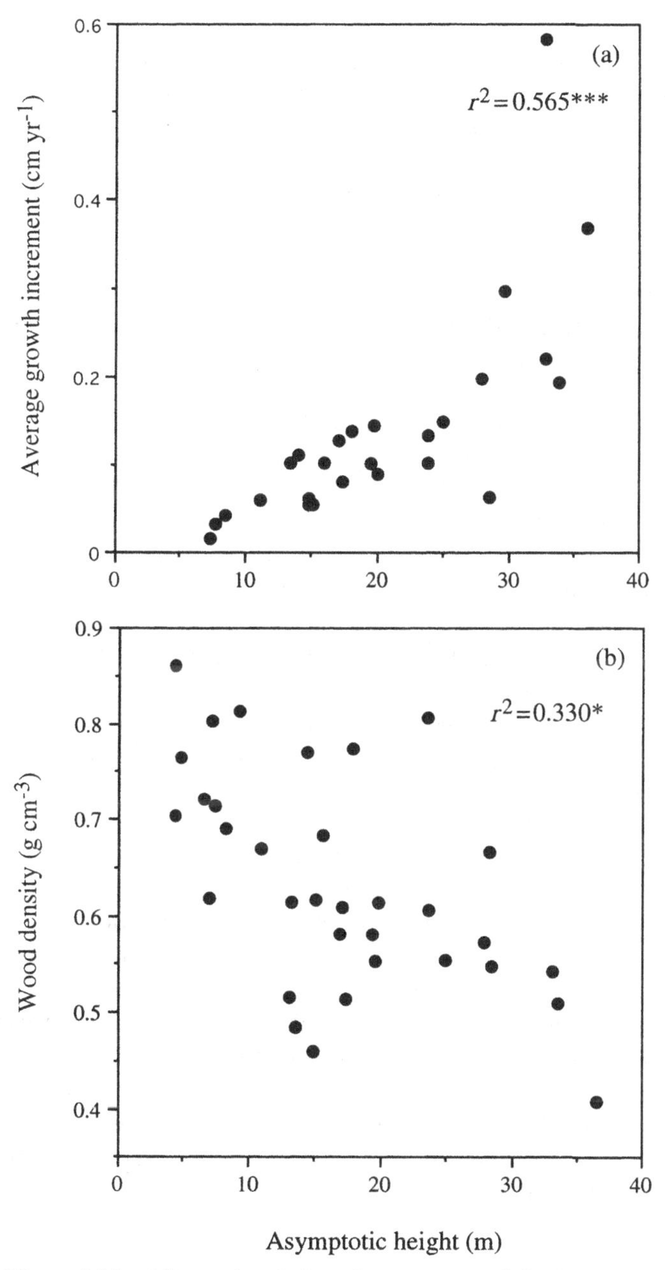


Figure 2.15 Allometric relations between growth increments and species' asymptotic maximal height for adult trees (a) and between their peripheral wood density and asymptotic maximal height (b). After Thomas (1996a).

references therein) has perhaps been the leading student of tropical tree root system form, although most of his papers have been concerned with aerial roots rather than with subterranean ones. Aerial roots are to be found in most tropical forests, and are particularly abundant in forests with periodic or continuous inundation by fresh or salt water, but they are far from common in most well-drained forests, unless buttresses are considered as aerial roots. A preliminary analysis of some tropical species showed there were possibilities that a Hallé-and-Oldeman-type approach might be feasible for classifying root system architecture, and that processes such as reiteration do appear below ground (Atger & Edelin 1994). Hallé (1995) has even argued that tree trunks and branches may be better considered as roots.

Another major means of classifying root systems is by their relative depth. Some trees have much of the root system close to the soil surface, sometimes with major roots running over the top of the soil. Other species have greater development of roots at depth. Tree roots may descend a long way into the soil. Nepstad *et al.* (1994) report roots at depths of up to 18 m in the southern, more seasonal, part of the Amazon Basin in Brazil. These deep roots reach to the permanent water table, allowing rain forest to exist in an area where the dry season is both prolonged and severe, averaging 5 months with less than 250 mm of rain. A considerable portion of the Amazon rain forests probably requires very deep roots to survive.

Becker & Castillo (1990) compared the root systems of three species of shrub or treelet with saplings of three tree species in the forest on Barro Colorado Island, Panama (BCI). They found that the shrub or treelets had shallower root systems (Fig. 2.16), with a greater proportion of the root system in the top 20 cm of the soil and more root area absolutely and per unit leaf area. The authors argued that this reflected a greater requirement for nutrients by the small-statured species because of reproduction. However, the relative scarcity of deep roots in the shrubs may make them more susceptible to drought than the saplings. Using stable hydrogen isotope analysis ($\delta$D) to fingerprint water supply, Jackson *et al.* (1995) found considerable variation in $\delta$D for shrubs on BCI, indicating that contrary to Becker & Castillo's observations, some species were probably deep-rooted. A repeat of the shrub versus sapling comparison for species at Andulau in Brunei (Becker *et al.* 1999a) failed to find a significant difference in dry mass allocation between the two life-form groups because some of the shrub or treelet species were deep-rooted.

### Root hemi-parasites

A relatively few tropical rain-forest tree species, restricted to the order Santalales, produce haustoria from their roots (Fineran 1991). These

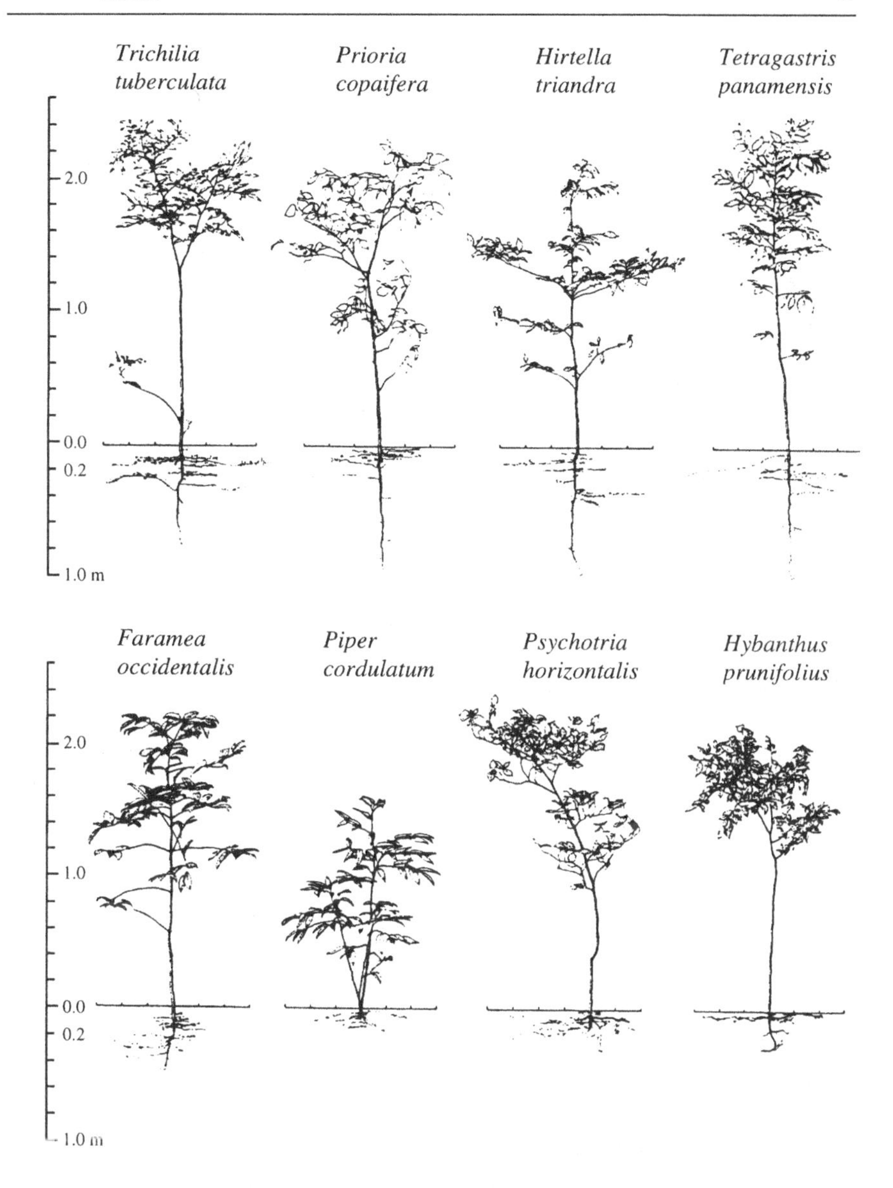


Figure 2.16 Above- and below-ground architecture of saplings of canopy and subcanopy species (upper panel) and a treelet and shrubs (lower panel). After Becker & Castillo (1990). Copyrighted 1990 by the Association for Tropical Biology, P.O. Box 1897, Lawrence KS 66044-8897. Reprinted by permission.

Table 2.2. *Multi-species surveys of mycorrhizal types in tropical rain forests*

| Site (study) | No. of species | Percentage of species with VAM infection | Percentage of species with EM infection |
|---|---|---|---|
| Java (Janse 1896) | 46 | 100 | 0 |
| Sri Lanka (de Alwis & Abeynayake 1980) | 63 | 86 | 8 |
| Nigeria (Redhead 1968) | 51 | 86 | 6 |
| Puerto Rico (Lodge 1987) | 48 | 96 | 6 |
| Trinidad (Johnston 1949) | 30 | 100 | 0 |
| Venezuela (St. John & Uhl 1983) | 22 | 86 | 14 |
| Venezuela: heath forest (Moyersoen 1993) | 42 | 100 | 12 |
| French Guiana (Béreau *et al.* 1997) | 73[a] | 100 | 4 |
| Brazil (St. John 1980) | 83 | 67 | 3 |

[a]Excluding two species only recorded from seasonally flooded coastal savannah.

tap into the xylem of other roots in the soil. It is likely that the hemi-parasites extract water and other contents (e.g. mineral nutrients) of the host xylem but there have been no detailed physiological studies of the parasitic process in arborescent species of the tropical rain forest. Self-parasitism (haustorial connections to other roots of the same plant) is readily observed in such species that occur in the families Santalaceae, Olacaceae and Opiliaceae. They are mostly shrubby, often scandent, but some, such as *Scleropyrum*, are quite large trees.

### Mineral nutrition

It is dangerous to generalise, but typically tropical rain forests grow on strongly acidic soils, low in concentrations of available nutrients required by the trees. The low pH and lack of buffering from other cations often mean that free aluminium ions ($Al^{3+}$) are abundantly present in the soil water at concentrations that would be toxic to many plants from higher latitudes. Tropical trees need to be tolerant of high free aluminium concentrations, although there has been relatively little research on this in wild species. On sites with infertile mineral soil, it is often found that tree fine roots are abundant close to the mineral soil surface, or even found on top of the mineral soil. The nutrient inputs to the soil from the decomposition of the forest litter are likely to be large in comparison to the amounts being released by the mineral soil. The tree fine roots may therefore proliferate where the important nutrients are being released by decomposition. However, toxicity due to free aluminium or free protons may also deter root growth in the mineral soil.

Table 2.3. *Features of the symbioses between microbes and the roots of tropical trees*

| | Vesicular–arbuscular mycorrhiza (VAM) | Ectomycorrhiza (EM) | Rhizobial nodulation | Actinorhizal association | Coralloid roots |
|---|---|---|---|---|---|
| **Taxa** | | | | | |
| Microbes involved | fungi<br>Zygomycotina: Endogonales, Glomales | fungi<br>Basidiomycotina (many spp.)<br>Ascomycotina (many spp.) | bacteria<br>*Rhizobium sensu lato* | bacteria<br>Actinomycete: *Frankia* | bacteria<br>Blue-green algae<br>*Nostoc*, *Anabaena*, ?*Calothrix* |
| Tropical trees involved | majority of species, wide taxonomic range | relatively few including Dipterocarpaceae, Fagaceae, Caesalpinioideae, Myrtaceae – Leptospermoideae, *Uapaca*, *Neea* | Leguminosae, *Parasponia* (Ulmaceae) | few species, Myricaceae, Casuarinaceae | Cycadopsida |
| **Morphology** | | | | | |
| Features | vesicles<br>hyphal coils<br>arbuscules | Hartig net<br>sheath | nodules with peripheral vasculature (except *Parasponia*) | nodules with central vasculature | coralloid, apogeotropic roots<br>green band in cross section |
| **Functionality** | | | | | |
| P acquisition | important | important | — | — | — |
| N acquisition | marginal | important | important | important | important |
| Water uptake | marginal | ?important | — | — | — |
| Protection from pathogens | ?important | ?important | — | — | — |
| Others | ?micronutrient uptake | storage of nutrients<br>?protection from toxic ions<br>?carbon transfer between plants | — | — | — |

In part, after Fitter & Moyersoen (1996).

### Mycorrhizas

A mycorrhiza is a sustainable, non-pathogenic, biotrophic interaction between a fungus and a root (Fitter & Moyersoen 1996). Most tropical tree species habitually possess vesicular–arbuscular mycorrhizas (VAM) (Alexander 1989), as can be seen from a survey of reports on mycorrhizal infection in different tropical rain-forest sites (Table 2.2). VAM involve primitive fungi from the Zygomycotina orders Endogonales and Glomales (Smith & Read 1997) (Table 2.3). There are probably about 200 species of VAM fungi, although as they are primitive and without sexual stages there are few characters available for detailed systematic analysis (Allen *et al.* 1995). VAM fungi are cosmopolitan and seem to have a wide range of hosts. In tropical roots, VAM fungi rarely form arbuscules, but coils of hyphae are common in host cells (Alexander 1989).

Ectomycorrhizas (EM) involve higher fungi from the Basidiomycotina or Ascomycotina (Table 2.3). There are probably over 5000 species of EM fungi world-wide (Allen *et al.* 1995), including both epigeous (above-) and hypogeous (below-ground-fruiting) species. The EM fungi produce a sheath or mantle of hyphae over the surface of the infected root, and a network of hyphae running between cells within the epidermis of the root called the Hartig net. Both VAM and EM can have extensive mycelia in the soil, although those of EM are generally better developed. Relatively few tropical tree species have EM. Notable groups that do, include all Dipterocarpaceae and Fagaceae, many Leguminosae subfamily Caesalpinioideae and Myrtaceae subfamily Leptospermoideae. EM fungi vary in host specificity, with some fungi apparently being very limited in range whereas others infect many different species of host. In general, EM fungi are more vagile than VAM species because of aerial spore release and much smaller spore size. It is interesting to note the distinction between the relatively few species of VAM fungi forming symbioses with many species of vascular plant and the high diversity of EM fungi restricted to a much smaller number of plant hosts. Plants can be infected with VAM and EM simultaneously.

The transfer of fixed carbon from the root to the fungus is a common feature of VAM and EM, although the latter are probably more energy demanding. In return, the plant probably gains a number of benefits from the fungus; the best studied of these is a transfer of phosphorus. Lowland tropical soils frequently have very low concentrations of available phosphorus and it appears that mycorrhizas improve the supply of P to the root or are cheaper ways of obtaining it. There are several potential mechanisms for increased rates of P uptake by mycorrhizal plants (Bolan 1991). These include:

1. The exploration of greater soil volumes by the mycelia of mycorrhizal fungi.
2. The faster movement of P into fungal hyphae than roots.
3. The solubilisation of soil P not available to roots.

P is not only scarce in solution in most tropical soils, it is also relatively immobile, diffusing effectively over very short distances. Fungal hyphae can ramify through large volumes of soil and in this way probably are more efficient than roots at scavenging for P in the bulk soil because they reduce the distance the P has to diffuse before being absorbed and provide a large surface area for absorption. Faster P movement through the soil could be achieved by fungi if they have a greater affinity for P ions than roots and can take up P at a lower threshold concentration than roots. There is however, little evidence for this. Mycorrhizal fungi may be able to solubilise P by excreting organic acids and phosphatases. Unequivocal evidence that mycorrhizal fungi do actually contribute plant-unavailable P to host plants is still being sought.

Mycorrhizas may also be able to assist the host plant by transferring nutrients other than P. Mycorrhizal fungi are often associated with the litter layer and so-called direct cycling has been proposed as an advantage of possessing mycorrhizal symbioses. Direct cycling is the transfer of nutrients from the decaying litter to the root by the action of the mycorrhizal fungus breaking down the organic matter. It is difficult to demonstrate conclusively that nutrient transfer by a mycorrhiza came only from its own breakdown of organic matter, when there will be many other micro-organisms in the soil. However, one interesting, largely anecdotal, observation is that in forest areas dominated by EM tree species there is a greater accumulation of forest-floor litter (Alexander 1989). This has been attributed to the EM fungi outcompeting the decay fungi for important nutrients, making it difficult for them to break down the rest of the litter (Högberg 1986). Mycorrhizas may also assist in water uptake, tolerance of the adverse conditions in the soil, and resistance to root pathogens (Table 2.3).

EM have the advantage over VAM in that the fungi generally penetrate into the soil even further and are more long-lived. In fact, more than one plant can be infected by the same EM mycelium and the fungus may become a conduit for carbon transfer between plants. It has been demonstrated experimentally that EM linkages allowed considerable transfer of fixed carbon between seedlings and mature trees in a Canadian forest (Simard *et al.* 1997). This raises the possibility that deeply shaded seedlings might be able to survive by tapping into the EM network of surrounding trees. EM, because of their greater bulk, also have the ability to store absorbed nutrients which may

Table 2.4. *Examples of lowland tropical rain forests on moderately to freely draining soils dominated by a single species that regenerates* in situ

| Location | Dominant species (family/subfamily) | Dominance | Dispersal | Mycorrhizas |
|---|---|---|---|---|
| Malay Peninsula, Sumatra | *Dryobalanops aromatica* (Dipterocarpaceae) | 60–90% of timber trees | large winged fruit | EM |
| Borneo, Sumatra | *Eusideroxylon zwageri* (Lauraceae) | pure stands | large, heavy seeds | ? |
| India | *Poeciloneuron pauciflorum* (Guttiferae) | almost pure stands | ? | ? |
| Uganda | *Cynometra alexandri* (Caesalpinioideae) | > 75% of canopy | poor dispersal | VAM (Torti & Coley 1999) |
| Central Africa | *Gilbertiodendron dewevrei* (Caesalpinioideae) | > 90% of canopy | very large seeds with low dispersal distance | EM with some VAM infection (Torti & Coley 1999) |
| Central Africa | *Julbernardia seretii* (Caesalpinioideae) | > 90% of canopy | large seeds | EM with some VAM infection (Torti & Coley 1999) |
| Trinidad, Guyana | *Mora excelsa* (Caesalpinioideae) | > 80% of canopy | very large seeds | VAM (Torti *et al.* 1997) |
| Costa Rica | *Pentaclethra macroloba* (Mimosoideae) | > 50% of canopy | large seeds | ? |

Based on the works of Hart *et al.* (1989) and Connell & Lowman (1989).

make them advantageous in areas with seasonality of nutrient availability (Högberg 1986). A greater drain of fixed carbon is the price of EM compared with VAM, and it appears in general there is little to decide between the two in terms of relative merits.

In most tropical rain forests the general background is VAM species with a minority forming EM. Two groups (Hart *et al.* 1989; Connell & Lowman 1989) working independently observed a correlation between low-diversity tropical lowland forests (Table 2.4) and the reported possession of EM. These forests were generally dominated by a single species of caesalpinoid legume, e.g. *Gilbertiodendron dewevrei* in the Congo Basin and *Mora excelsa* on Trinidad and in Guyana. Both of these species can have stands exceeding 100 $km^2$, where more than 50% of the canopy trees and 80% of the total basal area are made up of the one species. These monodominant forests are probably not just early stages in succession. They generally occur on soils low in nutrients, although not extremely so. Does the possession of EM make a species competitively superior over VAM species on low-nutrient sites? There does not seem to be any very strong evidence for this. Investigations of several of the monodominant species could not find EM, but VAM were abundant (Torti *et al.* 1997; Torti & Coley 1999) (Table 2.4). Moyersoen (1993) found no consistent increase in the frequency of EM species down a soil fertility gradient in Venezuela; nor did Béreau *et al.* (1997) in French Guiana. Dipterocarpaceae, the most obligate of EM tropical trees, are represented by many species in rain forests across a range of soil fertilities in Sri Lanka and West Malesia.

In a very detailed study of the P relationships of forest trees in Cameroon, Newbery *et al.* (1997) have concluded that EM may allow species of caesalpinoid legume (*Microberlinia bisulcata*, *Tetraberlinia bifoliolata* and *T. moreliana*) to persist in certain sites in a landscape generally too poor in P for them. The EM-tree-dominated groves have much higher total P concentrations (309 compared with 186 $\mu g\, g^{-1}$) in the soil than the matrix forest, which Newbery *et al.* (1997) hypothesised was due to the more efficient P cycling by the EM trees. The EM also supposedly allowed the trees to mast fruit by buffering their P requirements through the P stored in the fungal hyphae. It is notable that many of the EM monodominants are masting species, and it may be that masting and the production of large seeds with limited powers of dispersal are what allow the species to dominate the landscape patches they occupy.

In conclusion, the possession of EM may be involved in the dominance of certain tropical forests by one or a few species, but more concrete evidence is needed before it can be accepted that EM are competitively superior to VAM. The fact that VAM and EM are mixed together in most tropical forests, and may simultaneously infect the same tree, indicates that dominance of one over the other is unlikely.

**Nitrogen fixation**

A symbiotic relationship that allows an increased availability of N to the plant will be beneficial. However, it may be simplistic to assume that mechanisms of supplementing nitrogen income are an evolutionary response to low nitrogen availability. McKey (1994) has argued that nitrogen fixation evolved in legumes as a mechanism to maintain a high-N internal economy, not as a means of facilitating invasion of low-N sites in the landscape.

Nitrogen-related symbioses of plants generally work by harnessing microorganisms with the ability to fix atmospheric nitrogen. The commonest such relationship is the legume nodule containing *Rhizobium*. Some plants are actinorhizal; that is, they have a symbiotic relationship with an Actinomycete for nitrogen fixation. Some cycads have cyanobacteria associated with their roots that can reduce nitrogen from the air.

Legumes are by far the commonest nitrogen-fixing tropical trees, but it is a mistake to assume that all legumes fix nitrogen. Nodulation is rare in the Caesalpinioideae (23% of species investigated according to Sprent (1994)) and commonest among the Papilionoideae (97%). We still lack strong evidence of significant rates of $N_2$-fixation in tropical woody legumes in natural ecosystems, although it is available from agroforestry systems. It is possible that the low P-availability, acidity and abundant $Al^{3+}$ make leguminous $N_2$-fixation difficult in the tropics. Stable nitrogen isotope studies do, however, indicate active nitrogen fixation in tropical tree legumes, species of *Casuarina* and cycads (Yoneyama *et al.* 1993). In legumes, the nitrogen is fixed by bacteria formerly all placed in the genus *Rhizobium*. This has now been split into several smaller genera. *Parasponia* (Ulmaceae), a genus of small trees from New Guinea and Australia, is the only non-legume known to be naturally nodulated with rhizobia. Interestingly, the structure of the *Parasponia* nodule is more similar to the *Frankia* nodule of actinorhizal plants than to the legume nodule (Soltis *et al.* 1995).

*Frankia* is an actinomycete: a filamentous prokaryote. Strains, rather than species, of *Frankia* have been recognised (Baker & Schwintzer 1990). Among vascular plants world-wide, 25 genera in eight families are known to be actinorhizal (Moiroud 1996; Huss-Danell 1997). These include *Alnus* (Betulaceae), *Allocasuarina*, *Casuarina*, *Ceuthostoma*, *Gymnostoma* (all Casuarinaceae), *Coriaria* (Coriariaceae), *Elaeagnus* (Elaeagnaceae) and *Myrica* (Myricaceae) that are found in the humid tropics, of which only *Myrica* and the Casuarinaceae are represented as trees in the lowlands, where they are restricted to the Indo-Pacific region. Thus actinorhizal species are rarely found in lowland tropical rain forest, and do not play a major role in its nutrient economy.

Cycads often have cyanobacteria (blue-green algae) associated with their coralloid roots (Johnson & Wilson 1990; Jones 1993). These are apogeotropic (gravity-insensitive) roots usually located near the soil surface. The bacteria (from the genera *Nostoc*, *Anabaena* or *Calothrix*) fix nitrogen that may be made available to the cycad plant. Cycads are found throughout the tropics and are common in some lowland rain forests.

It has generally been believed that, with the exception of the legume–rhizobia relationship, vascular plant–micro-organism symbioses are somewhat scattered through the higher-plant phylogeny with the implication that symbioses evolved many times. However, recent use of DNA sequences to generate phylogenetic trees has shown a relatively narrow phylogenetic base for the actinorhizal symbiosis (Soltis *et al.* 1995) and also possibly the ectomycorrhizal symbiosis (Fitter & Moyersoen 1996). Both symbioses and the legumes are centred within the rosid clade, a fairly discrete group of the higher eudicots (see Fig. 1.1). This has led to the speculation that the 'gene for advanced symbiosis' had a single evolutionary origin in the angiosperms, with of course another EM origin among the gymnosperms. The scattered presence of EM in groups of angiosperms outside the rosid line needs explaining under this hypothesis, however.

### Other methods of obtaining extra nutrients

There are many free-living micro-organisms capable of fixing atmospheric nitrogen. Free-living algae or lichens may grow as tiny epiphylls on the leaves of trees. It has been shown that nitrogen fixed by epiphylls can be transferred to the host plant (Bentley & Carpenter 1984). However, epiphylls shade the underlying mesophyll cells of the host leaf, making it more difficult for them to photosynthesise rapidly. Therefore, it is unlikely that plants will evolve leaf forms that encourage growth of nitrogen-fixing epiphylls.

Certain species of tropical Rubiaceae and Myrsinaceae characteristically possess bacteria-filled nodules in their leaves. It has been hypothesised that these might act in a nitrogen-fixing capacity, but there is little evidence to support this conjecture (Miller 1990). See page 82 for more details on the bacterial leaf-nodule symbiosis.

Plants that are habitually inhabited by ants may absorb nutrients from the waste of the ant colony. However, examples of so-called myrmecotrophy are very largely confined to epiphytic species (Beattie 1989). There is one report of uptake of nutrients from domatia of a tropical tree. Nickol (1993) found that the leaf domatia of *Tococa guianensis* took up $^{14}$C-labelled glutamine, whereas the leaf surface did not. The ants on the *Tococa* plants appear to be sustained entirely from honeydew secretions of scale insects in the leaf domatia and by eating glands on the leaf surface. Therefore, any absorption of nutrients by the leaf domatia will very largely be recycling of material originating from the plant. Labelling experiments failed to demonstrate any uptake by stem domatia of *Macaranga* (Fiala *et al.* 1991).

In forests on poor soils, the litter falling to the forest floor represents a major flux of mineral nutrients. Plants in such sites that can obtain sole access to the nutrients released by decomposition from an amount of litter may be at an advantage. This has been the argument put forward to provide a functional explanation of litter trapping by plants (Raich 1983). A basket arrangement of leaves or branches will catch falling leaves where they can decompose. *Agrostistachys* is a genus of understorey treelets that often has the litter-catching form. It can develop roots to absorb nutrients from the decomposition of 'arrested' litter in the nest of leaves on the top of the stem. Canopy trees may also develop adventitious roots to exploit nutrients in the debris under epiphytes (Nadkarni 1981). The presence of damp epiphytes and humus or sponges soaked in nutrient solution stimulated the development of adventitious roots from the branches of the neotropical montane tree *Senecio cooperi* (Nadkarni 1994).

Tropical trees can also develop adventitious roots into debris-filled cavities within their own hollow trunks (Dickinson & Tanner 1978; de Foresta & Kahn 1984). Janzen (1976a) conjectured that being hollow might be an advantage to big, old trees if it allows some recycling of nutrients locked away in the heartwood and exploitation of waste products from roosting or nesting vertebrates.

Bruijnzeel (1989) hypothesised that the multi-stemmed form seen quite commonly in species from heath and upper montane forests might have an advantage in nutrient capture. Stemflow is considerably enriched in nutrients and the multi-stemmed form may funnel more stemflow to the tree base than a single stem. There has been no experimental demonstration that this is the case.

### Comparative use of nutrients

We know very little about the comparative nutrient demand or ability to conserve nutrients of different tropical tree species. Most studies of nutrients and their cycling in tropical forests have concerned multi-species stands. This allows us to draw some preliminary conclusions about differences among species of contrasting forest types, but tells us little about differences among species within the same community. Comparisons of biomass accumulation and nutrient content in plantation stands growing on the same soil type demonstrate that species differ appreciably in ability to take up and use nutrients (see, for example, Montagnini & Sancho 1990; Wang *et al.* 1991; Stanley & Montagnini 1999).

A question of considerable interest is which mineral nutrient is in shortest supply to the trees and hence limits growth if light and water are available in adequate amounts. Several approaches have been employed in an attempt to answer this question:

1. *Bioassays.* These generally involve growing a test plant in pots of forest soil under a variety of nutrient addition treatments. The test plant may be herbaceous (often a crop plant), or it may be a woody species. Seedlings (or cuttings) have to be the subjects of such studies for logistic reasons. These will be referred to in more detail in a later chapter. Suffice it to say here that although bioassay experiments are relatively easy to perform, they have many problems in interpretation. Soil removal from the forest may affect its nutrient availability. Nitrogen in particular will be influenced by the activities of soil micro-organisms, which can be altered considerably by soil disturbance and the increased abundance of dead roots. The test species, particularly crop plants, but also tree seedlings, may behave in a different way from mature trees in the forest. For instance, the likelihood of mycorrhizal infection may be reduced in pots. The degree of infection can influence the outcome of fertilisation experiments. The absence of competing roots from surrounding trees might also be a critical difference.

2. *Litter production and nutrient concentration.* Vitousek (1984) found a tighter correlation between litter production and litter phosphorus (P) concentration than that with litter nitrogen or calcium concentrations across many tropical forest sites. Correcting for climatic variables maintained the correlation with P. Silver (1994) reported that litterfall mass per unit of P contained (inverse of P concentration) was negatively correlated with extractable (mostly montane sites) and total (montane and lowland sites) soil P in a similar study. Again N and Ca were much weaker correlates. These results indicate greater stand-level P-use efficiency at low P availability and imply that P is the limiting nutrient in most tropical forest systems.

3. *Stand fertilisation experiments.* Forests plots can be given treatments of different nutrients and growth can be estimated from litterfall and stem increment data. Positive growth effects to additions of either N or P have been demonstrated for montane forest in Venezuela (Tanner *et al.* 1992) and 140-year-old stands in Hawaii (Raich *et al.* 1996). In an unreplicated treatment application, montane forest in Jamaica showed a similar response (Tanner *et al.* 1990). However, old growth stands at 1220 m in Hawaii did not show a growth improvement after N addition (Vitousek *et al.* 1993). In the only study of the effects of fertilisation on lowland rain-forest stands, N and P

addition, separately and together, increased fine litterfall and its P concentration, but did not have a significant effect on stem diameter growth at Barito Ulu in Borneo (Mirmanto *et al.* 1999).

### Element accumulation

Surveys of the elemental concentrations within tropical trees have shown that some species accumulate various elements to much higher concentrations than others do. For instance, in a survey of the bark concentrations of 15 elements among 113 species in a forest in Sumatra (Masunaga *et al.* 1997), 22 species were found to be accumulators (to be more than two standard deviations above the log mean) of one or more elements. Aluminium was the commonest element to be accumulated. A similar pattern was observed for leaf samples from the same forest (Masunaga *et al.* 1998a), although not all leaf accumulators were bark accumulators and vice versa. At the family level, aluminium accumulation ($>$ 1000 ppm in leaf tissue) is well represented in the tropics (Chenery & Sporne 1976), which is probably not surprising given the high free aluminium concentrations in many lowland rain-forest soils. Aluminium-accumulating shrubs from southeast Asia (*Melastoma malabathricum*, *Urophyllum* spp.) have been shown to have high Al concentrations in their young leaves (Osaki *et al.* 1997; Masunaga *et al.* 1998b), a pattern different from the classic aluminium accumulator, the tea plant (*Camellia sinensis*), which has high Al concentrations in mature leaves only.

## Leaf form and physiology

In this book, I employ a functional definition of leaf. Leaves are essentially the organs of photosynthesis. To a plant morphologist, leaves can be present in many modified, non-photosynthetic roles. However, from a functional viewpoint leaves can include those derived from a modified rachis (phyllodes) or stem segment (cladodes), although among tropical rain-forest trees neither are particularly common.

With extremely few exceptions, tropical rain-forest trees perform C3 photosynthesis (Medina 1996). There are two tree species of *Euphorbia* (*E. forbesii* and *E. rockii*) from the understorey of Hawaiian rain forest that are known to be C4 plants. Some hemi-epiphytic species of *Clusia* from tropical America utilise CAM or CAM–C3 intermediate metabolism, at least while they are truly epiphytic.

Despite the uniformity in the basic biochemistry of photosynthesis among trees in the tropical rain forest, there is generally a large diversity of leaf form represented. Compound leaves are common, including some quite bizarre forms such as those found in members of the Araliaceae. Leaves or leaflets vary in size, shape, texture and in a whole suite of anatomical, chemical and physiological ways. They vary between positions on the same tree at any one time, with age of the tree, with the growth environment of the tree, and among individuals of the same species due to both genetic and environmental factors. There are also characteristic trends in average leaf form among different groups of species within a forest and among forest types along environmental gradients such as those of increasing altitude or decreasing precipitation. Leaf form is one of the important physiognomic characters used to define the different tropical forest formations. The description of this variation is an important activity for plant ecologists, but elucidating why leaf form and performance vary with environment in a manner that is, to some extent, predictable must be the ultimate goal.

An important approach to understanding the relative merits of different leaf designs has been the use of economic analogies in ecological theory. First put forward by Bloom *et al.* (1985), the consideration of resource acquisition in terms of profit, loss and costs of investment and maintenance has proved to be a valuable conceptual tool. Using this approach, Givnish (1987) has identified three main evolutionary areas of constraint on leaf design. These are the constraints to do with gas-exchange, those concerning support and supply and finally those involved with protection of the leaf. The uncertainty ecologists face in understanding the evolutionary optimisation of leaf design is knowing how to measure the profit and loss and the investment. Photosynthetic rates of leaves are usually measured as the net rate of carbon uptake (or oxygen evolution) per unit area or per unit mass (fresh or dry) of leaf. Area-based rates are useful for comparing leaves in their apparent efficiency of light interception. Mass-based rates express photosynthetic efficiency per unit investment in leaf mass. Dry mass is often used as a measure of investment in ecological studies as it probably reflects the amount of photosynthate involved in construction and maintenance. A leaf of low leaf mass per unit area (LMA) may have low area-based photosynthetic rates, but its mass-based rate may be high. However, it may not just be instantaneous rates of return that are important: a relatively low profit margin sustained over a long period can bring in large total profits. Therefore, the length of the active period of a leaf's photosynthesis will also be important in determining its profitability.

Material or photosynthate may not be the correct currency for the profit-and-loss equation. In sites where nutrient supply is limiting growth, the investment of that nutrient in photosynthesis may be more important than dry matter as a whole. For instance, the efficient use of nitrogen or phosphorus for photosynthesis may be important on infertile sites.

Photosynthetic rates in the field are generally measured by using infra-red gas analysis equipment with the carbon dioxide concentration measured in a cuvette containing the leaf, or clamped to it. The cuvette has a forced ventilation system that disrupts the leaf's boundary layer. However, it is possible that in still-air conditions boundary-layer resistance might limit photosynthetic rates. Meinzer *et al.* (1995) found that porometry substantially over-estimated evapotranspiration in small trees and shrubs in tropical forest gaps when compared with measurements made using the heat-pulse technique and potometry of whole plants. Similarly, studies of canopy leaves in Panama found that under conditions of high stomatal conductance and low wind speeds transpiration becomes decoupled from stomatal conductance (Meinzer *et al.* 1997). The implication from these findings is that gas exchange measurements may often over-estimate maximal photosynthetic rates of individual leaves.

**Size, shape and other structural characteristics**

Leaf size has classically been dealt with in descriptive vegetation ecology by reference to leaf size classes. The Danish plant ecologist Raunkiaer developed an arbitrary classification of leaf sizes based upon a geometric series $9^n$ with a base leaf size of 25 $mm^2$ (Raunkiaer 1916, 1934). He found that a series based on 9 produced a system that lent itself more meaningfully to the description of vegetation types across the globe than one based on 10. Each of the six size classes originally recognised was given a name derived from Ancient Greek for leaf (φυλλον, *phyllon*) with a prefix indicating relative size. Webb (1959), while investigating the tropical vegetation of Queensland, Australia, found that many of the forests he was studying fell into the mesophyll-dominated class but there was still a distinction in types. To provide a means of doing so, he divided the mesophyll class into two, recognising the notophyll class (22.5–45 $cm^2$) as the designation for the small mesophylls, with large mesophylls becoming the mesophyll class *sensu* Webb.

The mesophyll *sensu* Raunkiaer leaf size is the commonest among tree species in lowland tropical rain forests (Table 2.5). Usually at least three quarters of species belong to this group. Another feature of tropical lowland forests is the presence of at least a few species that reach the megaphyll class ($> 1640.25$ $cm^2$), even if palms are omitted from the analysis.

Rain-forest tree leaves are typically entire-margined (Bailey & Sinnott 1916; Rollet 1990) with an ovate–lanceolate shape, roughly three times as long as broad (Bongers & Popma 1990a) and frequently with a pronounced acuminate tip. Compound leaves are usually quite common; for example, of 183 species surveyed in Venezuela, 57 (31%) had compound leaves (Rollet 1990).

Table 2.5. *Leaf sizes in various lowland tropical rain forests*

Values given are the percentage of the species sampled in each of the leaf size classes of Raunkiaer.

| Site | lepto | nano | micro | meso | macro | mega |
|---|---|---|---|---|---|---|
| Mexico[1] | 0 | 3 | 9 | 79 | 8 | 1 |
| Ecuador[2] | 0 | 0 | 9 | 64 | 27 | 0 |
| Brazil[3] | 2 | 2 | 13 | 74 | 7 | 0 |
| Brazil[3] | 3 | 1 | 12 | 75 | 9 | 0 |
| Brazil[4] | 1 | 0 | 15 | 78 | 5 | 1 |
| Nigeria[5] | 0 | 0 | 10 | 84 | 6 | 0 |
| Gabon[6] | 2 | 6 | 11 | 80 | 1 | 0 |
| Gabon[6] | 1 | 3 | 8 | 84 | 4 | 0 |
| Gabon[6a] | 0 | 0 | 9 | 82 | 5 | 0 |
| Gabon[6] | 1 | 4 | 8 | 86 | 1 | 0 |
| Philippines[7] | 0 | 0 | 4 | 86 | 10 | 0 |

Sources: [1]Bongers *et al.* (1988); [2]Grubb *et al.* (1963); [3]Cain *et al.* (1956); [4]Mori *et al.* (1983); [5]Richards (1996); [6]Reitsma (1988); [7]Brown (1919).
[a]Data, as given in source, do not sum to 100%.

The relative rarity of toothed leaves and other non-entire margin forms in the tropical rain forest is well enough established for palaeobotanists to use the frequency of toothing in leaf fossil assemblages, together with leaf size, to infer the type of climate under which the vegetation represented by the assemblage grew (Wiemann *et al.* 1998). There are two basic angiosperm leaf venation patterns: craspedodromous, where secondary veins form a pinnate arrangement and run more or less parallel to each other and terminate at, or near, the leaf margin, and brochidodromous, where the secondaries loop and join within the leaf margin (Fig. 2.17). Leaves of temperate deciduous trees are frequently craspedodromous with a toothed margin, but those of tropical rain-forest trees are more likely to be brochidodromous with an entire margin. Roth *et al.* (1995) employed a hydrodynamic model to develop hypotheses concerning the relative merits and constraints of the different leaf designs. The advantage of the craspedodromous pattern over the brochidodromous one is that it uses a shorter length of vein per unit leaf area and therefore would make leaves cheaper to construct if 'plumbing costs' are high and should make leaves more efficient at intercepting light as a greater proportion of the lamina area can be devoted to photosynthetic tissues. The hydrodynamic model also highlighted the limitations of the craspedodromous design. The delivery of water to cells farthest away from the terminal ends of the veins, i.e. areas on the leaf margin between the secondaries, was poor. This probably explains the toothed leaf margins of many craspedodromous leaves. The tissues on the leaf margin that would be vulnerable to dehydration or over-heating because of the low rate of water supply are simply not formed,

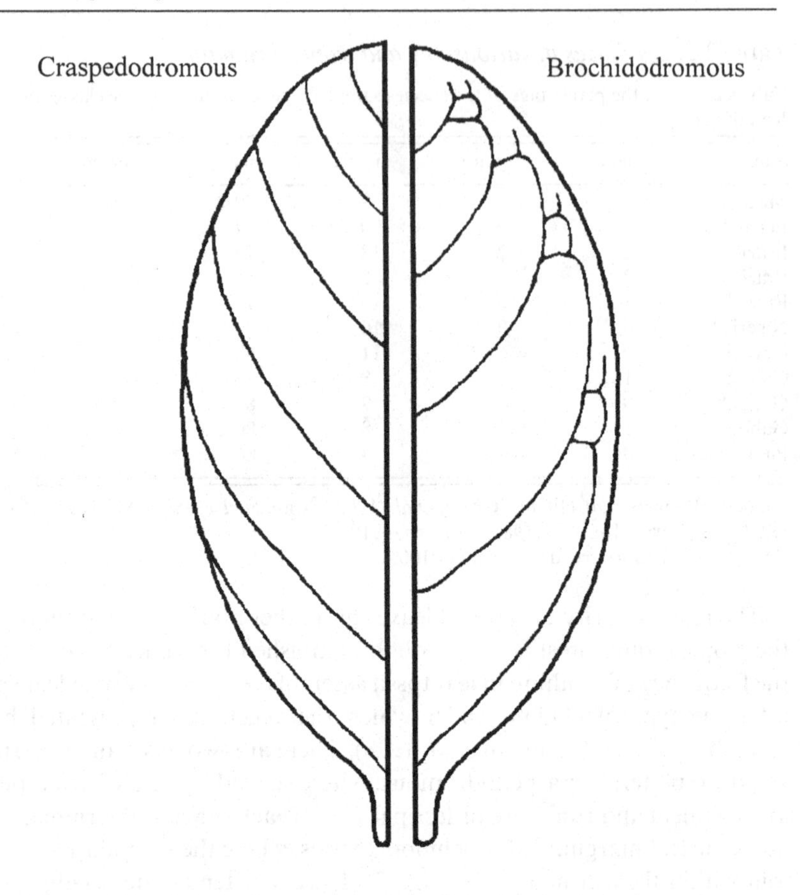


Figure 2.17 Schematic representation of the two basic leaf venation patterns of dicotyledons.

resulting in a toothed edge. Rain-forest leaves generally live longer than do those of temperate deciduous forest (Coley & Barone 1996), so therefore it is probably necessary to invest in a more reliable water-supply system for the mesophyll tissues, represented by brochidodromous venation.

The acumen of tropical leaves is often referred to as a drip-tip because of the tendency of water to accumulate there, forming an ever larger drop until it drips off. There has been considerable speculation about the advantages of having a means of increasing the rate of drainage of water from the surface of a leaf (Richards 1996). These include reduced periods of a water layer obscuring the incoming radiation and leaching minerals, particularly potassium, from the leaf tissues, and the likelihood of water encouraging the growth of epiphylls. A shrubby species of *Piper* in the montane forests of Costa Rica was shown to drain drops of water from leaves at a slower rate when the drip-tip was removed (Lightbody 1985). However, no allowance was made for drop volume. Drip-tips appear most pronounced on the leaves of tree saplings growing in the shade (Roth 1996) and in species with pinnate leaves (Rollet 1990).

Along gradients of increasing seasonality of rainfall there is usually an increased proportion of drought-deciduous species and leaf size tends to decrease (Gentry 1969). Increasing altitude is generally reflected in a decrease in relative abundance of large-leaved species (Dolph & Dilcher 1980). Tree leaf size, frequency of compound leaves and frequency of drip-tips was found to decrease with elevation in New Guinea (Grubb & Stevens 1985).

There is considerable variation both within and among tropical forests and forest types in leaf structural characteristics such as leaf mass per unit area (LMA), lamina thickness, anatomical features and concentrations of nutrients (Figs. 2.18 and 2.19). The collation of data from published reports on leaf properties does not show such a clear trend as is generally described in the literature. The expectation is for the leaves of species from higher or more infertile sites to be smaller and thicker, with more dry mass per unit area, and probably volume, and to have lower concentrations of important nutrients (Grubb 1977; Turner 1994).

The typical leaves of lowland heath and upper montane forests in the tropics can be referred to as sclerophylls. The name means 'hard leaf', and is an allusion to the textural properties of the leaf in comparison to the soft and flexible leaves of more mesic sites. Leaf texture includes properties such as hardness, stiffness, strength and toughness. The first two, resistance to impression and bending, do not have any successful methods for quantification in leaves. Tensile strength and work of fracture (toughness) can be measured by using calibrated load cells in the appropriate equipment (Lucas & Pereira 1990; Lucas *et al.* 1991a). The highly anisotropic nature of leaf material causes difficulties. Frequently the midrib and veins are tougher than the intercostal region of the lamina (Lucas *et al.* 1991a; Choong *et al.* 1992). It is not easy to derive an average toughness or strength for the leaf, and in practice pre-determined fracture paths have to be chosen for comparative purposes.

The average LMA of different forest types varies considerably among sites (Fig. 2.18), with a large degree of overlap among the different forest formations. A more direct study of leaf form on one mountain, Gunung Silam in Borneo, has shown a trend of increasing LMA with altitude (Bruijnzeel *et al.* 1993). LMA also increased down a soil fertility gradient at San Carlos de Rio Negro in Venezuela (Medina *et al.* 1990). Grubb (1998a) reported a positive correlation between LMA and leaf size for species from caatinga in Venezuela

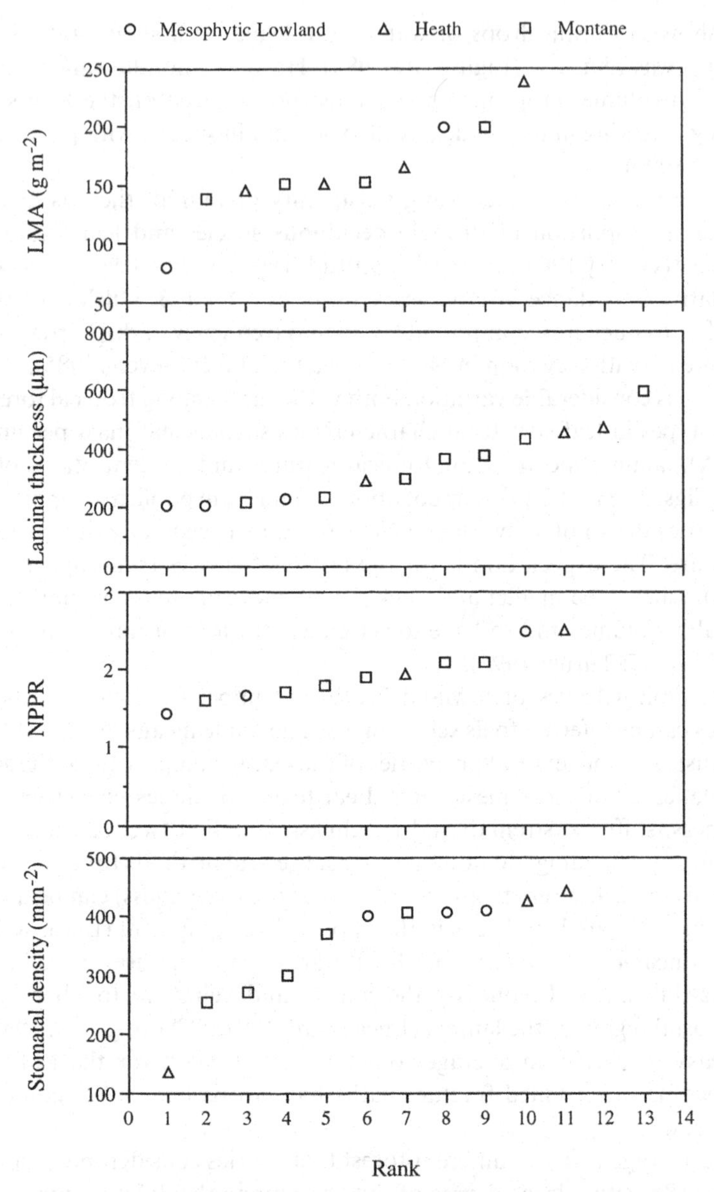


Figure 2.18 Species average leaf characteristics for different forest sites. Circle, mesophytic lowland forest; triangles, lowland heath forest; squares, montane forest. NPPR, non-palisade to palisade tissue thickness ratio in the mesophyll. Data from Bongers & Popma (1990a), Choong *et al.* (1992), Turner *et al.* (2000), Sobrado & Medina (1980), Tanner & Kapos (1982), Kapelle & Leal (1996), Sugden (1985), Cavelier & Goldstein (1989), Coomes & Grubb (1996), Grubb (1974) and Peace & MacDonald (1981).

and argued that this reflected a requirement for proportionally more material for support in larger leaves. However, this correlation is not evident in data provided by Rollet (1990) or Turner & Tan (1991). At Los Tuxtlas, Mexico, there was a positive correlation between leaf area and mass-based concentrations of N, P and K (Bongers & Popma 1990a) across the species in the lowland forest.

Lamina thickness more convincingly distinguishes the supposedly scleromorphic types from more mesophytic lowland forest (Fig. 2.18). However, non-palisade to palisade tissue thickness ratios in the mesophyll (NPPR) do not appear to show any consistency in distinguishing forest types. Montane forests tend to have a lower density of stomata than lowland ones (Fig. 2.18). The canopy leaves of mesophytic lowland forests can be as tough as those of heath forests (Turner *et al.* 1993b, 2000). Heath and montane forest species do appear to have generally lower foliar concentrations of major nutrients (Fig. 2.19). This is most noticeable for nitrogen where the ranges of community averages for mesophytic lowland and heath forests barely overlap, but for most of the other elements the heath and montane forest averages rarely approach 60% of the maximum value of a mesic lowland site. There was relatively little difference in mass-based foliar nutrient concentrations among a small sample of arborescent dicot and monocot species at La Selva (Bigelow 1993). Area-based concentrations were higher in the monocots; this reflected the higher LMA of these species.

**Leaf protection**

Rates of leaf herbivory are relatively high in the wet tropics (Coley & Barone 1996), with insects being the most important herbivores. Immature leaves lose area to herbivores at 5–25 times the rate of fully enlarged, mature leaves (Table 2.6), with most leaf damage being sustained in the expanding stage. Pathogens are also a major cause of leaf damage. On Barro Colorado Island, 29% of leaf damage across 25 tree species was attributable to pathogens (Coley & Barone 1996). Leaves are attractive to herbivores and pathogens because they are more nutritious than other, regularly available, parts of the plant. In particular, they have higher protein concentrations than other plant organs.

Leaf area loss to herbivores will reduce the rate of carbon fixation achieved by the plant, and also represents a material drain on the plant's resources. In a study (Marquis 1984, 1987) of the effects of defoliation on the growth and reproduction of the understorey shrub *Piper arieianum* at La Selva, Costa Rica, it was found that heavy defoliation (50% loss of area or more) caused significant reduction in seed output in plants for two years subsequent to the leaf loss (Fig. 2.20). This must provide strong selection pressure in favour of

Table 2.6. *Rates of herbivory in tropical rain-forest tree species*

Value in parentheses after each entry is the number of species included.

| | Mature leaves % area per day | Young leaves % area per day | Proportion (%) of leaf lifetime herbivory on young leaves |
|---|---|---|---|
| Shade-tolerant species | 0.03 (105) | 0.71 (150) | 68.3 (31) |
| Gap specialists | 0.18 (37) | 0.65 (37) | 47.3 (30) |

Data from a literature compilation by Coley & Barone (1996).

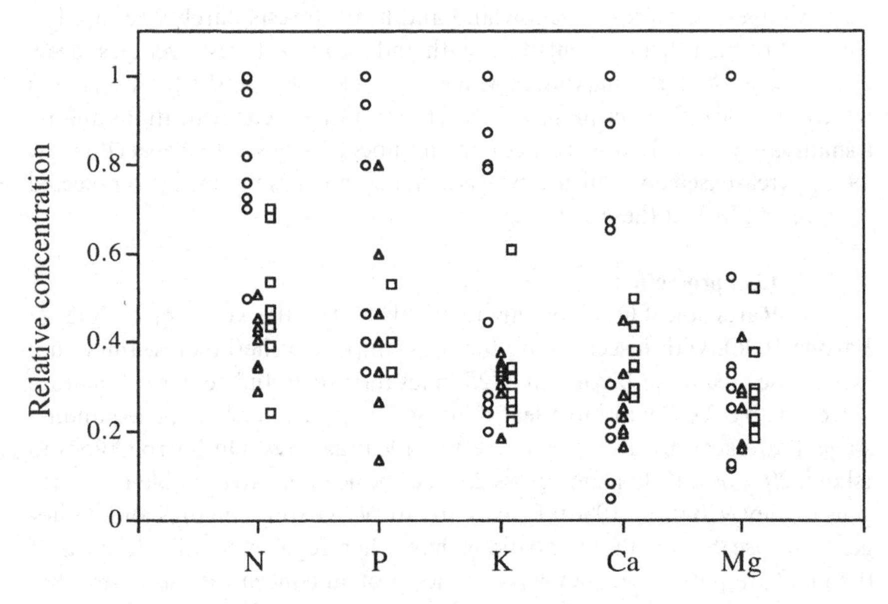


Figure 2.19 Foliar concentrations of important elements in different tropical-forest sites averaged across species, expressed as a proportion of the highest recorded mean value. Data from Vitousek & Sanford (1986).

plant-based mechanisms of defence. However, there may be an alternative strategy if there is a cost to defence in terms of reduced growth rates. If the production of defences uses resources that would otherwise go to growing more plant tissue, then in certain circumstances there may be selection in

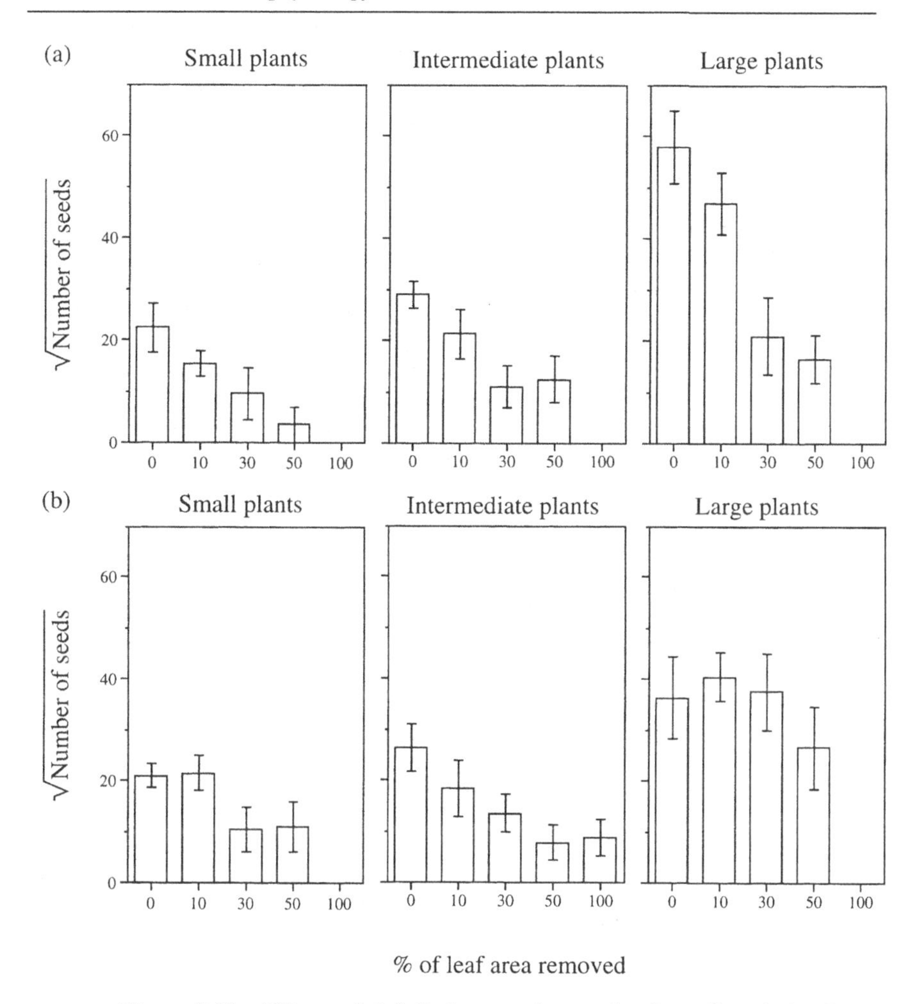


Figure 2.20 Effects of defoliation on the production of seeds in *Piper arieianum* during the first (a) and second (b) years after treatment. Bars represent standard error. After Marquis (1987).

favour of maximising growth, even at the cost of being susceptible to herbivores or pathogens. However, most plants do have some mechanisms of defence.

## The defences of tropical forest trees

Plants are attacked by a wide range of herbivores and pathogens. These vary in size, specificity and many other characteristics. It is therefore not surprising that plants normally possess many different sorts of defence.

Some of these are outlined with examples from tropical forest trees below. More examples can be found in Chapter 5. Harborne (1993) and Bennett & Wallsgrove (1994) provide good reviews of plant defences.

### *Toughness/Fibre*

DEFENCE

In order to ingest and digest plant material, herbivores need to comminute leaves and other parts that they eat. Materials that increase the strength and toughness of the plant body increase the work to be done by the herbivore during feeding. Cell wall materials such as cellulose and lignin are the major structural components of plant tissues. The disposition of cells with thick walls is the major determinant of the mechanical properties of plant tissues (Lucas *et al.* 1995). In addition, structural compounds provide the fibrous portion of the plant material, a component that most herbivores are unable to digest. Therefore toughening also reduces food quality because the nutritious material is diluted by indigestible fibre.

MODE OF ACTION

In most leaves it is the veins that are the toughest part (Lucas *et al.* 1991a; Choong *et al.* 1992; Choong 1996). The fibres in the vascular bundles provide toughening to the lamina, making it more difficult for herbivores to cut or tear pieces out of the blade. Sclerenchymatous bundle sheaths may protect veins from phloem-sucking insects and also effectively compartmentalise the lamina, preventing a chewing insect from having an unhindered path in its grazing and reducing the digestibility to larger animals.

DISTRIBUTION AMONG TROPICAL TREES

In general, tropical forest trees have leaves of high toughness and fibre concentration (Coley & Barone 1996).

CASE STUDIES

A range of generalist folivores from tropical rain forests have been shown to select low fibre/toughness leaves of relatively high protein content from among those available in their habitat; examples are howler monkeys (Milton 1979), leaf monkeys (McKey *et al.* 1981; Davies *et al.* 1988) and lowland gorillas (Rogers *et al.* 1990). Caterpillars avoid eating the tough veins of *Castanopsis fissa* (Choong 1996).

### *Spines*

DEFENCE

The presence of hard, sharp parts among the foliage may deter animals from coming too near the plant. The terms spine, thorn and prickle and others variously used for spiny plant parts await a universally agreed system of application.

MODE OF ACTION

An animal risks injury when contacting a spiny plant. To be effective, the spine needs to be sharp and relatively hard in order to penetrate animal tissue. Plant spines are often hardened by accumulations of inorganic crystalline substances such as silica or calcium salts. Small animals such as insects may easily move between the spines, rendering them ineffective as defences against such organisms. Spines can also serve as anchorage mechanisms in climbing plants, and as modifiers of a plant's energy and gas exchange with the environment, but in tropical trees it seems likely that defence is the main role of armature.

DISTRIBUTION AMONG TROPICAL TREES

Not many trees in the tropical rain forest are spiny. Perhaps the most notable group is the palms.

CASE STUDIES

The prickles found on young shoots of Hawaiian treelets in the genus *Cyanea* have perplexed ecologists because the Hawaiian archipelago is devoid of indigenous mammalian herbivores. Flightless grazing ducks and geese, exterminated after Polynesian settlement of the islands, have been invoked as the selection pressure for these defences (Givnish *et al.* 1994).

### *Hairs*

DEFENCE

Plant trichomes can play a defensive role (Levin 1973).

MODE OF ACTION

Hairs can be likened to spines at a reduced scale. A layer of hairs may physically obstruct insects and other tiny herbivores from attacking a plant. They may be hard and sharp through inclusion of crystalline materials, strongly hooked, sticky, stinging or contain toxins.

DISTRIBUTION AMONG TROPICAL TREES

Densely pubescent leaves are not common among tropical rain-forest tree species. Of 205 species from Venezuela studied by Roth (1984), only 15 had a dense covering of hairs, and this was restricted to the leaf undersurface in all cases. Tropical trees possessing stinging hairs are confined to the Urticaceae tribe Urticeae and possibly a few Euphorbiaceae (Thurston & Lersten 1969).

CASE STUDIES

*Streblus elongatus* has short razor-sharp silicified hairs on the undersides of the leaves (Lucas & Teaford 1995). The urticaceous genera *Dendrocnide* and *Urera* are the largest groups of stinging species among tropical rain-forest trees, with species of the former, notably *Dendrocnide stimulans*, inflicting stings of tremendous ferocity.

An interesting twist to the defensive role of pubescence was reported by Letourneau *et al.* (1993). They investigated the ant-protected euphorb *Endospermum labios* in New Guinea. This species occurs as what were believed to be genetically determined glabrous and hairy morphs. The hairless morph suffered less herbivory that the pubescent one because glabrous plants were more likely to be inhabited by ants. Colony foundress queens did not appear to make plant-host choice based on pubescence, but they were more successful on the glabrous morph. Letourneau *et al.* (1993) hypothesised that parasitic flies had more chance of egg laying on queens that were delayed by the presence of indumentum in making their nest chamber in the stem. Later research revealed that the hairy phenotype was actually an induced response to attack by stem-boring insects (Letourneau & Barbosa 1999).

### *Crystal inclusions*

DEFENCE

Plant tissues can contain crystalline inclusions, generally of calcium oxalate, calcium carbonate or silica. These can vary in shape from rounded to sharply pointed structures.

MODE OF ACTION

Sharp or pointed crystals can pierce animal tissues and membranes, causing injury. Otherwise the hardness of crystalline inclusions can wear teeth or mandibles of herbivores.

DISTRIBUTION AMONG TROPICAL TREES

Many plants in the rain forest possess crystalline inclusions; for instance, the leaves of 20 out of 22 species investigated in Venezuela possessed crystals (Pyykkö 1979). Approximately 300 out of 1300 species examined from South America (mostly Surinam) had silica grains in the wood (ter Welle 1976).

CASE STUDIES

Lucas & Teaford (1995) proposed that siliceous deposits in leaves could cause microwear on the teeth of long-tailed macaques.

### *Gums, latexes and resins*

DEFENCE

Many plants secrete sticky fluids when parts are broken. These may be distasteful or toxic, but they can also deter herbivores through the physical property of stickiness or by hardening rapidly on exposure to air.

MODE OF ACTION

Many plant species possess canal systems running through their tissues that contain resins, latexes or gums. When the plant is damaged the contents of the canals pour into the wound. They may even be under pressure and squirt out rapidly. Gluey secretions can gum up insect mouthparts. The latexes may also dry and harden, making removal difficult. Herbivores may waste time and energy cleaning themselves of sticky exudates from wounds that they create in the plant.

DISTRIBUTION AMONG TROPICAL TREES

Exudates are relatively common among tropical trees. For instance, at four sites in Gabon the proportion of species producing latex or resin from the trunk ranged from 16 to 35% (Reitsma 1988). Families such as the Apocynaceae, Sapotaceae, Guttiferae and Moraceae, and Euphorbiaceae subfamily Crotonoideae are well known for latex production. The Burseraceae and Dipterocarpaceae are highly resinous. Farrell *et al.* (1991) noted that latex- or resin-canal-bearing taxa tend to be larger than sister groups without canals. This, they argued, reflected the evolutionary success of groups that evolved laticiferous defence.

CASE STUDIES

Six out of seven tree species growing on Barro Colorado Island, Panama, inflicted with wounds to the trunk responded by latex or resin production which led to the formation of a scab of dried exudate over the wound, protecting it from infection (Guariguata & Gilbert 1996).

### *Non-protein amino acids*

DEFENCE

Plants may contain amino acids other than those usually incorporated into proteins by animals. These can be extremely toxic (D'Mello 1995).

MODE OF ACTION

The non-protein amino acids when ingested by herbivores often act as analogues to specific amino acids used in protein synthesis. The plant amino acid molecules become incorporated into proteins, but the substitution leads to proteins that cannot fulfil their normal function, thus disrupting, sometimes irreparably, cell biochemical machinery.

DISTRIBUTION AMONG TROPICAL TREES

The legumes are the major group that employs non-protein amino acids as a defence. These defences are often found in seeds, but may also occur in leaves and other plant parts.

CASE STUDIES

Canavanine is a non-protein amino acid quite widespread in the legumes. It is an analogue of arginine. Hypoglycin A is found in unripe fruits and the funicle of the sapindaceous tree *Blighia sapida*. The seeds of *Cycas circinalis* contain a β-*N*-methyl amino acid that is associated with degeneration of the brain in Pacific Islanders who eat seeds improperly prepared. Seeds of *Lecythis ollaria* contain seleno-cystathionine, a toxic analogue of cys-

Table 2.7. *Presence of alkaloids in mature leaves among tropical rain-forest tree floras*

| Site | Number of species tested | Percentage alkaloid-positive | Reference |
|---|---|---|---|
| Douala-Edea, Cameroon | 70 | 16 | Waterman & McKey (1989) |
| Kibale, Uganda | 20 | 50 | ditto |
| Makokou, Gabon | 160 | 7 | Hladik & Hladik (1977) |
| New Guinea | 108 | 6 | Hartley *et al.* (1973) (as cited by Hladik & Hladik (1977)) |

tathionine where selenium replaces sulphur in the amino acid molecule (Aronow & Kerdel Vegas 1965).

### *Alkaloids*

#### DEFENCE

Alkaloids are *N*-heterocyclic molecules, generally with a biosynthetic origin in amino acids (Hegnauer 1988; Waterman 1996). They are complex and varied in structure and biochemical properties, but tend to be bitter to taste and are frequently toxic to animals.

#### MODE OF ACTION

Toxic effects of alkaloids on animals include inhibition of DNA and RNA synthesis, inhibition of mitosis, breakdown of ribosomes and cell membranes and interference with nerve transmission.

#### DISTRIBUTION AMONG TROPICAL TREES

Some surveys for alkaloid-bearing species have been conducted in tropical rain forests (Table 2.7). These have mostly shown quite low frequency of alkaloid-positive species. The one exception, although this has a smaller sample size than the rest, is the rain forest at Kibale, Uganda, where half the species tested gave positive results. Despite the general rarity of alkaloidal trees, lowland tropical rain forests have the highest frequency and mean concentration of alkaloids of any major vegetation type (Levin 1976). Alkaloids are found in many evolutionary lines and occur in about 20% of higher plant families, but abundant alkaloids are particularly associated with the Annonaceae, Apocynaceae, Lauraceae, Leguminosae, Rubiaceae, Rutaceae and Solanaceae among tropical trees (Hartley *et al.* 1973; Levin 1976). On the other hand, Guttiferae, Melastomataceae, Myrsinaceae, Myrtaceae and Sapindaceae are notable for the presence of few, if any, alkaloidal species.

CASE STUDIES

Humanity has found uses for many plant alkaloids as medicines, stimulants and poisons. The arrow poison curare is derived largely from menispermaceous vines, but the arborescent legume genus *Erythrina* contains alkaloids with a similar toxicology (Chawla & Kapoor 1997). The main ordeal poisons of Africa are another group of legume alkaloids from the genus *Erythrophleum* (Neuwinger 1996). Tropical trees have been domesticated for their alkaloidal properties among other things, for example caffeine (*Coffea*, *Cola*, *Theobroma*) and quinine (*Cinchona*).

### *Cyanogenesis*

DEFENCE

Cyanide is extremely toxic to nearly all living organisms. Cyanogenic glycosides generate cyanide when the CN group they contain is cleaved from the sugar moiety of the glycoside molecule. They provide the commonest form of cyanogenesis in plants.

MODE OF ACTION

When a herbivore attacks a cyanogenic plant either enzymes that are released by the plant, or the digestive enzymes of the herbivore, break down the glycoside, releasing the poisonous cyanide. Cyanide interferes with the cytochrome system thus inhibiting the terminal part of the main respiration pathway in cells.

DISTRIBUTION AMONG TROPICAL TREES

Cyanogenesis is not common among tropical trees. In a survey of 430 tree and shrub species from Costa Rica only 20 (4.7%) tested positive for cyanide production (Thomsen & Brimer 1997). Woody plants probably show a lower frequency of cyanogenesis than herbaceous species. Taxonomically, cyanogenesis seems quite widely spread with families such as Annonaceae, Elaeocarpaceae, Euphorbiaceae, Flacourtiaceae, Leguminosae, Proteaceae, Rosaceae, Rubiaceae, Sapindaceae and Sapotaceae containing cyanogenic members (Hegnauer 1977; Thomsen & Brimer 1997).

CASE STUDIES

Leaves of two Central American *Acacia* species were found to be toxic to the southern armyworm (*Podenia eridania*, Noctuidae) owing to the presence of cyanogenic glycosides (Rehr *et al.* 1973).

### *Phenolic compounds*

DEFENCE

Phenolic compounds are substances that possess one or more hydroxyl group (OH) bonded into an aromatic ring structure. A wide range of complex molecules, mostly synthesised via the shikimic acid pathway, can be included in the class of phenolic compounds. The simpler molecules include lignans, coumarins and flavanoids. Alkaloids and terpenes can also be phenolics. Tannins are a complex set of polyphenolic compounds. They cause the astringency found in many plant products and are the group of phenolic compounds most often associated with plant defence.

#### MODE OF ACTION

Tannins possess the ability to cross-link protein molecules. This property is employed in their traditional use of tanning leather. Tanning makes the leather much less susceptible to microbial decay by binding the proteins in the animal skin. In plant leaves, the tannins, particularly the condensed-tannin group, are believed to help deter herbivores by cross-linking plant proteins in the herbivore's mouth and gut and making them less accessible for digestion, and also by linking to the gut wall proteins and interfering with the process of nutrient uptake.

#### DISTRIBUTION AMONG TROPICAL TREES

Tannins are commonly found in tropical trees. The leaves of tropical trees generally have higher concentrations of condensed tannins than those of temperate ones (Coley & Aide 1991), but assays of total phenolics do not show any major latitudinal difference. Woody plants are more likely to contain tannins than herbaceous ones (Mole 1993), possibly because of a biochemical link to lignin synthesis. This may also explain the association of tannins with more primitive angiosperm groups and their relative rarity in advanced clades such as the Asteridae. A few families, such as the Araliaceae and Moraceae, appear to have relatively few members that synthesise them.

#### CASE STUDIES

There have been a number of studies of the role of tannins in the defence of juvenile stages of tropical trees, but little concerning adults.

### *Terpenes*

#### DEFENCE

Terpenes are perhaps the largest and most diverse group of plant secondary chemicals. They share a common biosynthetic origin in mevalonate and are based on a five-carbon molecule, which can be repeated from a few to many times to produce different chemicals. $C_{10}$ molecules are referred to as monoterpenoids, $C_{20}$ as diterpenoids and so on. The larger polymers are the basis of plant resins and essential oils. Terpenoid glycosides, including cardiac glycosides, and saponins are among the most toxic of the terpenes found in plants.

#### MODE OF ACTION

Terpenoids probably have a wide range of functions, but there is evidence that some are toxic, deterrent or inhibitory to herbivores and pathogens. The modes of action are varied, but include an evidently unpleasant bitter taste and irritation of animal tissues. Pyrethroids are a group of monoterpenoids that are effective 'knock-down' insecticides.

DISTRIBUTION AMONG TROPICAL TREES

A number of families are particularly rich in highly resinous tropical trees, notably the Burseraceae, Dipterocarpaceae and Leguminosae. Tropical conifers are also very resinous. Essential oils are abundant in the Rutaceae.

CASE STUDIES

The South American legume genus *Hymenaea* contains the terpenes caryophyllene and caryophyllene oxide. The former has been demonstrated to be an effective deterrent to insect herbivores, whilst the latter inhibits the growth of leaf-spotting fungi (Langenheim 1994) and makes leaves repellent to leaf-cutting ants (Hubbell *et al.* 1983) presumably because of its anti-fungal properties.

### *Ants*

DEFENCE

Myrmecophytic plants have a mutualistic relationship with ants. Ants are often highly aggressive to invaders of their territories, and myrmecophytic plants harness ant colonies to attack and deter potential herbivores.

MODE OF ACTION

The trees elicit protection from ants by providing food and/or accommodation. Food is presented in the form of sugary secretions from extra-floral nectaries (EFNs) and food bodies (Fig. 2.21), typically found on young leaves and shoots. Many myrmecophytes provide domatia (little houses) for ants. These are usually pouch-like leaves or stipules, or hollow petioles or twigs. The ants seem particularly aggressive protectors around their nest site and can be effective defenders of host trees.

DISTRIBUTION AMONG TROPICAL TREES

EFNs are quite common in tropical trees: 12% of species (91/741) at Pasoh, Peninsular Malaysia (Fiala & Linsenmair 1995), with most of the EFNs (70%) occurring on the leaf blades. The frequency of EFNs increased with tree stature at Pasoh (Table 2.8), possibly indicating that canopy trees can better spare the photosynthate delivered to the nectaries. An even higher proportion of species (33%) on Barro Colorado Island were found to possess EFNs (Schupp & Feener 1991), but the sample was strongly biased towards light-demanding species. Both studies found EFNs to be particularly common in advanced angiosperm clades, notably Dilleniidae, Rosidae and Asteridae. McKey & Davidson (1993) record more than 100 genera of tropical trees from 38 families that are regularly symbiotic with plant ants. Most of these provide domatia.

CASE STUDIES

The benefit of ant-defence is clear from the data of Vasconcelos (1991) who demonstrated a 45-fold higher rate of fruit production in ant-inhabited individuals of the Amazonian melastome shrub *Maieta guianensis* compared with those from which the resident ant colony was removed.

Small ants may be better defenders against insect herbivores than large ones if reduced individual size translates into a higher density of defenders. This was hypothesised by Gaume *et al.* (1997), who showed that the African arborescent myrmecophyte *Leonardoxa africana* was well defended by the tiny ant *Petalomyrmex phylax*. Paired leaves, one with ants, one without, showed that the ants were very effective at keeping off marauding insect herbivores, including sap suckers that are usually not quantified in such studies (Fig. 2.22).

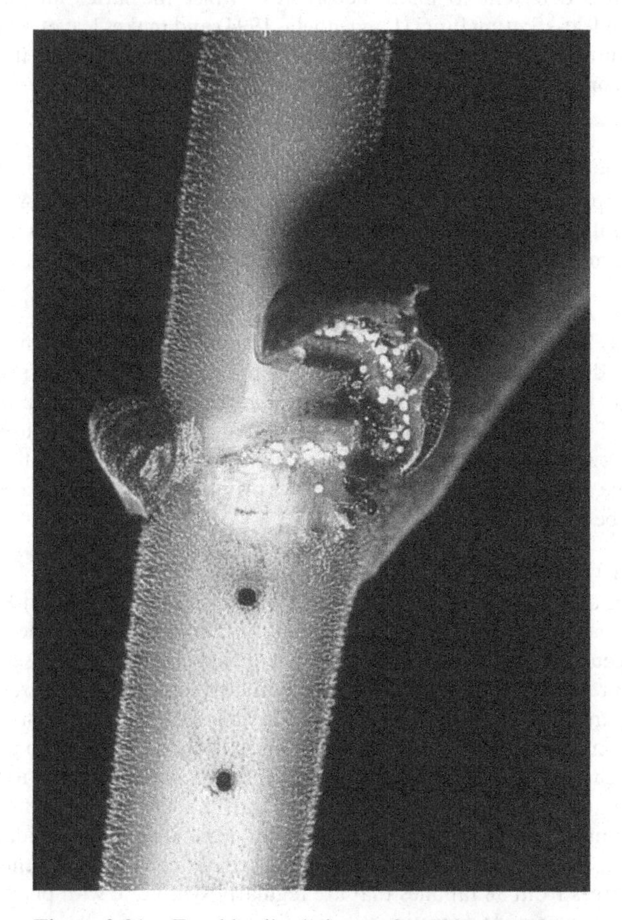

Figure 2.21 Food bodies being gathered by *Crematogaster borneensis* ants from under the stipule of a *Macaranga trachyphylla* plant. The holes in the stem are entrances cut by the ants to the hollow twig which acts as a domatium. Photo: Tamiji Inoue.

Table 2.8. *Occurrence of woody plants with extra-floral nectaries (EFN) in Pasoh Forest Reserve, Peninsular Malaysia*

See Table 2.1 for a definition of the stature classes.

| Stature class | Percentage of species with EFN (no. EFN spp./total no. spp.) | |
|---|---|---|
| 'shrub' | 2.3 | (1/44) |
| 'treelet' | 8.8 | (9/102) |
| understorey | 13.2 | (34/257) |
| canopy | 10.3 | (29/286) |
| emergent | 21.4 | (12/56) |

Data from Fiala & Linsenmair (1995).

It is not just animals that ants will attack if they come into contact with their host plant. Ant exclusion from trees of *Stryphnodendron microstachyum* resulted in more leaf damage from pathogens (de la Fuente & Marquis 1999). Ants may also defend the host against other plants. Perhaps the most extreme example of this ant-mediated competition between plants has been reported in some neotropical spreading shrubs belonging to the Melastomataceae (Morawetz *et al.* 1992; Renner & Ricklefs 1998). Morawetz *et al.* (1992) described how *Tococa occidentalis* (which is actually *Tococa guianensis* according to Renner & Ricklefs (1998)) is found growing as clones in gaps in lowland forest in Peru. These clones are generally surrounded by clear space with no living plants. The ants (*Myrmelachista* sp.) that inhabit the hollow stems and leaf domatia of the *Tococa* actively attack living plants growing near the host. They chew the shoot tips and main nerves of the leaves and spray a chemical from their abdomens into the wounds that they have made (Morawetz *et al.* 1992). Plants treated in this way, including small trees, die rapidly. Morawetz *et al.* (1992) concluded that the ants were injecting a potent herbicide into the plants that they attacked. Renner & Ricklefs (1998) report similar observations for clumps of *Tococa* mixed with another myrmecophytic melastome (*Clidemia heterophylla*) in Ecuador.

It should not be assumed that ants are always beneficial. Very aggressive ant defenders could deter animals needed by the plant such as pollinators or seed-dispersers (Thomas 1988). Ant defence of *Psychotria limonensis* on Barro Colorado Island resulted in a 50% increase in fruit set but the amount of mature fruit removed was reduced by 10–20% (Altshuler 1999). In the South American myrmecophytic shrub *Cordia nodosa*, the common ant inhabitant of the hollow stems actively sterilises the tree by biting off flower buds (Yu & Pierce 1998). The apparent advantage to the ants of doing this is to maintain vegetative growth that allows more domatia and hence a bigger ant colony to develop. Fortunately for the *Cordia*, not all individuals are inhabited by the 'bad' ant.

*Mites*

DEFENCE

Tufts of hairs in the axils of veins, often associated with shallow pits or pockets, are common on leaves (Walter 1996). These domatia are believed to be important as sites in which mites can hide.

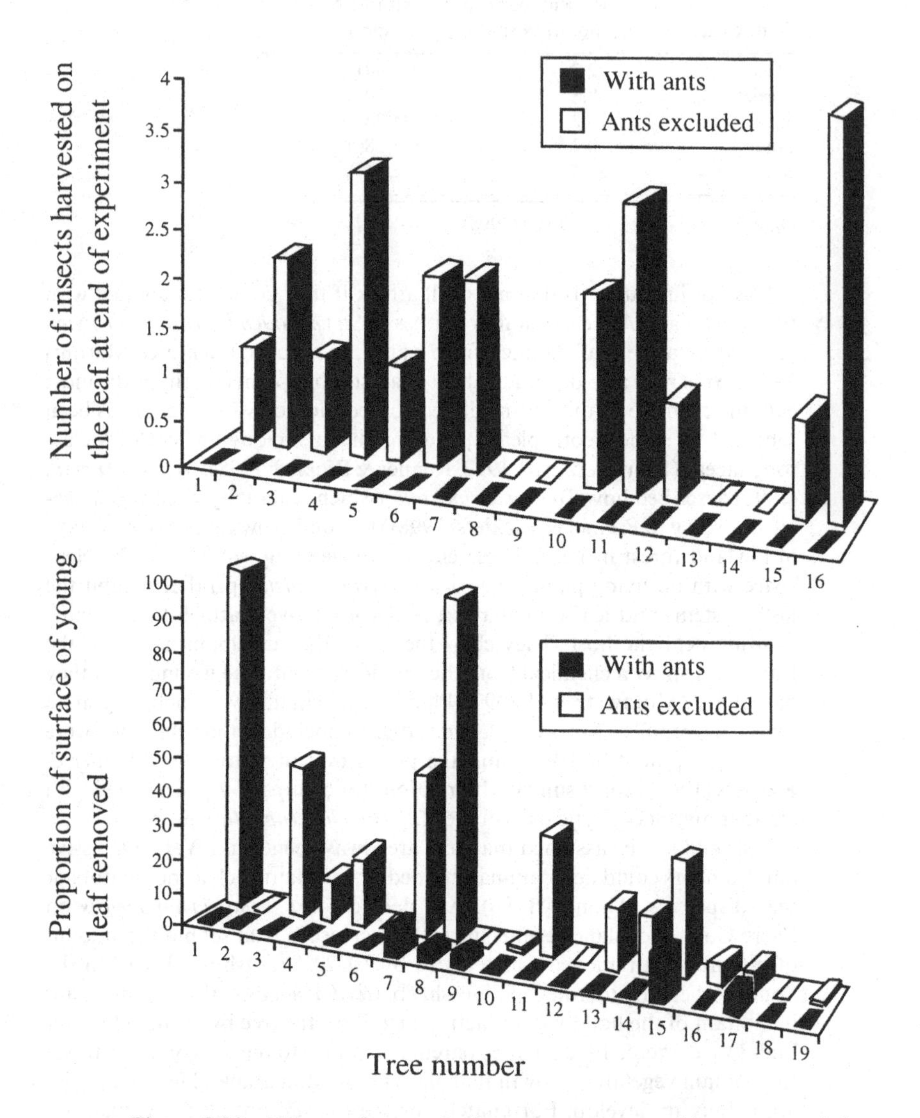


Figure 2.22 Effects of ant exclusion on occurrence of insects over 4 days (upper) and on herbivory caused by chewing insects over 14 days (lower) on young leaves of *Leonardoxa africana*. After Gaume *et al.* (1997).

MODE OF ACTION

Many mite species are major pests of plants, sucking the contents out of individual leaf cells. They can also act as disease vectors. However, studies have shown that the mite inhabitants of leaf domatia rarely include plant parasites. Instead, they are used by fungivorous and predatory species (Walter 1996). These mites may act like ant colonies, but at a further reduction of scale. Some may eat fungal spores and hyphae on the leaf surface before they have a chance to attack the leaf. The predatory species will devour the parasitic mites that would otherwise infest the plant. The presence of mite domatia has been shown to influence numbers of predatory mites (Walter 1996).

DISTRIBUTION AMONG TROPICAL TREES

Domatia are frequently encountered on the leaves of trees in the tropical rain forest.

CASE STUDIES

Agrawal (1997) compared two cultivars of avocado pear (*Persea americana*), one of which had domatia and one which lacked them. Herbivore numbers on the plants with domatia were consistently lower than those without, or where they were experimentally blocked. However, the differences failed to achieve statistical significance.

## Classification of plant defences

Plant defences can be classified in several ways. Defences can be classified as physical or chemical. The former alter the mechanical properties of the foliage, making it more difficult to attack. The latter involve molecules that deter or poison the attacker. The distinction between the two is not always clear. For instance, the irritant latexes and resins produced by some plants may act both physically in gumming up the herbivore and chemically in containing a toxic substance. Stinging and glandular hairs are physical defences, but may deliver noxious chemicals.

It is the physical defences of mature leaves, notably their high toughness, that makes immature leaves the most attractive to herbivores. The low fibre concentration and relatively high protein and water contents make young, expanding leaves easier to digest and more rewarding food.

Chemical defences consist of a wide range of molecules produced by secondary metabolism in plant cells. It has been estimated that there are over 100 000 plant secondary metabolites (Waterman 1996), of which we know the probable function of a tiny proportion only. Some may be involved with defence, but we cannot assume that they all are. Apparently defensive chemicals may play additional roles. For instance, high concentrations of tannins in leaves may interfere with the decomposition process. The tannins bind the protein in the decaying leaves and reduce nitrogen availability to the micro-organisms responsible for decomposition (Kuiters 1990; Bruijnzeel *et al.* 1993; Northup *et al.* 1995). This results in a build-up of litter and a reduced supply rate of nitrogen to the plants. There is a general inverse correlation between soil nitrogen availability and foliar tannin concentrations, so there tends to be a positive feedback increasing the litter and soil concentration of polyphenolic compounds. It has been speculated that this is important in determining distribution patterns of forest formations and individual species (Bruijnzeel *et al.* 1993; Northup *et al.* 1995). The production of tannin-rich litter may give tree species an advantage if, by reducing nutrient availability and increasing soil acidity and toxic phenol concentrations, they compromise the growth potential of competitors. Alternatively, on highly infertile soils the tannins may help prevent loss of nitrogen from the system, improve the cation exchange capacity and provide a thick litter layer more hospitable to tree roots than the mineral soil (Northup *et al.* 1995). They may also bind potentially toxic free aluminium ions, though these are unlikely to be present in white sand soils.

Chemical defences may be concentrated at the sites poorly protected physically. For instance, the mesophyll tissues of leaves may store relatively high concentrations of secondary compounds as they are low in toughness (Choong 1996). Increasing wall thickness of cells in the mesophyll would reduce their effectiveness at photosynthesis, so chemical defence is better for these tissues.

Another classification of defences contrasts constitutive defences, which are always present, with inducible ones that are produced in response to attack. Most research on inducible defences has been conducted on species from outside the tropics. However, Agrawal & Rutter (1998) compiled published and unpublished data that provided support for ant-based defences sometimes being inducible. These are largely reports of increases in volume or sugar concentrations in secretions from extra-floral nectaries in response to real or simulated herbivory. Pubescence in *Endospermum labios* from New Guinea may be induced by attack from stem-boring insects (Letourneau & Barbosa 1999). The extra hairs on the plant's surface may interfere with oviposition by more borers. Other inducible defences, such as the production of proteinase inhibitors, seem likely to occur in at least some tropical tree species.

### Foliar bacterial nodules

Miller (1990) has reviewed our knowledge of an intriguing symbiosis between plants and bacteria found in a few genera of the Rubiaceae (*Pavetta*, *Psychotria* and *Neorosea*) and Myrsinaceae (*Ardisia*, *Amblyanthus* and *Amblyanthopsis*) that are mostly pygmy trees of the tropical rain forest. These plants possess bacterial nodules in their leaves, usually detectable as dark spots or indentations in the leaf margin. The symbiosis is probably obligate to both partners. The bacteria are restricted to the plants and without the bacteria the plants stop development and die. It seems probable that the bacteria are needed by the hosts because they produce a growth regulator, most likely to be cytokinin, that the plant cannot synthesise. A complex system of morphological adaptations for ensuring bacterial colonisation of all shoots and of the seeds so that offspring plants are able to grow clearly indicates the closeness of the association involved. However, only a minority of species in the genera listed above have foliar bacterial nodules. One wonders whether more tree–microbe symbioses await discovery in the rain forest.

### Leaf development: coloured young leaves

A common feature of tropical rain forest trees is the production of new leaves of distinctive colours other than green. Young leaves are often a shade of red, varying from pale pink to bright scarlet, but white or even purple or blue can be observed. In a survey of 250 species across the tropics (Coley & Kursar 1996), 33% of species were found to have non-green young leaves. These included representatives of 61% of the families in the sample. Among tropical trees, coloured young leaves are particularly common in the Sapindaceae and the caesalpinoid legumes. The proximate reason for the young leaves not being green is that they lack chlorophyll (Fig. 2.23). Greening is delayed in these species in comparison with those that do have young leaves containing chlorophyll. Anthocyanins, generally located in the cell vacuoles, are the pigments responsible for the red, purple or blue hues of the flush leaves. Physiological reasons for delayed greening have been put forward. Possibly young leaves are susceptible to damage from high light intensities and chlorophyll would be destroyed if it were put in sooner, and maybe the anthocyanin pigments are protectants against bright light. However, the observation that coloured young leaves are commoner in understorey species than elsewhere (Kursar & Coley 1992a) seems to discount this argument.

Other theories to explain delayed greening involve selection due to herbivore pressure. Chlorophyll is a nutritious molecule and will add to the attractiveness of leaf tissue to herbivores. The cost of not synthesising chlorophyll and beginning photosynthesis in young leaves is a loss of photosynthate. But young leaves with chlorophyll would be more likely to be eaten by herbivores. It is already well established that young leaves are particularly susceptible to herbivory (Coley & Barone 1996). This is probably because of their low fibre concentrations, and despite the fact that young leaves are often relatively well protected chemically (Fig. 2.24), often with twice the

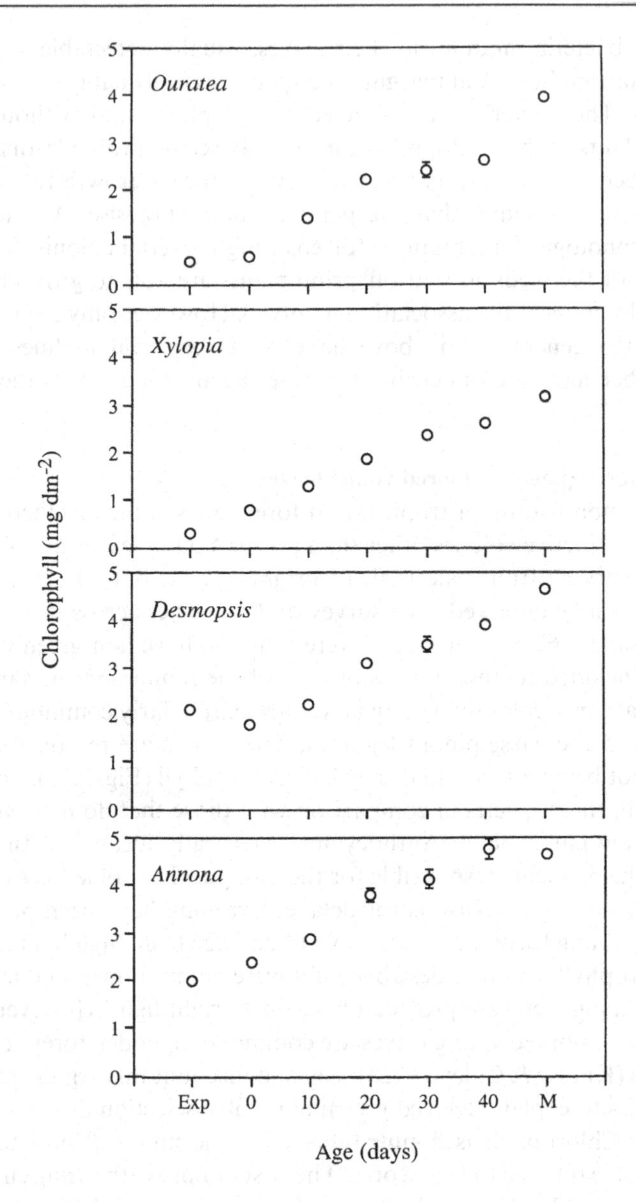


Figure 2.23 Chlorophyll per area as a function of leaf age in *Ouratea lucens*, *Xylopia micrantha*, *Desmopsis panamensis* and *Annona spraguei*. The error bars indicate plus or minus one standard error. In some cases, the error bars are smaller than the point and therefore are not shown. Exp is the average time taken for leaves from 10% of full leaf expansion until 4 d prior to full leaf expansion. The data at 0 d are the averages for leaves three days prior to full leaf expansion until +4d. The data at 10 d are the averages for leaves at +5 d to +15 d, etc. After Kursar & Coley (1992b).

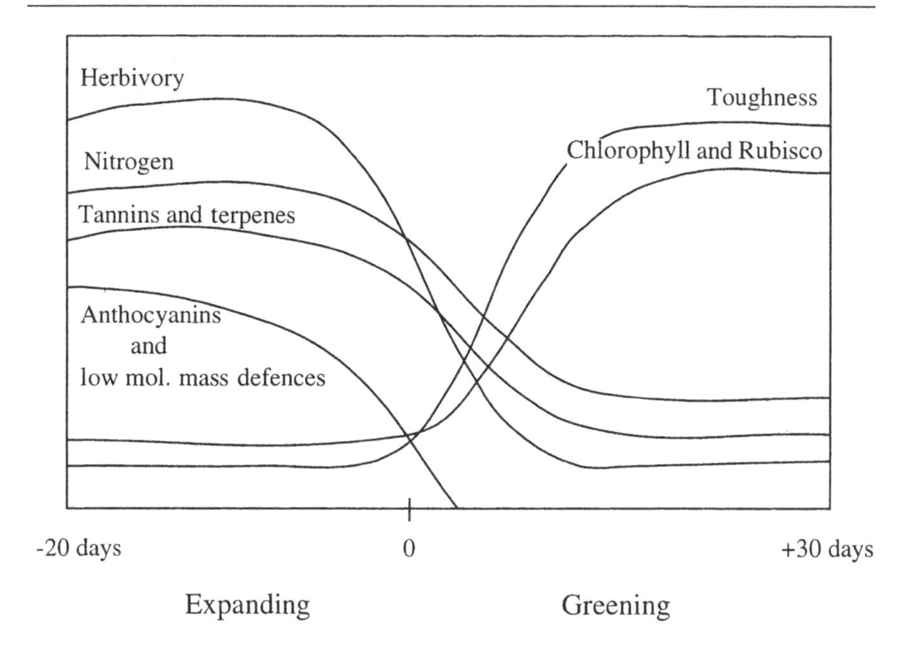


Figure 2.24 Schematic representation of changes occurring during leaf development for species with delayed greening. Age 0 indicates full leaf expansion. Negative ages are for expanding leaves, and positive ages are for days since full expansion. Low molecular mass secondary compounds decrease during expansion, but no data exist to test this. After Coley & Kursar (1996).

concentrations of tannins and other polyphenols of mature leaves (Coley 1983; Turner 1995a). The reduction in nitrogen and defence chemical concentrations as leaves mature may be due more to dilution by cell wall material than an actual reduction of the total amount of the substances in the leaf. Anthocyanins are possibly anti-fungal compounds (Coley & Aide 1989). Pathogens, as well as herbivores, find young leaves more tempting. Therefore, delayed greening may be a plant strategy to reduce the attractiveness of its already vulnerable young leaves to herbivores.

Cost–benefit analysis shows that delayed greening should be more favoured where the foregone benefits of greening are low. This will be in sites where potential photosynthetic rates are only moderate, and these include the forest understorey where, as already mentioned, delayed greening is commoner. Delayed greening may also be a corollary of very rapid leaf expansion (Coley & Kursar 1996). If a tree develops its leaves quickly it will reduce the period when it is particularly at risk from herbivory. It may be that the degree of physiological activity required both to expand rapidly and import and synthesise the pigments and other photosynthetic apparatus is not feasible, so the latter is delayed to allow greater speed in leaf expansion (Coley & Kursar 1996). It is possible that coloured young leaves are rarer in the temperate region than the tropics because lower temperatures slow down leaf expansion rates and allow the importation and synthesis of chlorophyll to keep pace with expansion. Coley & Kursar (1996) found that the amount of herbivory on expanding leaves was positively correlated with expansion rate in a wide-ranging survey. Aide (1993) also failed to find the expected negative correlation between herbivory and expansion rate. Clearly the hypothesised advantage of rapid expansion reducing the likelihood of herbivore attack is not evident in cross-species comparisons. Coley & Kursar (1996) suggest that slow-expanders are better protected and protection slows expansion. This may represent another 'grow or defend' trade-off. The evolutionary advantages of rapid leaf expansion, presumably in terms of reducing the period of susceptibility to herbivores and the earlier implementation of full photosynthetic potential, may outweigh the risk of losing leaf material to herbivores during expansion.

Juniper (1993) has taken a somewhat different line in the debate about coloured young leaves, concentrating more on why young leaves have the non-green and limp form. He hypothesised that the main advantage of flush leaves is that they do not appear to be leaves at all. Thus they do not provide a strong signal to herbivores, particularly insects, that they are potential food. He noted the apparently large intrapopulation, and even intragenotype, variation in flush-leaf form and proposed that this may be a way to avoid reinforcing the cue to insect attackers that new leaves are available.

Lucas *et al.* (1998) proposed that leaf colour may be used as a signal of food quality by some herbivores. In tropical rain-forest leaves, low toughness and probably peak protein concentrations correspond to a phase where leaves are light and predominantly yellow in hue, sometimes dappled red. Increasing greenness and darkness reflect increasing toughness and fibre concentration as leaves mature, which would reduce food quality. The authors proposed that trichromacy (possession of three colour receptors on the retina) in primates may have evolved to assist in foraging for leaves, as colour vision greatly facilitates distinction of the best quality food.

### Leaf longevity

Data on the leaf longevity of tropical trees are quite sparse, but it is clear that a large range in values is involved. Among seven species of *Piper* (mostly shrubs) at Los Tuxtlas, Mexico, mean leaf life span varied from less than three months to more than two years (Williams *et al.* 1989). Even longer lived are the leaves of the cycad *Zamia skinneri*. The average longevity of a

*Zamia* leaf at La Selva, Costa Rica, was 4.6 years, with some lasting nearly 9 years (Clark *et al.* 1992).

Species across a soil fertility gradient in Venezuela were chosen for a range of leaf life spans in order to investigate how longevity influenced form and function (Reich *et al.* 1991). Among the 23 species studied, leaf life length varied 30-fold from 1.5 to 50 months. There were strong correlations between several structural and physiological characteristics measured and longevity. Stomatal conductance to water and both mass- and area-based photosynthetic rates decreased with increasing leaf life span (Fig. 2.25). LMA and leaf 'toughness' increased with life span also. Mass-based net photosynthetic rate was negatively correlated with LMA and with leaf nitrogen concentration. The study indicated that leaf life span was a major determinant of leaf characteristics with short-lived leaves tending to be designed for high performance with high maximal rates of photosynthesis whereas long-lived leaves had generally lower rates of carbon assimilation.

To last over a long period, leaves need to be built with greater levels of protection against herbivores, pathogens and the physical environment (Turner 1994). A greater allocation of resources to these functions within the leaf reduces the allocation to photosynthetic machinery, which leads to reductions in maximal photosynthetic rates. Increased leaf life span is selected for in resource-poor conditions, such as forests on infertile soils, because the shortage of resources makes the possession of high performance leaves impracticable. Long leaf life length will increase the payback on the initial investment in leaf construction, particularly in terms of mineral nutrients on poor soils. There may be a positive feedback involved. When greater leaf longevity is favoured more investment in protection is required, which in turn will require greater longevity to pay back for the protective investment. Williams *et al.* (1989) did not find a positive correlation between leaf longevity and construction cost (estimated from calorific and nitrogen contents) among seven species of *Piper* in Mexico. The species with the long-lived leaves were forest understorey shrubs with generally lower construction costs than the shade-intolerant species. But the lower potential carbon gains of the forest understorey conditions meant that the ratio of construction cost to daily carbon gain (the amortisation rate) increased with leaf longevity (Fig. 2.26).

Leaf defence type has also been linked to leaf longevity. Coley *et al.* (1985) proposed the resource availability hypothesis of plant defence. They defined two main defence types. Qualitative defences, such as alkaloids or glycosides, operate by being toxic at low concentrations. They are small molecules, comparatively cheap for the plant to synthesise, though often containing nitrogen, but with relatively short turnover times. Quantitative defences are effective in proportion to the amount present and can include cellulose and lignin that contribute the fibrous portions of plant tissue, tannins and terpenes. These molecules are relatively expensive to manufacture but require much less maintenance, as shown for terpenes by Gershenzon (1994). The resource availability hypothesis predicts a switch between qualitative defences in short-lived leaves and quantitative ones in long-lived leaves because the lower maintenance costs of the latter become more relevant the longer the leaf lasts. Lebreton (1982), in coming to conclusions similar to those of Coley *et al.* (1985), noted the strong negative correlation between frequency of con-

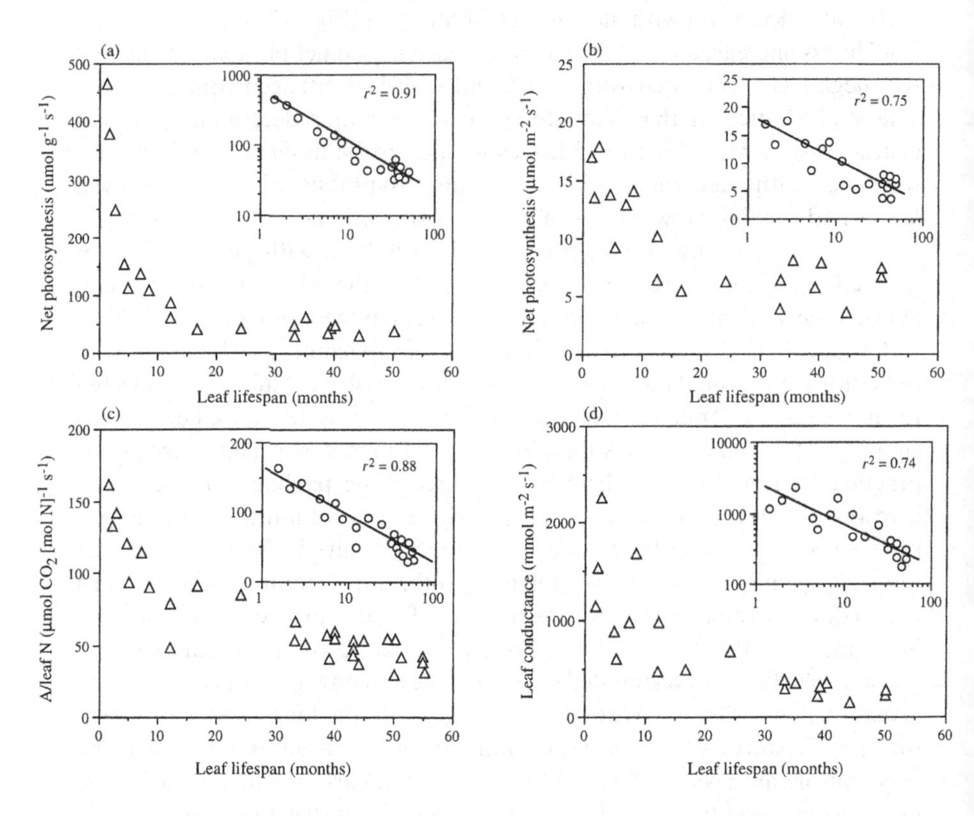


Figure 2.25 (a) Mass-based net photosynthetic rate (nmol $g^{-1} s^{-1}$) in relation to leaf life span (months) for sun leaves of 21 Amazonian tree species. The insert shows the same variables both on a $\log_{10}$-transformed scale. (b) Area-based net photosynthetic rate (μmol $m^{-2} s^{-1}$) in relation to leaf lifespan for sun leaves of 21 Amazonian tree species. (c) Net photosynthesis per unit leaf N (μmol $CO_2$ [mol N] $^{-1} s^{-1}$) in relation to leaf lifespan (months) for sun and shade leaves of 21 and 9 species, respectively (relation not affected by inclusion of shade leaves). (d) Leaf diffusive conductance to water vapour (mmol $m^{-2} s^{-1}$) in relation to leaf life span (months) (data as in a). After Reich *et al.* (1991).

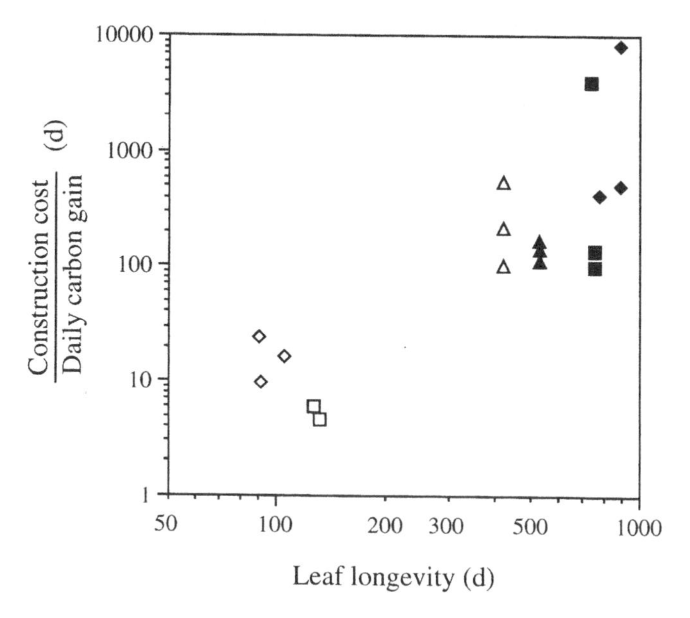


Figure 2.26 Relationship between leaf longevity and the ratio of construction cost (g $CO_2$ g leaf$^{-1}$) to daily carbon gain [g $CO_2$ g leaf$^{-1}$ d$^{-1}$]. This ratio is an estimate of the time required for a leaf to amortise its construction cost. Different symbols represent different individuals of several species of *Piper*. After Williams *et al.* (1989).

densed-tannin-containing species and of alkaloid-bearing ones in major taxa of plants. Waterman & McKey (1989) reported a more direct finding of the negative correlation between condensed tannins and alkaloids in samples of rain-forest trees from Africa. This they describe as 'intuitively pleasing' because tannins should bind to alkaloids, making the dual possession of both defences self-defeating.

Average foliar chemistry values for multi-species samples show, with a few exceptions, that fibre and tannin concentrations increase as nitrogen concentration decreases (Fig. 2.27), as would be predicted by the resource availability hypothesis. However, not all studies provide support for the hypothesis. Nascimento & Langenheim (1986) found no significant difference in foliar concentrations of terpenes and tannins between *Copaifera multijuga* trees growing on sites of different soil fertility in Central Amazonia. There was no significant difference in foliar tannin concentrations between secondary forest species from fertile and nutrient-poor sites in Singapore (Turner 1995a). However, the species typical of infertile sites had much tougher leaves (Choong *et al.* 1992).

**The influence of shade**

Bongers & Popma (1988) have conducted the most detailed study of leaf form variation within the crown of individual tropical trees. They compared leaves grown in exposed conditions receiving direct sunlight with well-shaded leaves lower in the canopy for 61 species growing in the rain forest at Los Tuxtlas, Mexico. There were significant differences between sun and shade leaves such that the sun leaves were smaller, had a higher leaf mass per unit area (LMA), had a lower mass-based potassium concentration but more nitrogen per unit area, with greater stomatal density and lower spongy mesophyll to palisade mesophyll thickness ratio than the shade leaves (Fig. 2.28). Sun leaves are probably smaller to allow more efficient convective

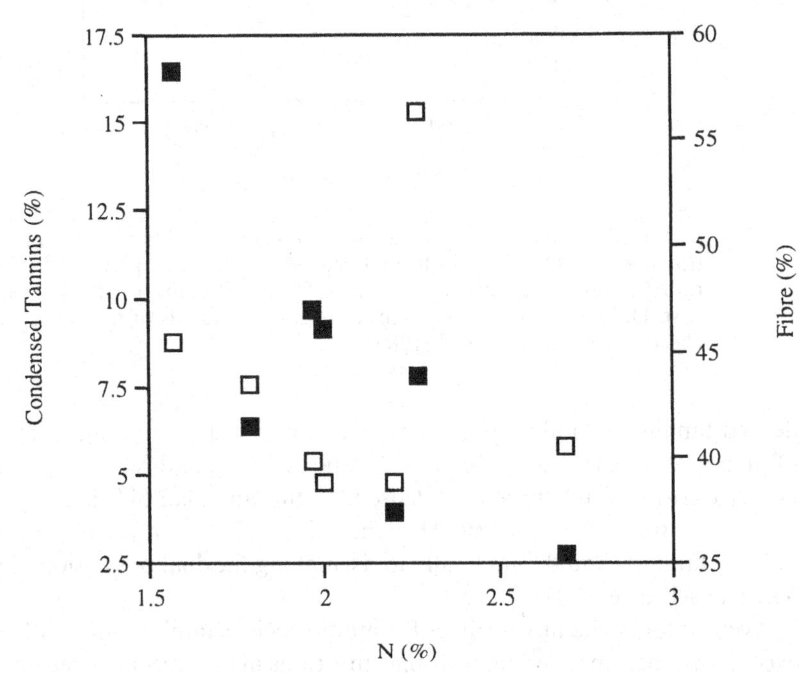


Figure 2.27 Average foliar condensed tannin (open symbols) and fibre (closed symbols) concentrations against foliar nitrogen concentration for different tropical rain-forest sites. Each point represents the mean value of an average of 28 species (range 14–44). The sites are Barro Colorado Island, Panama (saplings of persistent species growing in gaps) (Coley 1983), Douala-Edea, Cameroon (Waterman *et al.* 1988), Kibale, Uganda (Waterman *et al.* 1988), Kakachi, India (Oates *et al.* 1980), Korup, Cameroon (Waterman & Mole 1989), Kuala Lompat, Peninsular Malaysia (Waterman *et al.* 1988), Sepilok, Sabah (Waterman *et al.* 1988), Ranomafana, Madagascar (Ganzhorn 1992), and Tiwai, Sierra Leone (Oates *et al.* 1990). The last site seems anomalously high in condensed tannins.

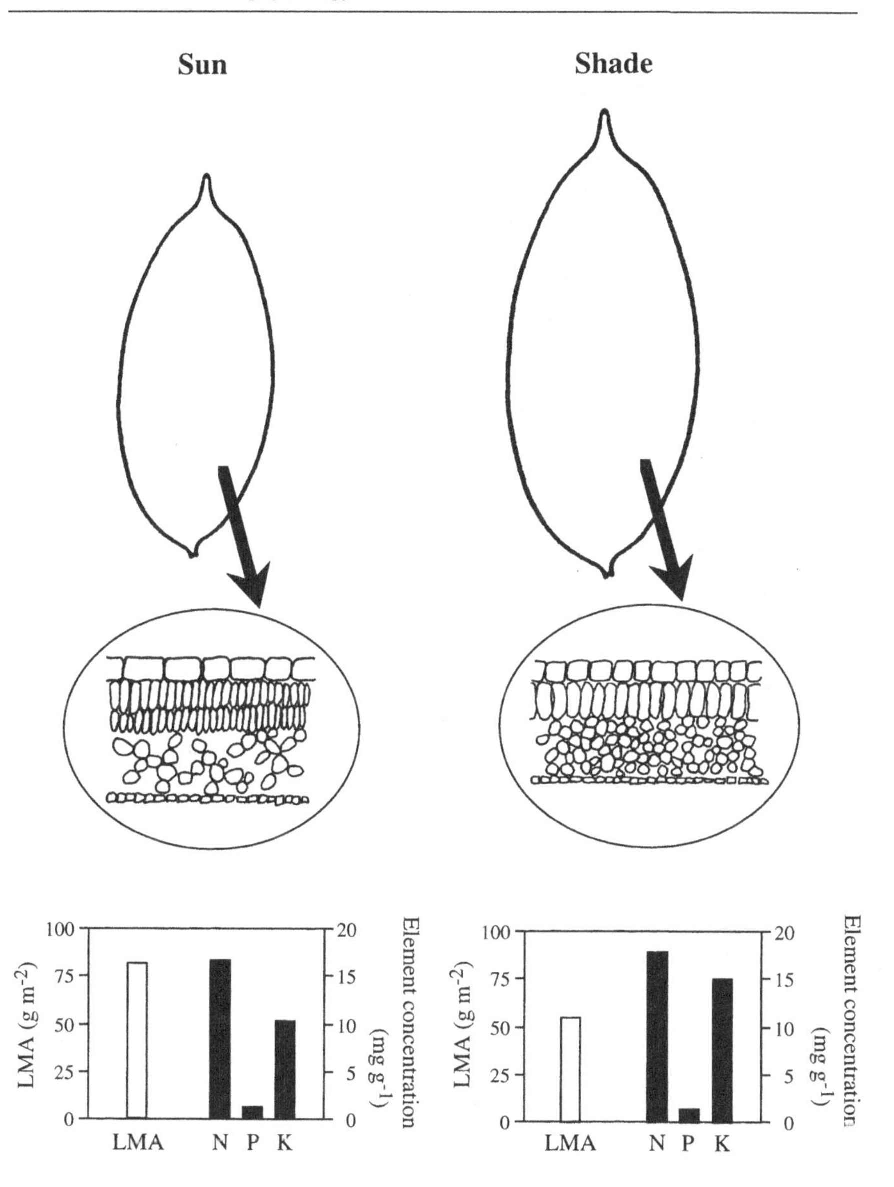


Figure 2.28 Schematic representation of the relative differences in leaf form between sun and shade conditions based on data from Bongers & Popma (1988).

cooling. Large leaves have a high boundary-layer resistance that reduces the rate of transfer of sensible heat between the leaf and the general atmosphere. Large leaves in full sun can reach lethally high temperatures. Therefore, small leaves are favoured in exposed conditions. Large-leaved species must be able to maintain rapid transpiration to prevent overheating. Leaf thickness may be greater in sun leaves because the high incident radiation means that sufficient light penetrates several layers of chloroplast-bearing mesophyll cells to allow them all to photosynthesise at high rates. Thicker leaves also tend to have better water-use efficiency because the carbon fixation rate rises more rapidly with thickness than does the rate of water loss.

Foliar defences may also vary with degree of shading. Mole *et al.* (1988) found that sun leaves generally had higher concentrations of phenolics, including condensed tannins, than shade leaves for four species in Sierra Leone. This may be more of a response to environmentally related changes in metabolism than alterations in plant investment in defence. Fibre concentrations were not influenced by shading.

A simplistic argument, often resorted to by students, when asked why shade leaves are bigger than sun ones, is to state that this allows the leaf in the shade to capture more light. Under the low irradiances of the forest interior light interception is probably directly proportional to leaf area, but a plant with many small leaves could still intercept as much light as one with few large ones. The operative feature under natural selection in the shade is probably the amount of light intercepted by the leaves per unit of investment (including that in terms of support and supply) per unit area of leaf. Low LMA and low N per unit area reflect low investment costs per unit area and more efficient shade leaves, although a lower limit on LMA may be set by susceptibility to damage. In the sun, where there is ample light, other factors such as efficient use of water, or the need to avoid overheating come to the fore in leaf design. The lower potassium concentrations of sun leaves have not yet received a physiological explanation.

Poorter *et al.* (1995) studied the variation in leaf optical properties in four canopy species in Costa Rica with leaf height in the canopy. They found small, but statistically significant, differences in leaf absorptance with height in the canopy. The more 'shade-tolerant' species (*Lecythis ampla* and *Minquartia guianensis*) showed higher absorptance in the shade than the more 'light-demanding' species (*Dipteryx panamensis* and *Simarouba amara*), but the position was reversed in the bright light at 20 m up. The absorptance efficiency per unit dry mass decreased with height in the canopy. The total chlorophyll concentration on a mass basis decreased with height in the canopy for the two species studied. The results reflect changing limitations on photosynthesis with irradiance regime. Under the shaded conditions of the forest interior there is a shortage of light for photosynthesis, and efficient light capture is required with a greater amount of chlorophyll. Under bright light conditions there is often more than enough light to saturate the photosynthetic system of the leaf, and so carbon dioxide uptake may limit photosynthesis. Efficiency of photon capture may be less important than use of water or other factors in sun leaves.

The generally expected differences in photosynthetic performance between sun- and shade-grown leaves are for the sun leaf to have a higher dark respiration rate per unit area, higher light compensation point, higher saturating irradiance and higher maximal assimilation rate per unit area. There have not been many studies of leaves on mature tropical trees to confirm these expected responses to shade. Individuals of three pygmy-tree species of *Psychotria* growing in the shaded understorey and in gaps on Barro Colorado Island, Panama, were compared for photosynthetic performance (Mulkey *et al.* 1993). The gap plants showed higher respiration and assimilation rates than the shaded ones. However, available data indicate that there may be relatively little difference in physiological potential among leaves within the crowns of large rain forest trees in the upper canopy. For example, *Argyrodendron peralatum* and *Castanospermum australe* in Queensland, Australia (Doley *et al.* 1987; Myers *et al.* 1987), *Dryobalanops lanceolata* in Brunei (Barker & Booth 1996) and *Pentaclethra macroloba* in Costa Rica (Oberbauer & Strain 1986) were found to show little or no consistent intra-crown variation. There were differences detected in leaf morphology and performance within the crown of a 20 m tree of *Macaranga conifera* in Borneo (Ishida *et al.* 1999b). These were largely between leaves in the upper canopy and the lowest layer more than 45 cm beneath the uppermost leaves. It is possible that the light gradients at the top of the forest canopy are not marked enough to lead to differentiation in leaf performance in most species, or the differences are too slight to distinguish from the large variation present among leaves on a branch.

The instantaneous irradiance impinging on a leaf in the rain forest varies rapidly and with a large magnitude. This short-term variation is particularly marked in the forest understorey where the dim foliage-filtered light is occasionally interspersed with brief periods of much higher irradiance when direct radiation finds its way through holes in the canopy (Chazdon 1988). Such sunflecks can make up 10–85% of the total irradiance received in the forest understorey (Chazdon *et al.* 1996). Technical improvements over the last decade have made it possible to study the efficiency of plant use of rapidly fluctuating irradiance. Among six Rubiaceae species of pygmy tree on Barro Colorado Island, three shade-tolerant species of *Psychotria* growing in the forest understorey exhibited more rapid induction of photosynthesis when exposed to a sunfleck and greater persistence of the induced state than three more light-demanding species growing in gaps and at the edge of clearings (Valladares *et al.* 1997). The understorey species made more efficient use of short sunflecks, but once sunfleck duration reached about 10 s there was little difference in performance between the species groups.

**High light conditions**

Leaves growing in full sun at the top of the canopy can achieve very high rates of photosynthesis. *Ficus insipida* in Panama was recorded with a maximal photosynthetic rate of 33.1 μmol $CO_2$ $m^{-2}s^{-1}$ (Zotz *et al.* 1995), approaching that of herbaceous crop plants. However, on sunny days very high irradiances (> 2000 μmol $m^{-2}s^{-1}$) may cause photoinhibition, and high temperatures may lead to large vapour pressure deficits (VPD) between the leaf and the atmosphere. Diurnal studies of leaf gas exchange of canopy leaves of tropical trees generally show that photosynthetic rates follow the rates of incident solar radiation and high levels of fixation are maintained throughout the day (Zotz & Winter 1994). During a dry spell, trees may respond to internal water shortage or the high vapour pressure deficit by closing stomata in the afternoon to conserve water. Such midday depressions in photosynthetic rates have been observed in *Ficus insipida* in the dry season where the photosynthetic rate dropped from over 30 μmol $m^{-2}s^{-1}$ at 0945 h to 1 μmol $m^{-2}s^{-1}$ at 1120 h. Canopy leaves of *Qualea rosea* in French Guiana showed low carbon fixation rates (1.4 μmol $m^{-2}s^{-1}$) on an afternoon at the start of the wet season (Roy & Salager 1992), apparently not because of low shoot water potential, but because of the large VPD. Midday depression in maximal photosynthetic rates is probably due to a combination of stomatal limitation and reduced photochemical activity caused by high irradiances and high temperatures in most cases (Ishida *et al.* 1999c).

Chlorophyll fluorescence techniques have allowed the easier detection of photoinhibition over recent years. Species growing in 60–90 $m^2$ gaps on Barro Colorado Island were shown to exhibit photoinhibition by reduced ratios of variable to maximum chlorophyll fluorescence emission ($F_V/F_M$) when irradiances of 1700–1800 μmol $m^{-2}s^{-2}$ were incident at around midday (Krause & Winter 1996; Thiele *et al.* 1998). Recovery took about two hours under conditions of low light. During very high irradiances the photosystems become saturated and potentially damaging excess energy is present. The xanthophyll-cycle pigments are believed to play a key role in dissipating some of the excess energy. Under high light, violaxanthin is converted to zeaxanthin. The conversion is correlated with increased rates of heat dissipation, but the exact mechanism remains unclear. Königer *et al.* (1995) measured the xanthophyll-pigment pool size in canopy leaves from Panama. They found a strong positive correlation between maximal photosynthetic rates and the xanthophyll-cycle pool size per mole of total chlorophyll. Understorey species were substantially lower in the size of their xanthophyll-pigment pool: 25 out of 41 tree species spot tested on Barro Colorado Island showed significantly higher anti-oxidant activities in leaves growing in sunny conditions compared to those in shade (Frankel & Berenbaum 1999).

Young leaves may be more susceptible to photoinhibition than mature ones if they have less chlorophyll to absorb the incident photons. Krause *et al.* (1995) did find greater reductions in $F_V/F_M$ for young leaves (approaching full expansion) than mature ones in canopy foliage in Panama, despite the young leaves having a 20% larger xanthophyll-cycle pool. Recovery from photoinhibition by the leaves was a two-phase phenomenon. The first phase took about 30 min of low irradiance, e.g. clouds over the sun. The second phase was much longer. It is believed that the second phase involves synthesis of replacement D1 protein for the photosystem II reaction centre. D1 appears to be a deliberate weak link in the photosystem that gives way under pressure of very high irradiance, dissipating some of the excess energy in the process. Plants growing in high light conditions, e.g. gaps, seem to rely more heavily on the xanthophyll pigment cycle for recovery from photoinhibition whereas shade plants use the D1 protein breakdown and resynthesis mechanism (Thiele *et al.* 1998).

In the understorey on Barro Colorado Island pygmy trees and saplings of large trees that had long-lived leaves showed less marked reductions in $F_V/F_M$ on exposure to bright light than those with leaves of short life span (Lovelock *et al.* 1998). Presumably species investing in durable leaves require them to have protection from photoinhibition, whereas short-lived leaves can be economically replaced if damaged by high light.

### Other factors influencing leaf performance

Little research has been conducted on the influence of age of an individual on the physiology of its leaves in tropical trees. Leaf age probably does have a major influence on leaf performance, even after full expansion has been reached. There is often a very high variance in performance among leaves on the same individual. For example, canopy leaves of *Dryobalanops aromatica* in Peninsular Malaysia were found to have an absolute maximum net assimilation rate of 12 $\mu$mol $m^{-2} s^{-1}$, but the large variance resulted in a mean assimilation rate of 6.6 $\mu$mol $m^{-2} s^{-1}$ (Ishida *et al.* 1996). Leaf age may be one of the factors influencing photosynthetic performance. The season of development may be another. A number of studies on trees growing in seasonally dry climates have shown differences in average leaf performance between different seasons (Hogan *et al.* 1995; Kitajima *et al.* 1997).

Table 2.9. *The mean values (± standard error) of various characteristics of sun leaves of trees in three species groups from Los Tuxtlas, Mexico*

| Characteristic | Obligate gap species | Gap-dependent species | Gap-independent species | Statistical significance[a] |
|---|---|---|---|---|
| Leaf area ($cm^2$) | 202 ± 90 | 58 ± 13 | 54 ± 10 | ** |
| LMA ($g\,m^{-2}$) | 61.1 ± 4.7 | 94.3 ± 3.9 | 69.2 ± 6.5 | **** |
| N/area ($g\,m^{-2}$) | 1.22 ± 0.08 | 1.51 ± 0.08 | 1.03 ± 0.07 | *** |
| P/area ($g\,m^{-2}$) | 0.09 ± 0.01 | 0.12 ± 0.01 | 0.08 ± 0.01 | ** |
| K/area ($g\,m^{-2}$) | 0.63 ± 0.05 | 0.85 ± 0.07 | 0.76 ± 0.06 | * |
| N/P | 13.9 ± 0.7 | 14.0 ± 0.9 | 14.3 ± 1.1 | — |
| Stomatal density ($mm^{-1}$) | 524 ± 105 | 462 ± 98 | 201 ± 36 | ** |
| Leaf thickness (ωm) | 183 ± 10 | 229 ± 13 | 188 ± 11 | * |
| NPPR[b] | 1.25 ± 0.20 | 1.44 ± 0.12 | 1.68 ± 0.19 | — |

[a]ANOVA results. * $p < 0.05$, ** $p < 0.01$, *** $p < 0.001$, **** $p < 0.0001$, — $p > 0.05$.
[b]Non-palisade to palisade mesophyll tissue thickness ratio.
Based on data of Popma *et al.* (1992).

Popma *et al.* (1988) studied the vertical variation in leaf form in the forest at Los Tuxtlas, Mexico. They divided the forest into 8 m height intervals and for each interval included all the species that occurred in it. There was a trend for an increased proportion of compound-leaved species in increasing height classes in the forest, rising to nearly 40% of species in the tallest height class (> 24 m). This concurs with work at other sites (Brown 1919; Beard 1946; Rollet 1990). They also found a weak trend for reduced leaf size with maximal tree stature.

### Species groups based on leaf characteristics

The 61 species studied at Los Tuxtlas (Popma *et al.* 1992), were divided into three ecological groups based on their apparent light requirement for regeneration. The groups were 'obligate gap' species that were only ever found in gaps, 'gap-dependent' species that can survive in canopy shade as juveniles but need gaps to grow to large size, and 'gap-independent' species that can complete their life cycle in the shade (Popma *et al.* 1992). The obligate gap species had larger leaves than the other two groups (Table 2.9). The gap-dependent species had the highest LMA and more N and P per unit area than the other two groups. They also showed the greatest plasticity in leaf form between sun and shade leaves. The gap-independent species had more elongated leaves with fewer, but larger, stomata per unit area.

Analysis of published data on leaf photosynthetic rates (Fig. 2.29) for different tree species shows a pattern of increased carbon dioxide uptake rates

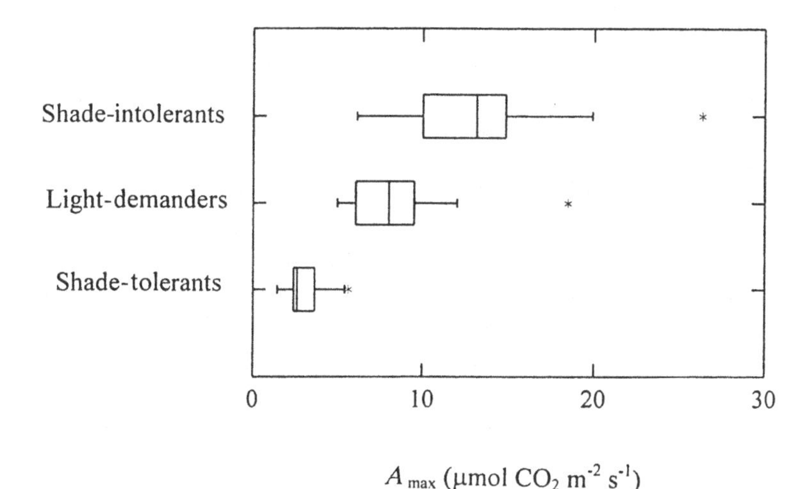


Figure 2.29 Box-and-whisker diagram of maximum photosynthetic rates ($A_{max}$) of leaves of tropical tree species divided into three ecological classes. Extreme outliers shown as asterisks. Species as follows: Shade-intolerants: *Adinandra dumosa*[12], *Annona spraguei*[1], *Antirrhoea trichantha*[2], *Bellucia grossularioides*[3], *Castilla elastica*[2], *Cecropia ficifolia*[3], *Cecropia longipes*[2,4], *Ceiba pentandra*[11], *Clidemia sericea*[3], *Dillenia suffruticosa*[12], *Ficus insipida*[5], *Ficus obtusifolia*[4], *Glochidion rubrum*[13], *Macaranga heynei*[12], *Macaranga hypoleuca*[13], *Mallotus paniculatus*[12], *Melastoma malabathricum*[12], *Piper umbellatum*[6], *Schefflera morototoni*[4], *Solanum straminifolium*[3], *Trema tomentosa*[12], *Urera caracasana*[2], *Vismia japurensis*[3], *Vismia lauriformis*[3]. Light demanders: *Anacardium excelsum*[2,4], *Chisocheton macranthus*[13], *Dialium pachyphyllum*[7], *Dipterocarpus caudiferus*[13], *Dryobalanops aromatica*[8], *Dryobalanops lanceolata*[13], *Licania heteromorpha*[3], *Luehea seemannii*[2,4], *Ocotea costulata*[3], *Parashorea tomentella*[13], *Pentace adenophora*[13], *Pentaclethra macroloba*[9], *Piper auritum*[6], *Protium* sp.[3], *Shorea seminis*[13], *Shorea xanthophylla*[13]. Shade-tolerant species (understorey conditions): *Desmopsis panamensis*[1], *Ouratea lucens*[1], *Piper aequale*[6], *Piper lapathifolium*[6], *Psychotria furcata*[10], *Psychotria limonensis*[10], *Psychotria marginata*[10], *Xylopia micrantha*[1].

Data from [1]Kursar & Coley (1992c), [2]Kitajima *et al.* (1997) (data for pre-dry-season leaves), [3]Reich *et al.* (1995), [4]Hogan *et al.* (1995), [5]Zotz *et al.* (1995) (dry season average values), [6]Chazdon & Field (1987), [7]Koch *et al.* (1994), [8]Ishida *et al.* (1996), [9]Oberbauer & Strain (1986), [10]Mulkcy *et al.* (1993), [11]Zotz & Winter (1994), [12]Tan *et al.* (1994), [13]Eschenbach *et al.* (1998).

for more light-demanding species. Fast-growing, shade-intolerant species are generally found to have maximal rates of more than 10 μmol $m^{-2}$ $s^{-1}$, often in the range 15–25 μmol $m^{-2}$ $s^{-1}$. Light-demanding species, which usually have relatively shade-tolerant juvenile stages, have mature leaves with a maximal net assimilation rate averaging 5–12 μmol $m^{-2}$ $s^{-1}$. Shade-tolerant species in the shade frequently photosynthesise at less than 5 μmol $m^{-2}s^{-1}$, but growing in the higher irradiances of canopy gaps they can probably improve to nearly 10 μmol $m^{-2}s^{-1}$. The degree of acclimation to light environment is likely to be less in the most shade-tolerant species than the more light-demanding ones. A similar relative pattern among the groups is seen for mass-based and N-based photosynthetic rates. Maximum instantaneous rates of carbon dioxide uptake can be related directly to total daily uptake across a wide range of leaf performance (Fig. 2.30).

The relative position of a species on the axis of shade tolerance is often referred to via its time of appearance in succession. Early-successional species are considered light-demanding, late-successional species are more shade

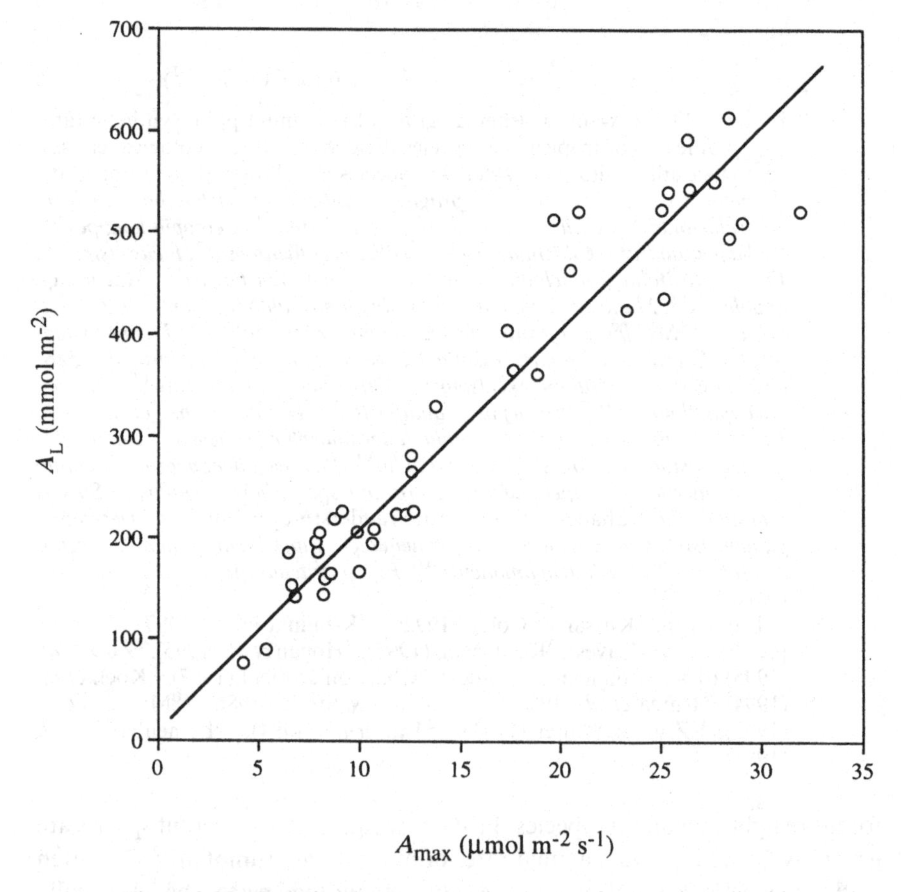


Figure 2.30 Relation between maximal rate of net carbon dioxide uptake ($A_{max}$) and diurnal carbon gain ($A_L$) for the leaves of five tropical tree species. After Zotz *et al.* (1995).

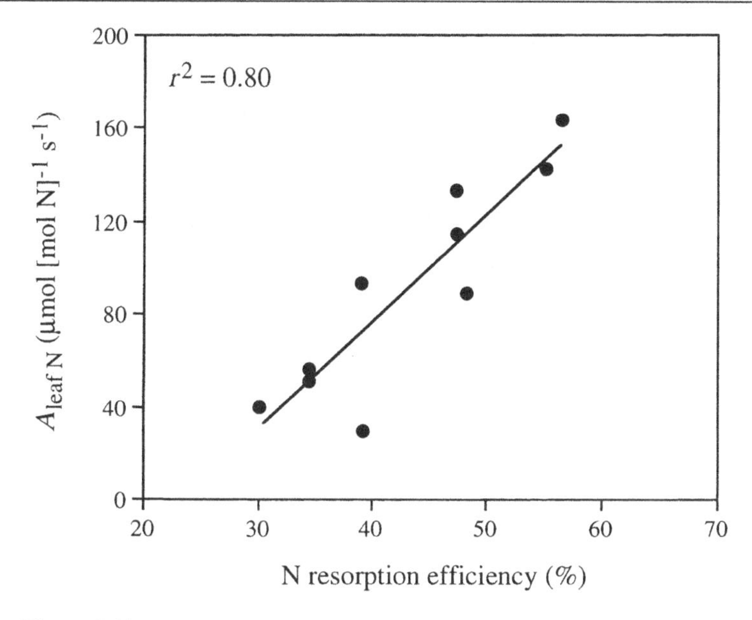


Figure 2.31 Potential photosynthetic N-use efficiency (defined as instantaneous net photosynthesis per unit leaf N) in relation to N resorption efficiency ($p < 0.001$, $r^2$=0.80) for ten Amazonian species. After Reich *et al.* (1995).

tolerant. Reich *et al.* (1995) analysed a subset of 13 of the 23 species studied in the San Carlos de Rio Negro region of Venezuela. There was a clear correlation between successional status and leaf longevity, with net assimilation rate and foliar nutrient concentrations (particularly Ca and Mg) decreasing and LMA increasing as leaf longevity increased with successional stage. Early-colonising species had higher foliar concentrations of nitrogen and phosphorus than late-successional and primary-forest species. All the species retranslocated more P (average of 62%) than N (43%). The early-successional species showed the greatest retranslocation rates. There was a positive correlation between N-resorption efficiency and the photosynthetic rate per unit foliar nitrogen across the species for which data were available (Fig. 2.31). For a given nitrogen concentration, early-successional species had a higher mass-based photosynthetic rate than late-successional species. Early-successional species showed greater variation in leaf N concentrations and a greater change in leaf photosynthetic rates for a given change in N concentration. Raaimakers *et al.* (1995) found a similar greater responsiveness in photosynthetic rates of early-successional species to changes in foliar nitrogen concentration at a site in Guyana. They also showed the same pattern

Table 2.10. *Assimilation rates in relation to foliar nutrient concentrations for species from Guyana*

Values are mean ± standard error.

| | $A_{sat}/N$ ($\mu$mol $CO_2$ $mol^{-1}$ N $s^{-1}$) | $A_{sat}/P$ (mmol $CO_2$ $mol^{-1}$ P $s^{-1}$) |
|---|---|---|
| All species | 84 ± 3 | 5.4 ± 0.2 |
| Pioneer species | 118 ± 5 | 7.8 ± 0.3 |
| *Cecropia obtusa* | 133 ± 9 | 8.1 ± 0.5 |
| *Tapirira marchandii* | 121 ± 6 | 7.2 ± 0.3 |
| *Goupia glabra* | 100 ± 7 | 8.1 ± 0.5 |
| Climax species | 66 ± 2 | 4.2 ± 0.2 |
| *Peltogyne venosa* | 80 ± 5 | 4.9 ± 0.3 |
| *Dicymbe altsonii* | 74 ± 6 | 4.7 ± 0.4 |
| *Mora excelsa* | 72 ± 4 | 3.3 ± 0.2 |
| *Eperua falcata* | 74 ± 6 | 4.5 ± 0.4 |
| *Eschweilera sagotiana* | 59 ± 3 | 4.6 ± 0.3 |
| *Chlorocardium rodiei* | 39 ± 3 | 3.1 ± 0.3 |

Data from Raaimakers *et al.* (1995).

with leaf phosphorus. At San Carlos de Rio Negro, leaf calcium concentrations were very low in the late-successional species. It is possible that a shortage of calcium was depressing photosynthetic rates in these species. There was no evidence of greater instantaneous nutrient use efficiency in the late-successional species. In Guyana, quite the opposite was the case, with the 'pioneer' species having significantly higher instantaneous nitrogen- and phosphorus-use efficiencies than the 'climax' species (Table 2.10). The only way in which late-successional species could have been using nutrients more efficiently was through longer residence times in the leaves.

Fredeen & Field (1991) measured the dark respiration rates of leaves of six species of the genus *Piper* growing at Los Tuxtlas, Mexico. The gap specialists *Piper auritum* and *P. umbellatum* had area-based respiration rates about twice those of the shade-tolerant species. Leaf dark respiration rate was positively correlated with daily total photosynthetic photon flux density and negatively to mean leaf longevity.

I believe it is possible to summarise the likely characteristics of species of differing shade tolerance, as given in Table 2.11. I have chosen to recognise three divisions of the shade-tolerance axis, referring to the least shade-tolerant species as shade intolerants, the most shade-tolerant as shade tolerants and the intermediate group as light-demanding. I would envisage the regeneration characteristics of the three groups to be as follows: shade intolerants require gaps to establish and grow to maturity and die relatively quickly if strongly shaded. Shade-tolerant species can persist readily in deep

Table 2.11. *Generalisations about leaf form among three ecological groups of tropical rain-forest tree species*

| | Shade intolerants | Light demanders | Shade tolerants |
|---|---|---|---|
| Leaf size | big | small | small |
| LMA | low | high | low |
| Lamina thickness | thin | thick | thin |
| NPPR | low | intermediate | high |
| Stomatal density | high | high | low |
| Nutrient/mass | high | low | low[a] |
| Nutrient/area | low | high | low[a] |
| Assimilation/mass | high | low | low |
| Assimilation/area | high | high | low |
| Leaf longevity | short | long | long |
| Leaf production | continuous | rhythmic | rhythmic |
| Leaf expansion | rapid | slow | rapid |
| Delayed greening | infrequent | infrequent | frequent |
| Chemical defence | weak | strong | strong |
| Leaf toughness | low | high | high |
| Ant defence | common | infrequent | rare |
| Plasticity of leaf form | low | high | low |

[a]Often high for K.

shade and may be able to complete their full life cycle in such conditions. However, most species probably do benefit from the increased irradiance in gaps, and may need such conditions for reproduction. The intermediate group can persist in shade, but probably not as successfully as the shade tolerants, but only make headway with growth if in, or near, a canopy gap. I do not set any formal boundaries on the groups, merely indicate that a three-way contrast results in a sufficient degree of repetition in suites of traits among species of similar regeneration ecology to allow recognition of certain generalisations. However, probably inevitably, there are species that do not conform to some, or even all, of these generalisations.

The leaves of shade intolerants can be summarised as short-term high-performance photosynthetic units. Fast growth is of paramount importance to these species, and this is achieved by a rapid turnover of large thin leaves with high assimilation rates, particularly per unit dry mass. Leaf respiration rates are high and an inability to down-regulate both respiration and leaf-production rate appears to be the proximate cause of the lack of shade tolerance in these species (see later chapters). Leaf defences, particularly physical ones, are poor, probably because their capital and maintenance costs would detract from fast growth. At the other end of the spectrum, shade-tolerant species also have thin leaves, but they have much lower metabolic rates. Conservation of resources and persistence are much more important in the resource-poor environment of the forest understorey. The intermediate group is the most flexible, producing high-investment, high-return leaves in favourable conditions.

# 3

# Tree performance

## Age, size and growth in tropical rain-forest trees

The growth rates of tropical trees are usually estimated by repeat measurements of dimensions, most often that of stem girth or diameter. If tropical trees could be aged easily and accurately then estimates of average growth rates would be possible from one-off measurements. The age of trees in the temperate zone can generally be estimated precisely from counting annual rings on cores taken from the trunk base. Tree rings are caused by periodic variation in the nature of the wood laid down at the cambium. Wood growth is often so uniform in tropical trees that rings are undetectable. Where they are present they may represent checks on normal growth that occurred at irregular intervals. However, there are certain tropical regions where dry seasons, or very wet ones with flooding, are strong enough annual signals to produce yearly rings (Martínez-Ramos & Alvarez-Buylla 1998).

Most tropical lowland forest trees do not have annual rings. Old trees can be aged by using $^{14}C$-dating techniques on wood samples from the heart of the trunk base. This method has recently been applied with some startling results on large trees from the Amazon basin (Chambers *et al.* 1998). Twenty large trees from near Manaus, Brazil, were found to have ages (± 80 years) of 200–1400 years. The oldest was an individual of *Cariniana micrantha*. There was a poor correlation between size and age, even within a species. These radio-carbon estimates of age are considerably longer than most earlier indirect ones (Chambers *et al.* 1998), and raise the possibility that a few big trees may exert a strong influence on the genetic make-up of populations over a very long period.

The commonest approach to ageing tropical trees has been to extrapolate from size measurements and known average increment data. This has severe limitations if an accurate estimate of a particular tree's age is required, because of the large variation in growth rates between individuals both within and between different sites. Lieberman *et al.* (1985a) developed a statistical technique to estimate the ages of large trees from known increment data. They used computerised random sampling of species-specific 13-year increment data for trees of certain size classes at La Selva in Costa Rica. From a sample of 1700 simulations, maximum, minimum and median growth trajectories were derived for each species with 12 or more individuals on the 12.4 ha enumeration plot. These could then be used to estimate the range of time it was likely to take a tree of a particular species to grow from the minimum size enumerated (10 cm dbh) to the maximum dbh occurring in the plot. Lieberman *et al.* (1985) proposed that the duration of the minimum growth trajectory was an estimate of the maximum longevity of the species. For the 45 species included in the study, the maximum projected lifespan ranged from 52 to 442 years (mean 190). Using similar techniques in Ecuador, Korning & Balslev (1994) estimated longevities for 22 species to range from 54 to 529 years (mean 254). Condit *et al.* (1993) also derived estimates for the average time taken by species to grow between particular stem diameters. They employed quadratic regression to obtain the relation between tree dbh and absolute diameter growth rate for 160 species on Barro Colorado Island, based on measurements made in the 50 ha plot. These equations could then be used to estimate passage times. Most species were found to take 5–120 years on average to grow from 1 to 10 cm dbh.

Tree longevities can also be estimated from population mortality rates (Martínez-Ramos & Alvarez-Buylla 1998), although these estimates tend to be very sensitive to variations in mortality rate and population size. Among 205 species on Barro Colorado Island, Panama, tree maximal lifespans (from 1 cm dbh) estimated in this way were as short as 35 years in *Solanum hayesii* and reached nearly 2000 years in *Swartzia simplex* (Condit *et al.* 1995).

In certain instances other features of trees can be related to age. Leaf or branch scars can be counted, particularly on species with a relatively simple architecture such as palms, and may provide a way of estimating individual age. Palms often produce new leaves at a very consistent rate, regardless of growth environment, as do some species of dicot, particularly as saplings (King 1993). Martínez-Ramos *et al.* (1988) found that the understorey palm *Astrocaryum mexicanum* grew so uniformly in stem length and was so common that tree-fall gaps in the rain forests of Los Tuxtlas, Mexico, could be aged by examining the palm stems for bends and kinks formed by the falling tree and measuring how far below the growing point of the palm they were. Few forests contain such amenable species.

Martínez-Ramos & Alvarez-Buylla (1998) compared the different indirect methods of assessing tree longevity with data for seven species from Los Tuxtlas, Mexico. Projected lifespan from mean growth rates was nearly twice as long as the other methods. Estimates based on maximum growth rates, mortality rates and forest-patch age from palm growth were quite consistent.

Does age matter in tropical trees? Because of the large variation between individuals in growth rate there is generally little correlation between age and size in tropical trees, with the possible exception of fast-growing shade-intolerant species where individuals not achieving fast-growth die. Size is therefore a better predictor of individual performance (e.g. growth rate, reproductive output, probability of survival) than age, and is also much easier to assess than age in tropical trees.

## The dynamics of tree populations in the rain forest

Trees are too long-lived to make it feasible to study cohorts of seedlings to maturity as a means of investigating the population dynamics of a given species at a particular site. The problems of ageing tropical trees make it easier to use a size-structured approach to population dynamics. However, relatively few species have been studied in sufficient detail to paramaterise the Lefkovitch matrix model of demography fully (Alvarez-Buylla *et al.* 1996). The sample is heavily biased towards the study of palms. Estimates of $\lambda$ (Table 3.1), the largest positive eigenvalue of the probability matrix, are close to unity for all the species studied except one. This indicates no long-term change in population size. The one exception, *Euterpe edulis*, was a population recovering from harvesting (Silva Matos *et al.* 1999). Sensitivity and elasticity analyses can be performed to identify at which growth stages the matrix probabilities most influential on adult population size occur. For long-lived tropical trees, the probabilities of remaining in pre-adult and adult stages are the most sensitive (Alvarez-Buylla *et al.* 1996). Probabilities for younger stages, changes between stages and those concerning fecundity are much less influential on population size. In contrast, in the fast-growing, short-lived *Cecropia obtusifolia* transitions from seedlings to juveniles and fecundity are the most important.

## Mortality in trees

Death can take many forms in tropical trees. Natural disasters including hurricanes, landslides, earthquakes, floods or droughts, lightning strikes, high winds, and diseases, as well as old age, can all be causes of mortality. Tropical rain forests occur at sites differing considerably in incidence of intense natural disturbances. For instance, Carey *et al.* (1994) reported that 64% of mortality in Venezuela involved trees dying standing, whereas in Costa Rica (Lieberman *et al.* 1985b) and Central Amazonia (Rankin-de-Merona *et al.* 1990) this figure was only one quarter of stems, and in the wet climate and rugged topography of the Rio Hoja Blanca Hills in Ecuador it was further reduced to 15% (Gale & Barfod 1999), indicating more death through treefall in the latter sites.

Repeated enumerations of plots of tropical rain forest have generally found

Table 3.1. *Estimates of population finite growth rates ($\lambda$) and elasticity analysis for tropical rain-forest tree species based on Lefkovitch matrix model*

Largest elasticities found for: $P_{ij}$, probabilities of remaining in the same stage (survival); $G_{ij}$, probabilities of advancing to following stages (growth); $F_{ij}$, fecundities of seedling recruitment. Life stages: s, seeds; j, juveniles; pa, pre-adults; a, adults.

| Species | Life history | $\lambda$ | Largest elasticity | Longevity (years) |
|---|---|---|---|---|
| Palms | | | | |
| *Astrocaryum mexicanum* | understorey, slow | 0.99–1.01 | $P_{ij(a)}$ | 125 |
| *Chamaedora tepejilota* | understorey, slow | 0.97–1.12 | $P_{ij(pa)}$ | 60 |
| *Coccothrinax readii* | understorey, slow | 1.01–1.10 | $P_{ij(pa)}$ | >145 |
| *Euterpe edulis* | understorey, slow | 1.19–1.28 | $P_{ij(a)}$ | n.d. |
| *Podococcus barteri* | understorey, slow | 1.01 | $P_{ij(pa)}$ | 75 |
| *Pseudophoenix sargentii* | understorey, slow | 1.01–1.20 | $P_{ij(pa)}$ | 80 |
| *Thrinax radiata* | understorey, slow | 0.99–1.01 | $P_{ij(pa)}$ | 120 |
| Others | | | | |
| *Araucaria hunsteinii* | canopy, slow | 0.99–1.09 | $P_{ij(a)}$ | 100 |
| *Araucaria cunninghamii* | canopy, slow | 1.01–1.02 | $P_{ij(a)}$ | 100 |
| *Brosimum alicastrum* | canopy, fast | 1.06 | $P_{ij(a)}$ | 120 |
| *Cecropia obtusifolia* | pioneer | 0.99–1.03 | $P_{ij(s)}$, $G_{ij(j)}$, $F_{ij}$ | 35 |
| *Omphalea oleifera* | canopy, medium | 1.01 | $P_{ij(a)}$ | 140 |
| *Pentaclethra macroloba* | canopy, slow | 1.00 | $P_{ij(pa)}$ | 150 |
| *Stryphnodendron excelsum* | canopy, slow | 1.05 | $P_{ij(pa)}$ | 150 |

From Alvarez-Buylla *et al.* (1996) and Silva Matos *et al.* (1999).

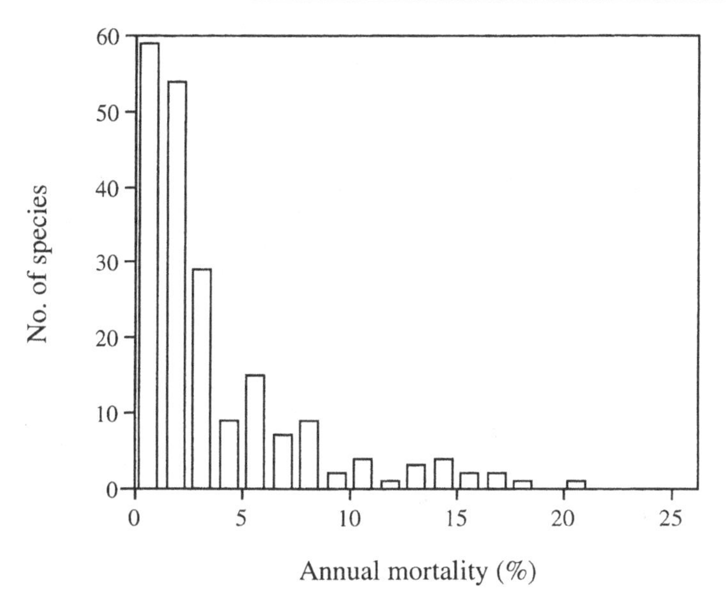


Figure 3.1 Frequency distribution of mean annual mortality rate for stems 10–99 mm dbh for 202 species on Barro Colorado Island, Panama. Data from Condit *et al.* (1995).

annual mortality rates for trees ($\geq$ 10 cm dbh) of 1–2% (Swaine *et al.* 1987). There tends to be some degree of size-dependence to mortality, at least when viewed across the full range of sizes, with seedlings having higher mortality rates than mature trees. However, once above about 10 cm dbh mortality levels off markedly.

Only in large plots are sample sizes for most species big enough to make meaningful comparisons between species. At Barro Colorado Island, Panama, it has been found that mortality rates vary considerably between species (Fig. 3.1). Over the period 1985–1990, annual mortality for stems of 10–99 mm dbh varied between zero in several species and 20% in *Solanum hayesii* (Condit *et al.* 1995). The modal mortality rates were 0.5–2.0 % (Fig. 3.1). The size dependence of mortality varied considerably between species (Fig. 3.2). Those with high mortality rates tended to show a strong increase in survivorship of individuals with size, whereas those with low overall mortality showed relatively little difference in survival between size classes.

Clark & Clark (1996) studied very large trees (> 70 cm dbh) on 150 ha at La Selva in Costa Rica. A total of 282 individuals in five species were monitored over six years. They represented only 2% of stems of 10 cm dbh or greater, but were 23% of stand basal area and 27% of its biomass. The mean annual mortality of the big trees was 0.6%, substantially lower than the average of 2.3% for all stems greater than 10 cm dbh.

## Tree growth in the forest

Most studies of the growth of tropical forest trees have found that the vast majority of trees grow very slowly. For instance, 64% of trees in plots at Bukit Lagong and Sungei Menyala in Peninsular Malaysia had diameter growth rates averaging around 1 mm $yr^{-1}$ (Fig. 3.3) over the 20–30-year

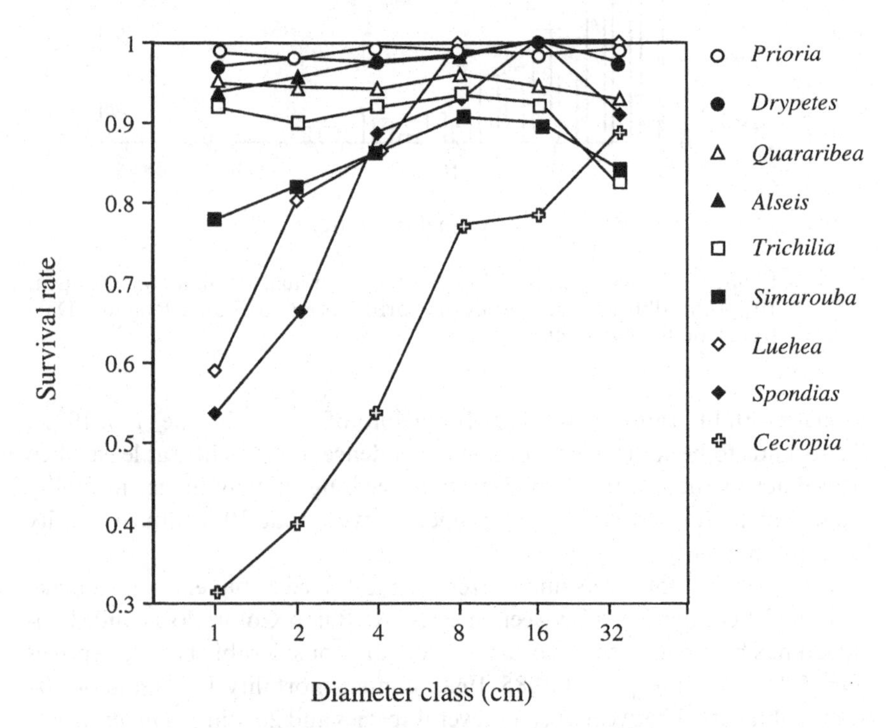


Figure 3.2 Fraction of plants surviving over the intercensus interval 1982–85 as a function of diameter class for sample tree species in the Barro Colorado Island plot. Many shade-tolerant species showed high and relatively constant rates of survival at all size classes except the largest. Examples of such species are *Prioria copaifera*, *Drypetes standleyi*, *Quararibea asterolepis*, *Alseis blackiana* and *Trichilia tuberculata*. In contrast, shade-intolerant species such as *Luehea seemannii*, *Spondias mombin* and *Cecropia insignis* all show increasing survival with increasing diameter, approaching the high survival rates of the shade-tolerant species once the shade-intolerant species reach the high-light environment of the canopy. After Hubbell & Foster (1990).

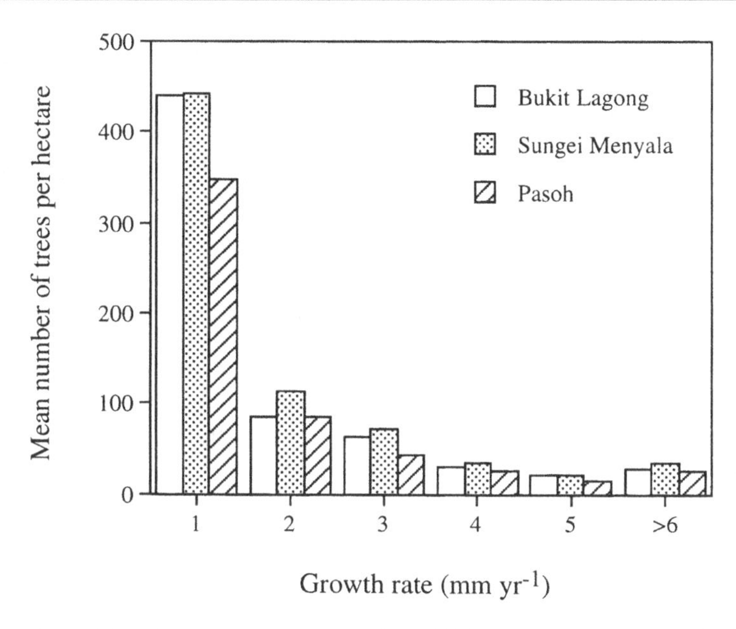


Figure 3.3 Frequency of trees (≥ 10 cm dbh) in growth rate classes in three primary dipterocarp forests in Peninsular Malaysia. After Manokaran & Kochummen (1994).

measurement period (Manokaran & Kochummen 1994). Similarly, saplings of 1–4 cm dbh on Barro Colorado Island showed a negative exponential distribution of growth rate frequency for three years of data (Welden *et al.* 1991). The modal class for growth was zero. In conclusion, most trees in the rain forest are hardly growing, and some even shrink. Being hit by fallen debris, broken off by large mammals or attacked by pests or disease can all cause leader shoot loss and a reduction in size, with young trees being the most susceptible (Clark & Clark 1991). Large trees generally grow faster (Fig. 3.4), even when growth is expressed on a basis relative to original size, probably because larger individuals are usually in more brightly lit conditions.

The growth rates of individual species in the rain forest vary considerably, with average diameter growth rates generally in the range 0.5–6 mm $yr^{-1}$, with maximal rates reaching 15 mm $yr^{-1}$ (Fig. 3.5). There tends to be strong autocorrelation between successive increment measurements for forest trees (Swaine *et al.* 1987). Slow growers remain stagnant in growth, whereas leading trees maintain high growth rates. There are several possible reasons for this. The most obvious is that the fast-growers are occupying particularly favourable sites in the forest whereas the slow-growers are in the generally unfavourable milieu of the forest interior. Trees need adequate supplies of light, water and nutrients to grow quickly. Soil water and nutrient availability do exhibit both temporal and spatial patchiness on the forest floor, but not with such a consistent and large variation in supply rate as light. In the dimly lit forest understorey where most tree individuals are situated there is insufficient light for fast growth. Only individuals with crowns near the top of the canopy, or located in gaps, are likely to receive sufficient light to achieve high rates of growth. Trees in the deep shade have strongly suppressed growth rates. It is possible that they lose the ability to respond if more light does become available, reinforcing the temporal autocorrelation of growth rates. There has been little research into long-term effects of shade suppression on tree physiology. Intrinsic growth rates of trees are also under a degree of genetic control and lead trees may be genetically predisposed to fast growth. Identifying such trees has long been the ambition of foresters. However, lead

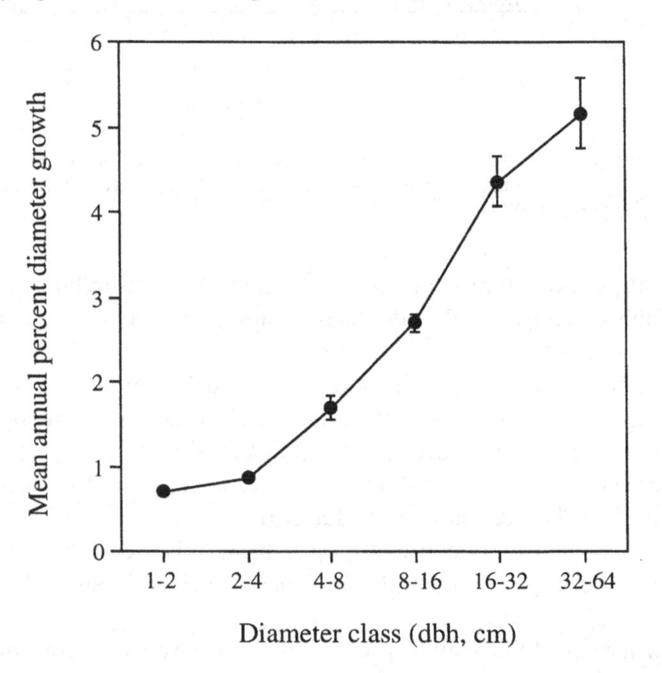


Figure 3.4 Mean annual percent growth rates for all individuals of canopy tree species in the 50 ha plot on Barro Colorado Island. These means are unweighted: the total growth increments divided by the total number of individuals. Some plants decreased in diameter due to stem breakage; these plants were excluded. Vertical lines indicate 95% confidence limits. Limits for the 1–2 and 2–4 cm dbh classes are too close to the circles to be visible (sample sizes were more than 17 000 individuals in these size classes). After Condit *et al.* (1992a).

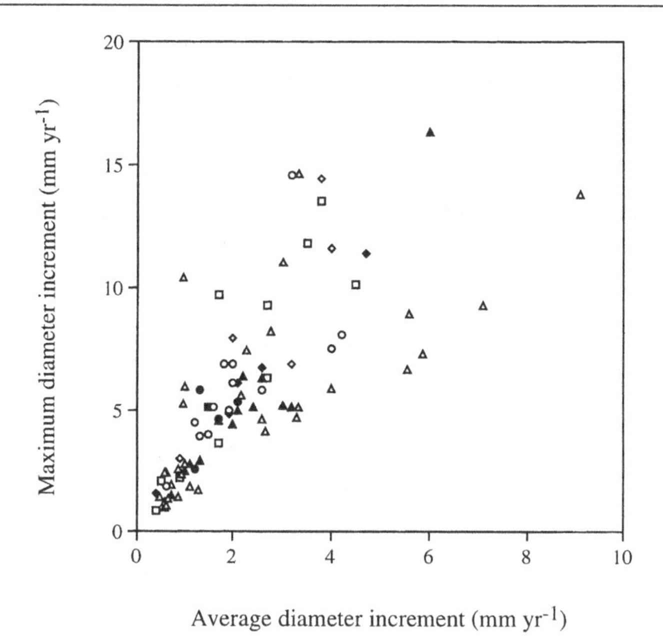


Figure 3.5 Average diameter increment against maximal increment for tree species from seven tropical-forest sites. Trees are $\geq 10$ cm dbh with $n \geq 20$. Data from Manokaran & Kochummen (1994), Ashton & Hall (1992), Lieberman *et al.* (1985a) and Korning & Balslev (1994).

trees need not necessarily be superior. Fast growth, for instance, may come at the cost of reduced resistance to pests and diseases.

**Tree performance in relation to light climate**

It is not easy to obtain quantitative data on the relationship between survival or growth rate and incident irradiance for large tropical trees because of the technical difficulties of obtaining long-term climatic data above tall trees. An approach used to get round this has been to employ semi-quantitative or qualitative measures of relative light availability to trees and use these in comparing individual growth rates. Various crown illumination indices are available (Clark & Clark 1992). These involve one or more fieldworkers in visually assessing the amount of light received by the crown of each tree in the study on a predetermined scale ranging from 'crown completely unobscured from the sun in all directions' to 'crown beneath continuous canopy'. Of the six large-tree species studied in detail by Clark & Clark (1992) at La Selva, all showed an increase in crown illumination index with increasing size of individual (Fig. 3.6). The biggest difference between the species was the failure of *Minquartia guianensis* to reach high crown-illumination-index scores. This species does not reach such large sizes as the others and hence cannot grow big enough to obtain a crown unobscured in all directions.

Lieberman *et al.* (1995) used a geometrical analysis of crown heights and distances from focal trees to generate an index of canopy closure. They regressed the closure index against 1000/dbh to obtain a linear relation, and

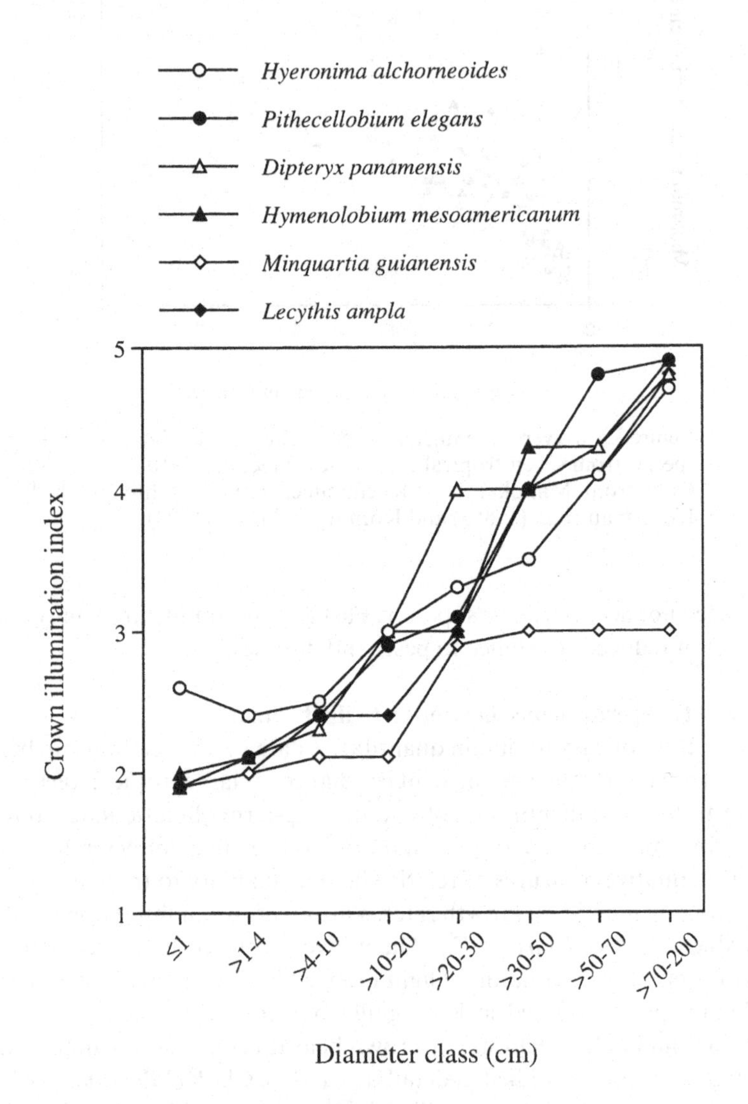


Figure 3.6 Crown illumination index for successive size classes for six large-tree species at La Selva, Costa Rica. Data from Clark & Clark (1992).

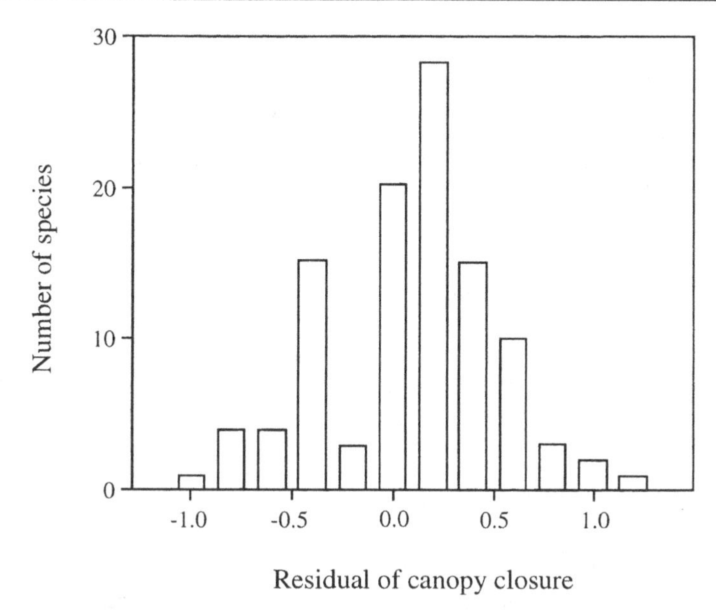


Figure 3.7 Frequency distribution of the mean residual of canopy closure for the 104 most abundant tree species in the study of Lieberman *et al.* (1995).

then used the residual from the linear regression as a size-compensated estimate of relative illumination, which could then be averaged for all the individuals of each species. The frequency distribution of the average residual for the 104 species studied is given in Fig. 3.7. Nine species, including *Hampea appendiculata* and *Cecropia obtusifolia*, had residuals significantly higher than those of all other species combined. In other words, they were found in sites significantly brighter than average. A smaller group of five species occupied sites shadier than average, with *Faramea terryae* being the most extreme shade-lover.

In the 50 ha plot on Barro Colorado Island, the forest was classified into high- and low-canopy (< 10 m tall) sites (12.7% of the total) on a 5 m × 5 m grid (Welden *et al.* 1991). This very rough classification of forest light climates has been employed to investigate the effects of shade on the survival, growth and recruitment of saplings of the commoner species (Table 3.2). Relatively few species showed a significant difference in survival between high- and low-canopy sites. Growth showed a much stronger effect, with all the species that displayed a significant difference between site types having faster growth in low-canopy sites. Recruitment to the 1 cm dbh class exhibited a similar pattern, with only two species (*Drypetes standleyi* and *Rheedia acuminata*) having

Table 3.2. *Sapling survival, growth and recruitment in relation to canopy height on Barro Colorado Island over the period 1982–85*

Tabulated are the number of species showing significantly greater survival, growth or recruitment in low-canopy (no foliage above 10 m height) sites, high-canopy sites, or with no difference between sites (indifferent). Saplings are 1–4 cm dbh, except in shrub species (adult height <4 m), where saplings were classed as 1–2 cm dbh.

| | Low-canopy sites | | High-canopy sites | | Indifferent | |
|---|---|---|---|---|---|---|
| | no. spp. | % of spp. | no. spp. | % of spp. | no. spp. | % of spp. |
| Survival | 6 | 4.0 | 19 | 12.8 | 123 | 83.1 |
| Growth | 66 | 57.4 | 0 | 0 | 49 | 42.6 |
| Recruitment | 70 | 44.9 | 2 | 1.3 | 84 | 53.8 |

Data from Welden *et al.* 1991.

better recruitment in high-canopy sites. A large proportion of the species was indifferent to canopy type in one or more of the demographic measures. Welden *et al.* (1991) argued that this implied most species were 'gap-neutral' in their regeneration ecology. The poor level of accuracy and discrimination in light climate possible from the method used for its estimation might be a readier explanation for the relative rarity of 'gap-positive' species.

## Mortality, growth and adult size

On Barro Colorado Island, the species were divided into four stature classes based on maximal height. These groups were 'shrubs' (< 4 m adult height), 'treelets' (4–< 10 m), mid-sized trees (10–< 20 m) and large trees (> 20 m adult height). The 'shrub' species showed mortality rates of about twice those on average of the trees and 'treelets' for individuals 1–9.9 cm dbh (Condit *et al.* 1995) (Table 3.3). 'Shrubs' tended to occupy the lower end of this size range, and being generally smaller may be more susceptible to death by falling debris. In addition, the reproductive activity of the 'shrubs' may deplete their resources and render them more liable to mortality. The saplings of the tree species were more likely to exhibit a positive growth and recruitment response to low-canopy sites than those of the two smallest size classes (Welden *et al.* 1991). For instance, 66% of the tree species showed a significant positive response in sapling growth to low-canopy sites, but only 41% of the 'treelet' and 'shrub' species did the same. The probable smaller average size of the 'treelet' and 'shrub' saplings may mean that the difference in light climate experienced between the two canopy-height classes is less marked. Another possible explanation is the tendency of understorey species to reproduce rather than grow more when in gaps.

Table 3.3. *Annual mortality rates (%) averaged across species for trees of different stature on 50 ha at Barro Colorado Island, Panama*

*n*, Number of species.

| Stature class | Stems 10–99 mm dbh | *n* | Stems ≥ 100 mm dbh | *n* |
|---|---|---|---|---|
| 20 m adult height | 3.2 ± 0.4 | 71 | 1.9 ± 0.2 | 63 |
| 10– < 20 m adult height | 2.7 ± 0.4 | 54 | 3.3 ± 0.2 | 49 |
| 4– < 10 m adult height | 2.9 ± 0.6 | 41 | 2.9 ± 0.8 | 16 |
| 4 m adult height | 6.3 ± 0.9 | 28 | — | — |
| all | 3.5 ± 0.3 | 194 | 2.6 ± 0.2 | 128 |

Data from Condit *et al.* (1995).

Table 3.4. *Annualised mortality rates for different stature classes for three sites in Malaysia*

| Stature class | Bukit Lagong | Sungei Menyala | Pasoh |
|---|---|---|---|
| emergent | 1.11 | 1.44 | 1.65 |
| main canopy | 1.33 | 1.84 | 1.90 |
| understorey | 1.45 | 2.58 | 2.35 |

Data from Manokaran & Swaine (1994).

Large-tree species showed significantly higher rates of survival among stems of 10 cm dbh or greater on Barro Colorado Island than those of species of smaller adult stature (Condit *et al.* 1995) (Table 3.3). At La Selva, the six emergent species studied by Clark & Clark (1992) showed low annual mortality for trees ≥ 10 cm dbh (0.44%) in comparison to the forest average (2.03%). The same pattern was seen in studies of lowland dipterocarp forests in Malaysia (Manokaran & Kochummen 1987) (Table 3.4), and trees in Ecuador (Korning & Balslev 1994). Emergent species also exhibit some of the highest maximal growth rates in forests (Manokaran & Kochummen 1994). The data from Figure 3.5 have been re-plotted indicating the stature of the species concerned (Fig. 3.8). Clearly, large trees tend to have higher average and maximal growth rates, particularly when compared with the understorey species. Subcanopy species tend to be more variable, with some having very rapid growth. Interestingly, the average rates for canopy-top species estimated from periodic increment data on individual trees in the forest are similar to the long-term averages (1–6 mm $yr^{-1}$) calculated for 20 large trees aged by using $^{14}C$-dating in the Amazon (Chambers *et al.* 1998). Maximal growth rates of large trees (5–15 mm $yr^{-1}$) in the forest still fall short of the rates of similar species in arboreta or plantation stands, where diameter increments greater than 10 mm $yr^{-1}$ are common, and over 20 mm $yr^{-1}$ are known (see for example, the data of Ng & Tang 1974).

The maximum projected lifespan estimates for trees from La Selva, Costa Rica, (Lieberman *et al.* 1985a) and Ecuador (Korning & Balslev 1994) show a trend of increasing longevity with increasing stature (Table 3.5). Only in the data from La Selva, are there statistically significant differences between stature classes. Here the understorey species have significantly shorter lifespans than the other two. However, the range of values is large for all the stature classes, with each height group possessing some fast-growing short-lived species.

## The use of growth and mortality data to recognise species groups

A frequent criticism of the functional classifications of tropical tree species is that they are largely based on *ad hoc* impressions of the species concerned rather than on detailed quantitative information. There have however, been several attempts to classify species by using growth and mortality data.

The 50 ha plot study on Barro Colorado Island is the largest data set

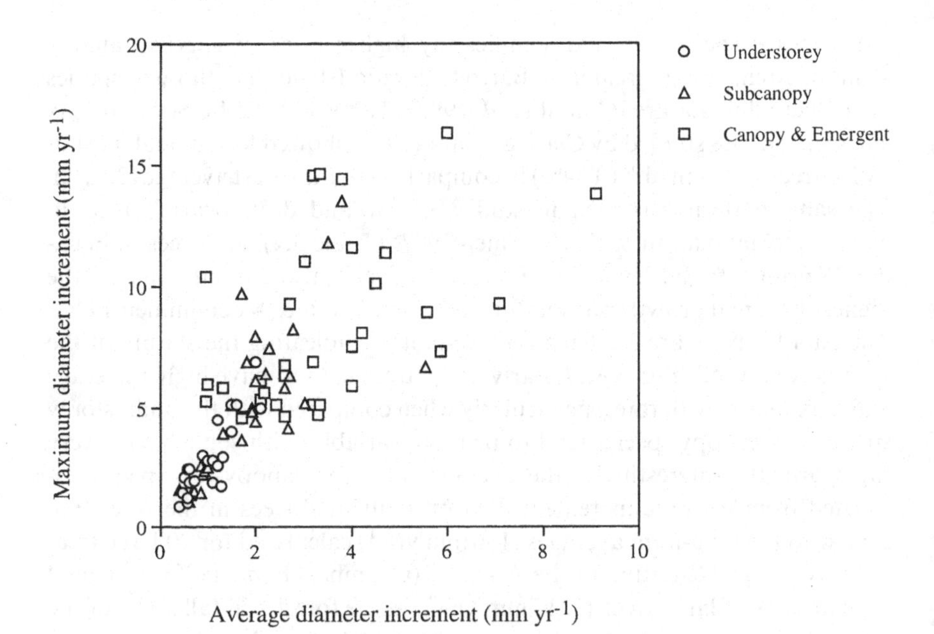


Figure 3.8 The same plot as Fig. 3.5, but here the symbols indicate the stature class of each species.

Table 3.5. *Maximum projected life span (10 cm dbh to maximum diameter) for tree species of different stature classes from Costa Rica and Ecuador*

Values given are mean ± 1 sd (range) *n*.

| Stature class | Costa Rica | Ecuador |
|---|---|---|
| understorey | 126 ± 56 | 199 ± 103 |
| | (52–221) | (69–348) |
| | 13 | 7 |
| subcanopy | 242 ± 81 | 270 ± 175 |
| | (78–338) | (54–529) |
| | 11 | 10 |
| canopy | 206 ± 102 | 299 ± 40 |
| | (78–442) | (250–353) |
| | 21 | 5 |

Data from Lieberman *et al.* (1985a) and Korning & Balslev (1994).

employed for such a purpose (Condit *et al.* 1996a). Five parameters were used to perform principal components analysis (PCA) for 142 species. The five vital statistics employed were: annual mortality (avoiding the drought years) for 1–9.9 cm dbh and ≥ 10 cm dbh classes; mean growth rate for 1–2 cm dbh and 10–20 cm dbh size classes; and the colonising index (the proportion of recruitment that occurred in low-canopy sites). The understorey species had to be dealt with separately from the others because they did not have sufficient data for the larger tree sizes. The two resulting PCAs produced similar results, particularly for the first axis. Most of the species occurred in a tight knot in the range -1 to 1 on the axis, with a scattering of species stretching from 1 to 6. These were species with a high mortality, high growth rates and tendency to recruit in gaps. This group was arbitrarily divided into 'pioneers' with first-axis scores of more than 3 and 'building-phase' species in the range 1–3. The pioneers included *Cecropia insignis* and *Zanthoxylum belizense* among the large-statured species and *Croton billbergianus* and *Palicourea guianensis* from the understorey group. An important point to note is that the species showed no marked discontinuity of distribution along the shade-tolerance axis, and that the groupings proposed by the authors were defined at arbitrary intervals.

At La Selva, Lieberman *et al.* (1985a) used their growth trajectory simulation technique to estimate maximum potential lifespan of tree species from 13 years of enumeration data on 12.4 ha of forest. They identified four species groups on a three-dimensional plot of maximum dbh (as measured on the plot), maximum potential lifespan (as projected from the slowest growth simulation) and maximum growth rate (as estimated from the fastest growth simulation). These were:

Group I: understorey species with slow maximum growth rates and short lifespans.
Group II: shade-tolerant subcanopy species that lived up to twice as long as the understorey species but had similar maximum growth rates.
Group III: canopy and subcanopy species that were shade tolerant but could grow fast in high light conditions and were long-lived.
Group IV: shade-intolerant species with fast growth and short life spans.

The longest-lived species were either relatively shade-tolerant mid-canopy trees or large canopy-top species. The distinction between groups I and II, and between III and IV, seems rather arbitrary to the external observer. That between the two pairs of groups appears clearer.

These two studies, and the more empirical observations of other tropical ecologists, lead to the conclusion that groups of species based on growth and survival, particularly in relation to irradiance regime, mature stature and probable longevity can be recognised among the tropical rain-forest tree community. These groups are not discrete, but better considered as regional references for locations within the character space available. The four groups of Lieberman *et al.* (1985a) seem to be reasonable and can be characterised as follows.

*Understorey species:* small-statured at maturity with relatively low maximal growth rates and quite high mortality (though not as high as that of pioneers). Generally shade tolerant with limited growth responses to increased irradiance.
*Subcanopy species:* intermediate in stature between understorey and canopy species. Often shade-tolerant with a relatively low mortality rate. Some species long-lived.
*Canopy species:* the largest trees, usually with high juvenile survival and a large increase in growth rates when exposed to high light.
*Pioneer species:* fast-growing, with high mortality rates, particularly in the shade and among juvenile stages. Relatively short-lived.

These groups, and possible subdivisions of them, are discussed in greater detail in Chapter 6.

## Relative performance of species of similar life history

Clark & Clark (1992) investigated the growth and mortality of six species in detail on a 150 ha area of forest at La Selva, Costa Rica. Each of the six would be placed in the canopy group of the scheme above. All individuals encountered greater than 50 cm tall were included in the survey. These were monitored annually for survival and growth and were assessed for crown illumination class and forest growth phase. Using these data, and some extra observations for *Simarouba amara* and two species of *Cecropia*, the authors identified four species groups as follows.

Group A: found in low crown illumination classes and with a high proportion of juveniles in the mature phase of the forest [*Lecythis ampla* and *Minquartia guianensis*].

Group B: steady increase in illumination class and proportion in early phases with increasing size [*Dipteryx panamensis* and *Hymenolobium mesoamericanum*].

Group C: a tendency to be found in low illumination classes as middle-sized juveniles, interpreted by Clark & Clark as a requirement for gaps as saplings but with a tendency to become overtopped in the building phase and then waiting for a new gap event to grow on [*Hyeronima alchorneoides* and *Pithecellobium elegans*].

Group D: occupying the brightest sites at all size classes [*Cecropia* spp.]

The Group A species showed no significant relationship between sapling mortality and crown illumination. The other species did show an improved chance of survival for saplings growing in brighter conditions. However, there was no great difference in the maximal growth rate of the six canopy-top/emergent species. This led Clark & Clark to question the view that there is a tradeoff between ability to persist in the shade and maximal growth rate in high light. We will return to this subject in a later section. In general, there was relatively little difference in the growth and survival responses to forest environment between the species, particularly the five emergents. As in the study of Welden *et al.* (1991), this might be attributable to methodological problems including low sample sizes for some diameter classes and the accuracy of the light regime estimates. However, it may also reflect reality. Tropical tree species of similar adult stature and shade tolerance may be very similar.

## What limits tree growth?

It is interesting to consider what factors actually limit the growth of individual trees in the forest. The answer, as already indicated, in most cases is light availability. Shading results in most trees in the forest growing at rates well below their potential maximum. Clark & Clark (1994) found large interyear variations in growth of six species at La Selva, Costa Rica. Over an eight-year period, diameter increments for individuals from more than 50 cm tall to 1 cm dbh were 3–10 times greater in the year of fastest growth compared with that of slowest (Fig. 3.9). The six species showed a strong concordance in performance over the observation period, i.e. stand productivity was varying relatively uniformly over time. The year of best growth was the one with the lowest rainfall. It seems improbable that poor growth in wet years was due to waterlogged soils. More likely was that wet years were much cloudier and had substantially lower total photosynthetically active radiation (PAR) received than dry years.

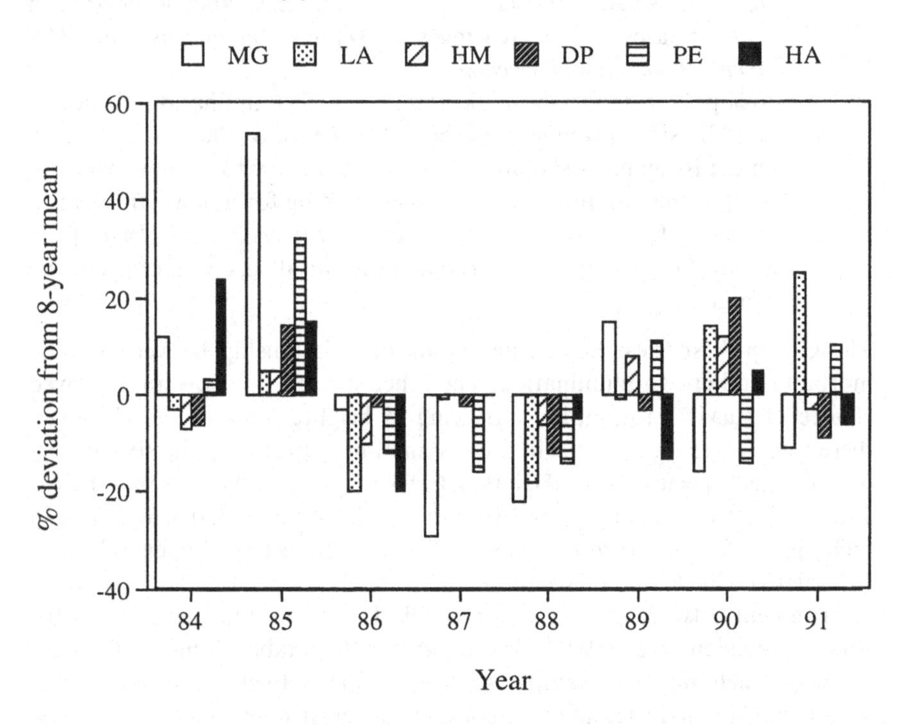


Figure 3.9 Annual variation in mean diameter growth rates of adult tree (> 30 cm in diameter) of six canopy and emergent species in primary forest at La Selva, Costa Rica (after Clark & Clark 1994). Data for each species are each year's mean annual adjusted increment (averaged over all individuals after detrending or subtracting the 8-year mean diameter increment from each tree's annual increment), given as a percentage deviation from the species' overall mean diameter increment over the 8-year period. Abscissa labels indicate the calendar year beginning each measurement period. MG, *Minquartia glabra*, LA, *Lecythis ampla*, HM, *Hymenolobium mesoamericanum*, DP, *Dipteryx panamensis*, PE, *Pithecellobium elegans*, HA, *Hyeronima alchorneoides*.

In seasonally dry forests, growth is probably reduced owing to water shortage during the dry season, unless the trees can tap water deep in the soil. At La Selva, Costa Rica, a detailed study of diameter growth patterns on a day-to-day basis (Breitsprecher & Bethel 1990) showed an annual periodicity in growth in a majority of species on well-drained soils. Most species showed reduced growth in the mild dry season at La Selva. During a severe drought in 1983 on Barro Colorado Island, there was considerably higher mortality among trees than normal, with large-diameter stems suffering the greatest increases in mortality.

## Herbivory

The loss of material to herbivores can be substantial, even in tropical trees. Complete defoliation of trees has been reported quite frequently. It is likely that such an event will influence the performance of a tree negatively. The studies by Marquis (1984, 1987) on *Piper arieianum* referred to earlier showed this to be the case. The best-studied group in terms of the influences of defoliation is understorey palms (Mendoza *et al.* 1987; Oyama & Mendoza 1990; Chazdon 1991; Cunningham 1997). Presumably their large, and relatively few, leaves make observations easier. All the neotropical palms studied proved to be remarkably resilient to leaf loss. Indeed, *Chamaedora tepejilote* showed increased reproduction after being defoliated (Oyama & Mendoza 1990). Neither leaf nor ramet removal showed any influence on leaf size or stem diameter in *Geonoma congesta* three years after defoliation (Chazdon 1991). It is probable that mobilisation of stored carbohydrate is important in buffering palms from premature leaf loss, but even so, Cunningham (1997) could not find any significant effect of defoliation on stored carbohydrate in *Calyptrogyne ghiesbreghtiana*.

# 4

# Reproductive biology

## Reproduction

The high species diversity of the tropical rain forest is inevitably associated with low population densities for a majority of species. Most tree species will exist at densities below ten mature individuals per hectare of forest, and many will live in much sparser populations. Tropical ecologists have long been intrigued by the possible ability of tropical tree populations of widely separated individuals to outbreed successfully. In recent years, new techniques for genetically fingerprinting individuals have provided strong evidence that many tropical tree species are strongly outbreeding despite large average distances between trees within a population.

## Vegetative reproduction

There have been relatively few studies of vegetative reproduction in tropical trees. However, a number of reports refer to the marked ability of some understorey species to root from plant fragments including stems and leaves (Gartner 1989; Kinsman 1990; Sagers 1993). Shrubs, such as species of *Psychotria* and *Piper*, may need to be resilient to damage as they stand a relatively high chance of being broken by falling tree parts or large animals. An ability to re-sprout and produce adventitious roots will allow broken fragments to establish as new plants. Even the large-tree species, *Tetramerista glabra*, from the peat swamp forests of Borneo has been reported to employ this method of propagation (Gavin & Peart 1997, 1999). Many *Tetramerista* 'seedlings' are actually sprouts from fallen branches or collapsed saplings.

A wide range of tropical trees are typically multi-stemmed. Palms, pandans and many dicot tree species spread laterally by sprouting from the trunk base or producing root suckers. A few species from tropical Africa (e.g. *Anthonotha macrophylla* and *Scaphopetalum amoenum*) are reported habitually to bend over as they grow and when the crown touches the ground to root and send up more shoots, which in turn arch and spread (Richards 1996). When root suckers emerge at a distance from the main trunk it can be difficult to distinguish clones from populations of genetically distinct individuals without close observation. It seems likely that some species in the rain forest are using root suckering to propagate themselves, but there have been few investigations into the phenomenon.

The opposite of one individual appearing as many, many individuals appearing as one, is also seen in tropical trees. Hemiepiphytic figs produce anastamosing aerial root systems by the fusion of separate roots. This union can occur between different fig individuals of the same species, leading to large stranglers that are genetic mosaics derived from two or more plants (Thomson *et al.* 1991). Somatic mutations within trees could possibly also lead to genetically heterogeneous individuals. Murawski (1998) has provided some preliminary evidence for genetic variation among branches of the same crown, for two trees from French Guiana.

## Sexual systems

Plants, unlike animals, show a wide diversity of sexual systems. Truly hermaphrodite, monoecious and dioecious species can be found in the tropical rain forest. Surprisingly similar proportions of the tree flora among these groups have been reported in different neotropical rain forests (Table 4.1). Dioecy – separate male and female trees – is quite common in tropical forests. Monoecy is rarer and is largely confined to palms, figs and members of the Euphorbiaceae.

There was no significant difference in the relative proportion of species among different sexual systems between 'pioneer' and 'persistent' groups at Los Tuxtlas, Mexico (Ibarra-Manríquez & Oyama 1992). In a comparison between dioecious and non-dioecious species, the dioecious species had significantly smaller flowers and more seeds per fruit and a higher proportion of simple as against showy flowers and fleshy rather than dry fruits.

Reviewing descriptions of pollination in tropical dioecious species, Renner & Feil (1993) concluded that vertebrate pollination and exploitation of long-tongued bees were rare in this group. Small insects were the main pollinators. About one third of the dioecious tropical species for which they had data were found not to offer floral rewards to visitors in pistillate flowers. Instead the female flowers mimic the reward-giving male flowers and dupe pollinators into visiting them. Papaya (*Carica papaya*), nutmeg (*Myristica fragrans*) and the vegetable ivory palm (*Phytelephas seemannii*) are examples. In the last named, the female inflorescences produce the same smell as the male ones and attract tiny staphylinid beetles, which normally lay eggs on the decaying male inflorescences (Bernal & Ervik 1996).

Dioecy is the most certain way of ensuring outcrossing. Many species also use temporal differences in the activity of male and female parts to avoid autogamy. Controlled crossing experiments have revealed quite high levels of

Table 4.1. *Relative frequency (%) of different sexual systems among the tree flora at various tropical-forest sites*

Some of the rows fail to add up to 100% because of a small number of species with mixed sexuality.

| Site | Forest type | Reference | Hermaphrodite | Monoecious | Dioecious |
|---|---|---|---|---|---|
| La Selva | lowland | Kress & Beach 1994[a] | 67 | 10 | 23 |
| Barro Colorado Island | lowland | Croat 1979[b] | 63 | 15 | 21 |
| Los Tuxtlas | lowland | Ibarra-Manríquez & Oyama 1992 | 63 | 9 | 27 |
| Altos de Pipe | montane | Sobrevila & Arroyo 1982 | n.d. | n.d. | 31 |
| Blue Mountains | montane | Tanner 1982 | 68 | 11 | 21 |

[a]Canopy and subcanopy classes combined, understorey omitted. The data may include a few non-tree species, but the results are not substantively different from those of Bawa *et al.* (1985b).

[b]Trees.

self-incompatibility in tropical tree species. Some 88% of hermaphrodite tree species ($n$=17) studied at La Selva, Costa Rica, were found to be self-incompatible (Kress & Beach 1994). Outbreeding was observed to predominate in the Malaysian rain forest also (Ha *et al.* 1988a).

An entirely female population of *Garcinia scortechinii* has been reported from Pasoh, Peninsular Malaysia (Thomas 1997). Agamospermy has been reported before in the genus (Ha *et al.* 1988b, Richards 1990a), and this and other forms of apomixis are probably present in some species in most tropical forests. Clues to the likely presence of apomixis include very consistent fruit crop size and polyembryony (multiple embryos from one ovule), often observable as more than one seedling germinating per seed.

## Genetic diversity

Allozyme analysis and, more recently, DNA fingerprinting techniques have revolutionised our understanding of the genetic structure of tropical forest tree populations. Hamrick (1994), reviewing the published data, concluded that tropical trees have a lower proportion of polymorphic loci and lower levels of heterozygosity per individual than tree species from colder climates. Possibly this is due to smaller effective population sizes in tropical trees because of the low population densities. Despite this, it is clear from allozyme analysis that most tropical canopy species have high outcrossing rates (Murawski 1995; Doligez & Joly 1997), with multi-locus outcrossing estimates normally above 0.9. Considerably lower values have been recorded, however, particularly in species of Bombacaceae and Dipterocarpaceae. In most species studied, there is more homozygosity in the progeny (seeds) than the adult population, but there is no evidence of a decrease in heterozygosity in the population over the successive generations. This implies a tendency to select against homozygotes among the new recruits, and the existence of inbreeding depression. Lower levels of outcrossing have been observed in years with lower flowering densities, for example in *Cavanillesia platanifolia* on Barro Colorado Island (Murawski & Hamrick 1992).

Genetic markers have been used to demonstrate successful long-distance pollen dispersal and to infer breeding areas for tropical trees (Table 4.2). Chase *et al.* (1996) used microsatellite DNA techniques to assign paternity to seeds of known mothers in a population of *Pithecellobium elegans* in Costa Rica. Mating distance between confirmed parents averaged 142 m, with pollination over 350 m proven. Interestingly, proximity had little influence on likelihood of paternity, the modal class of distance between parents was more than 10 individuals away from the focal tree. Stacy *et al.* (1996) conducted research on three tree species (*Calophyllum longifolium*, *Spondias mombin* and

Table 4.2. *Breeding unit parameters estimated for tropical tree species*

The breeding unit corresponds to the circular area about a female tree within which 99% of potential mates are expected to occur.

| Species | Reference | Pollen vector | Density ($ha^{-1}$) | Adults (no.) | Area ($km^2$) | Radius (km) |
|---|---|---|---|---|---|---|
| *Astrocaryum mexicanum* | 1 | beetle | 1364 | 1542 | 0.011 | 0.060 |
| *Calophyllum longifolium* | 2 | small insect | 0.28 | 35 | 1.241 | 0.629 |
| *Cordia alliodora* | 3 | small insect | 20.9 | 520 | 0.249 | 0.282 |
| *Ficus dugandii* | 4 | fig wasp | 0.004 | 252.7 | 631.7 | 14.2 |
| *Ficus obtusifolia* | 4 | fig wasp | 0.072 | 762.1 | 105.9 | 5.8 |
| *Ficus popenoei* | 4 | fig wasp | 0.013 | 393.1 | 294.8 | 9.7 |
| *Pithecellobium elegans* | 5 | hawkmoth | 0.88 | 45 | 0.636 | 0.450 |
| *Platypodium elegans* | 6 | small bees | 0.78 | 68 | 0.866 | 0.525 |
| *Spondias mombin* | 2 | small insect | 0.33 | 6 | 0.196 | 0.250 |
| *Turpinia occidentalis* | 2 | unknown | 1.27 | 5 | 0.040 | 0.113 |

References: 1, Eguiarte *et al.* (1992); 2, Stacy *et al.* (1996); 3, Boshier *et al.* (1995); 4, Nason *et al.* (1998); 5, Chase *et al.* (1996); 6, Nason & Hamrick (1997).

*Turpinia occidentalis* ssp. *brevifolia*) that occur at low population densities (one mature tree per 2.2 to 6 ha) on Barro Colorado Island and have diminutive flowers pollinated by small generalist insects. Allozyme studies showed that all three species were 100% outcrossed and that they were effectively pollinated over long distances. Multi-locus paternity exclusion analysis showed that for *Calophyllum longifolium* 62% of effective pollen moved at least 210 m. A small proportion (2.5–5.2%) of successful *Spondias mombin* pollen moved more than 300 m. Even when pollinated by diverse, small insects, tropical trees can be effectively outcrossed over large distances. However, figs (*Ficus* spp.) have by far the most effective long-distance dispersal of pollen yet reported from the tropical rain forest, with effective breeding areas of hundreds of square kilometres (Nason *et al.* 1998).

## When to flower?

It can only be favourable for a plant to delay reproduction if the benefits of waiting are greater than the advantages of going ahead at the present time. In other words, if total reproductive output as a larger plant will be greater than a small plant could achieve on average, then flowering should be delayed. Thomas (1996c) found that flower and fruit production follow allometric constants averaging 5.2 for $D$ (stem diameter) and 1.8–2.0 for $D^2H$ for trees at Pasoh in Peninsular Malaysia. These values are higher than would be predicted by biomechanical considerations of supporting the reproductive organs alone. They indicate a very considerable advantage of tree size in reproductive output. There seems little doubt that allocation of resources to reproduction does compromise growth rates in trees. Thomas (1996b) found evidence of a reduction in height growth with onset of reproductive activity in tropical trees, based on allometric studies (Fig. 4.1). Thus most species grow to relatively large size and then reproduce rather than combining both activities throughout development. But then, why don't all trees grow as big as possible before starting to flower? It is probably a question of risk of mortality. If there is a high chance of death there is no point in delaying reproduction. Fast-growing species that exploit transient high-resource sites such as canopy gaps are at risk of being overtopped by competing individuals. Species in the shade of the understorey risk trees or large branches falling on them from above.

In dioecious species, the cost of reproduction may differ between the sexes. Thomas & LaFrankie (1993) studied five dioecious species in the Euphorbiaceae at Pasoh. They found that the two species of smallest mature stature showed a consistent male bias in the flowering population with a greater variance in size among flowering males than among females. The male bias

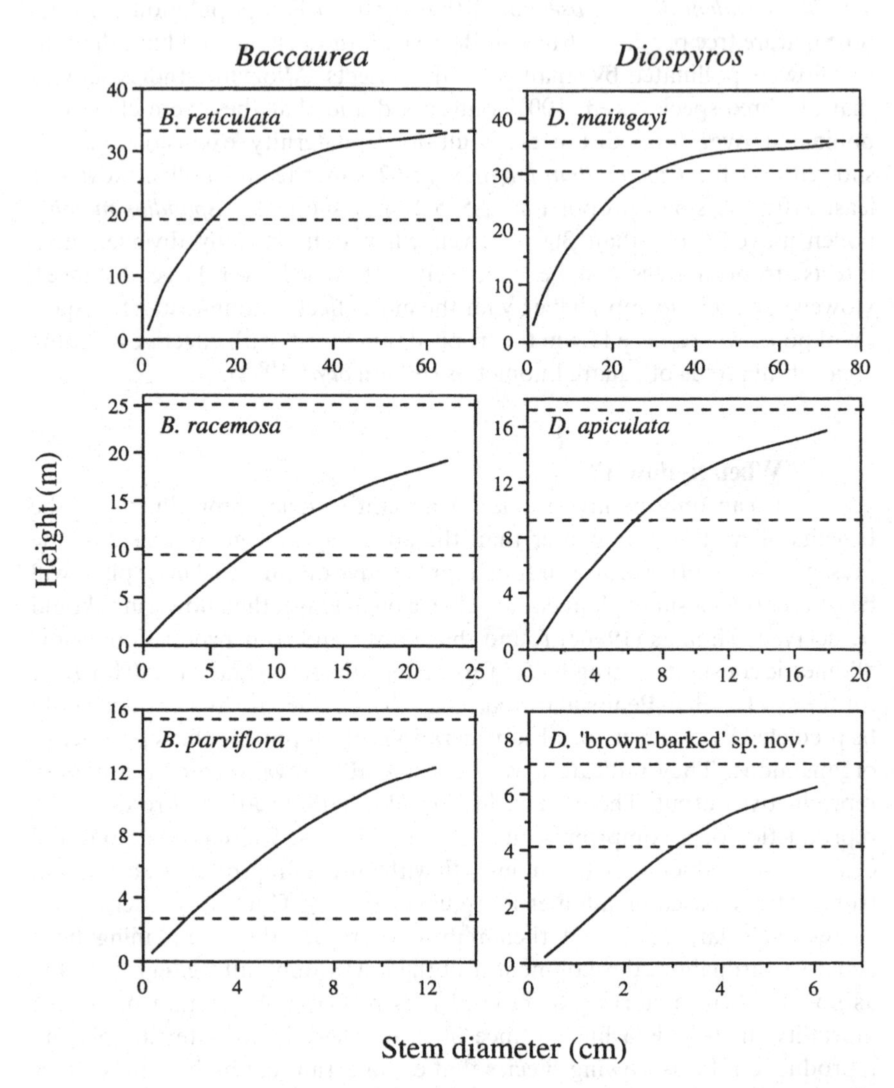


Figure 4.1 Representative height–diameter relations for six species of tree in two genera at Pasoh Forest Reserve, Peninsular Malaysia. Allometric curves illustrated for each species are least-squares fits of a generalised allometric function. On each graph, the upper dotted line represents the asymptotic maximal height ($H'_{max}$), and the lower dotted line indicates the estimated height at reproductive onset. After Thomas (1996a).

was particularly marked in the years when relatively few individuals flowered. A male bias in dioecious species is also evident in other published reports. It probably reflects the greater cost of reproduction in females that makes it less feasible for them to reproduce at small size, or in years when conditions do not favour reproduction. This gender difference will be more pronounced in the deep shade of the forest understorey where many of these small-statured species tend to occur.

Most trees reproduce over many years, but there are a few that do so only once. The only reports of monocarpic tropical dicotyledonous trees are some species in the neotropical genera *Tachigali* (Foster 1977; Richards 1996) and *Spathelia* (Richards 1996; Young & Augspurger 1991), and the tropical Asian *Harmsiopanax* (Philipson 1979). The genus *Strobilanthes* and some other species in the Acanthaceae are monocarpic and gregarious in flowering (Young & Augspurger 1991), events of which occur at long and irregular intervals (Janzen 1976b). These species are mostly herbaceous, but some are woody shrubs of the rain forest.

There is a link between architecture and monocarpy. Most monocarps utilise Holttum's model (Hallé *et al.* 1978): a terminal inflorescence on an unbranched shoot. When the apical meristem switches to producing flowers, vegetative growth must cease and the shoot module must die. This is clearly demonstrated in the palms, which represent the major group of rain forest monocarps. After death of the flowering shoot, suckers may grow from the base to replace the earlier one. Hapaxanthy is the term used to describe this behaviour, which is probably the most common form of monocarpy, at least in the rain forest (Young & Augspurger 1991). Bamboos are often monocarpic, frequently in synchrony across a population (Janzen 1976b), but it must not be assumed that all species of bamboo fit this pattern and there is no clear evidence of synchronous monocarpy being a common feature of tropical rain-forest bamboo species. The few species of *Tachigali* that are monocarpic are unique because they are highly branched trees that flower once and die.

The polycarpic majority of trees in the forest shows a wide range of temporal patterns of flowering activity. In a 12-year study of 254 trees from 173 species at La Selva it was found that 7% of species flowered continuously, 55% subannually, 29% annually and 9% supra-annually (Newstrom *et al.* 1993). Synchronous supra-annual flowering is a notable feature of the forests of West Malesia, which are dominated by the Dipterocarpaceae (Appanah 1985; Ashton *et al.* 1988). Many theories have been put forward to explain such masting or mass-flowering/fruiting behaviour (Kelly 1994). The predator-satiation hypothesis seems easily the most tenable for tropical examples of masting and will be dealt with later. However, synchronous flowering may also have the advantage of promoting the success of pollination, both in terms of number of pollination events and the likelihood of cross-pollination. Synchronously flowering individuals of the treelet *Hybanthus prunifolius* showed higher fruit and seed set (86% vs. 58% and 78% vs. 40%, respectively) than those induced to flower at a time different from the main population (Augspurger 1981).

## Pollination

The immobility of plants makes it necessary for them to have means of dispersing gametes between sedentary individuals in order to achieve sexual reproduction. In seed plants, pollen is the mobile, though not motile, vector of genetic material. Wind, and sometimes water, may carry the tiny pollen grains, and of the many millions liberated a few will come to land on the stigmas of flowers of the same species. Many plant species rely on animals, generally unwittingly, to carry pollen from flower to flower.

Notable features of the pollination biology of tropical forest trees are the rarity of wind pollination and the wide range of animal pollinators represented. There are some anemophilous species present. The best places to find them are on ridge tops, along seashores and riverbanks. This rarity is not surprising given that the general climatic conditions seem highly unfavourable to the dispersal of pollen on the wind. The high humidity dampens the pollen and causes it to stick together making it difficult for grains to remain airborne. The rain washes the pollen out of the air. The dense canopy filters out the pollen and makes the conditions inside the forest very still, not conducive to pollen dispersal by wind in the understorey. Linskens (1996) could not find any pollen in moss cushions examined in lowland and montane forest in Borneo. Adhesive-coated slide pollen traps also failed to detect any airborne pollen inside the rain forest. Not a single species of angiosperm in the rain forests of central French Guiana is known to be wind-pollinated (Mori & Brown 1994).

It has been estimated that 98–99% of tropical rain-forest plants are animal-pollinated (Bawa 1990). The most important groups of pollinating animals are insects, birds and mammals. The recent discovery of cockroach pollination in an annonaceous liana from Borneo (Nagamitsu & Inoue 1997) makes a sixth order of insects (Blattodea) reported to include pollinating species. The other five orders, Hymenoptera, Coleoptera, Lepidoptera, Diptera and Thysanoptera, are all known as pollinators of tropical trees. Mammals responsible for pollination in tropical trees include bats and to a lesser extent non-volant arboreal mammals. Birds are the final major animal group involved.

There has been an evolutionary tendency for plant species pollinated by the

Table 4.3. *Names of biotic pollination syndromes*

| Syndrome | Pollinators |
|---|---|
| cantharophily | beetles (Coleoptera) |
| chiropterophily | bats (Chiroptera) |
| entomophily | insects (Insecta) |
| lepidopterophily | Lepidoptera in general |
| melittophily | bees (Hymenoptera: Apoidea) |
| micromelittophily | small bees |
| myiophily | flies (Diptera) |
| ornithophily | birds (Aves) |
| phalaenophily | moths (Lepidoptera in part) |
| psychophily | day-flying Lepidoptera |
| sapromyiophily | saprozoic and coprozoic flies (Diptera) |
| sphingophily | hawkmoths (Lepidoptera: Sphingidae) |
| therophily | non-flying mammals |

same sort of animal to converge on a similar set of traits in inflorescence and flower morphology and presentation, attractants and rewards. These allow the recognition of a series of pollination syndromes, which in turn may permit the likely pollinating organism to be predicted from the floral traits of a species. However, caution is necessary in accepting the results of such extrapolations. Pollinators often visit flowers that are not 'typical' for them. Some species are visited by a wide range of organisms and careful observations are needed to confirm which species are the effective pollinators. For example, inflorescences of *Artocarpus* species are visited by many nocturnal insects including moths, flies, beetles and cockroaches (Momose *et al.* 1998a).

There follows a summary of the most important pollination syndromes found in tropical trees. These notes include information on the pollinators, examples of the plants they pollinate, common features of floral design and reference to case studies of pollination involving the pollinator group in question. The names of the pollination syndromes are given in Table 4.3. Important sources of general information about pollination syndromes not directly referred to include Bawa (1990), Prance (1985), Endress (1994) and Fægri & van der Pijl (1979).

## Pollination syndromes

### *Bats*

#### POLLINATOR

Bats comprise two suborders of the mammals. The Megachiroptera are restricted to the Old World where they are frequently agents of pollination and seed dispersal for tropical trees. Microchiropterans perform this role in the New World.

FLOWERS

Despite pollination being performed by different bat lineages in the palaeo- and neotropics, the bat-flower syndrome is quite uniform. Flowers open in the evening to coincide with the largely nocturnal foraging of the pollinators. They are usually large and pale with a distinctive musky, fermented odour and copious nectar. The flowers are mostly held away from the foliage to allow easy detection and access by the bats (Fig. 4.2). This is often acheived by having the flowers on the ends of branches held above, or dangling below the main crown. Alternatively, cauliflory or ramiflory are often associated with bat-pollination. The individual flowers are either large and robust or small and aggregated into many-flowered inflorescences that present a shaving brush-like array to the visiting bat (e.g. *Parkia*). Large bat-flowers are either the brush type with many long stamens, e.g. *Adansonia, Barringtonia,* or big, thick fleshy bowls, e.g. *Oroxylum indicum, Fagraea racemosa*[1]. The large size of bats in comparison to most other pollinators often means that bat-pollinated species have bigger flowers than most other members of their genus or even family. New World bat-flowers tend to produce hexose-rich nectar whereas the Old World equivalents secrete nectar with abundant sucrose (Baker & Baker 1983; Baker *et al.* 1998). Whether this reflects a difference between micro- and megachiropterans remains unclear. Bat-flowers are possibly the most nectariferous of all flowers, with copious, often mucilaginous, nectar and abundant pollen. This clearly makes bat-pollination an expensive undertaking for the plant. The benefits gained, however, are the long flights of bats, securing cross-pollination with distant individuals.

PLANTS

Many tropical rain-forest tree species that are bat-pollinated are from the families Bignoniaceae, Bombacaceae and Myrtaceae. The syndrome is also quite common in Leguminosae subfamily Caesalpinioideae. The Caryocaraceae are a small neotropical family of which all members are believed to be bat-pollinated.

CASE STUDIES

Bat pollination is the subject of a book (Dobat & Peikert-Holle 1985). Hopkins (1984) conducted research on the pollination of the pantropical caesalpinoid genus *Parkia*. The flowers of *Ceiba pentandra* were visited by a wide range of both diurnal and nocturnal animals in Central Amazonia, but only two species of phyllostomid bat actually pollinated them (Gribel *et al.* 1999). A bat-pollinated understorey palm was studied by Cunningham (1995). The Australian rain-forest tree *Syzygium cormiflorum* is largely bat pollinated (Crome & Irvine 1986). Several species of the genus *Lafoensia* (Lythraceae) were reported to be pollinated by bats in the Atlantic forests of

[1] Kato (1996) reports this species to be pollinated by anthophorid bees. This is probably a reflection of the taxonomic problems (Wong 1996) associated with this species or group of species.

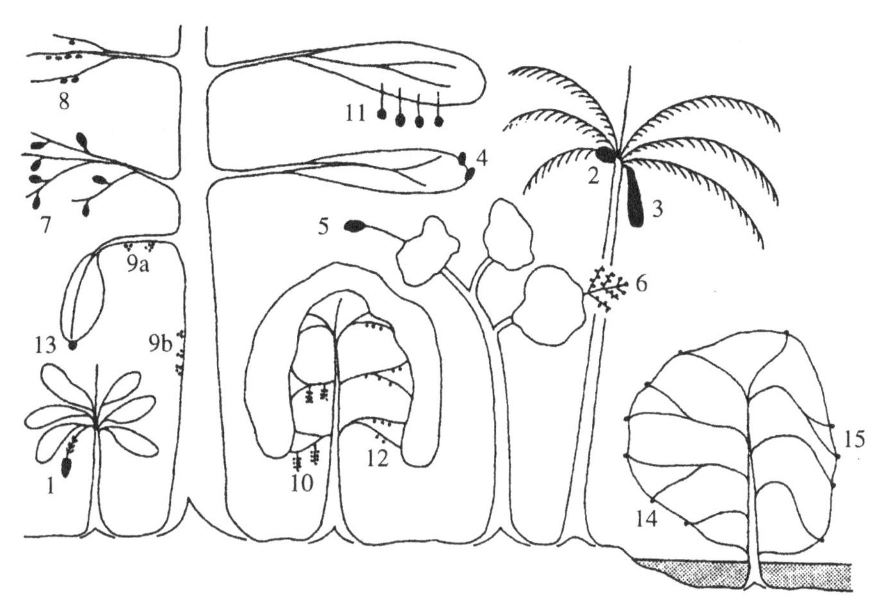


Figure 4.2 The positions of flowers on some Malaysian chiropterophilous trees. 1. Musaceae: *Musa* species. 2. Palmae: *Cocos nucifera*. 3. *Arenga pinnata*. 4. Anacardiaceae: *Mangifera indica*. 5. Bignoniaceae: *Oroxylum indicum*. 6. *Pajanelia longifolia*. 7. Bombacaceae: *Bombax valetonii*. 8. *Ceiba pentandra*. 9a. *Durio zibethinus*. 9b. *Durio graveolens*. 10. Lecythidaceae: *Barringtonia racemosa*. 11. Mimosaceae: *Parkia* species. 12. Myrtaceae: *Syzygium malaccense*. 13. Sonneratiaceae: *Duabanga grandiflora*. 14. *Sonneratia alba/ovata*. 15. *Sonneratia caseolaris*. After Marshall (1983).

Brazil (Sazima *et al.* 1999). Gould (1978) investigated the flower and inflorescence form and nectar production in *Oroxylum indicum* in relation to the behaviour of pollinating bats.

## *Non-volant mammals*

### POLLINATOR

Arboreal mammals including primates, marsupials and rodents have been suspected of being involved in pollination of tropical trees, but studies confirming pollen transfer, not just flower-visiting, are relatively few (Janson *et al.* 1981; Bawa 1990; Carthew & Goldingay 1997). The syndrome seems better defined in the shrublands of Australia and southern Africa. Primates will often eat flowers and in so doing may become liberally dusted with pollen but it seems probable that in relatively few cases are they effective pollinators, and are more likely to have a deleterious effect due to flower damage and removal.

FLOWERS

There are probably too few tree species designed for pollination by arboreal mammals to recognise a typical flower form. Diurnal and nocturnal pollinating species may even select for different floral traits (Carthew & Goldingay 1997), with dull-coloured but odoriferous blooms attracting night foragers and brightly hued and weakly scented flowers those species active during daylight hours.

PLANTS

A taxonomically diverse group of species has been associated with non-volant-mammal pollination (Carthew & Goldingay 1997). There is as yet no evidence of any families or genera of tropical trees being particularly specialised for pollination by arboreal mammals.

CASE STUDIES

The South American bombacaceous tree *Matisia cordata* produces bright yellow flowers on its branches that attracted a range of both diurnal and nocturnal primates and marsupials (Janson *et al.* 1981). Despite damage to the flowers by these visitors attracted to the copious nectar, fruit set was high. Birds and butterflies were also seen to attend flowering trees. Similarly, flowering individuals of the Central African *Daniellia pynaertii* (Leguminosae; Caesalpinioideae) attracted four species of diurnal primate (Gautier-Hion & Maisels 1994), yet still had large fruit crops. Few other potential pollinators were observed to visit the trees. A dormouse and two species of prosimian might act as pollinators of *Parkia bicolor* in Cameroon (Grünmeier 1990), although bats are probably the main agents of pollen dispersal. Momose *et al.* (1998d) report a squirrel-pollinated species of *Ganua* (Sapotaceae) from Sarawak. The sweet and fleshy berry-like corolla of the tree was the reward for the foraging squirrels. However, Fægri & van der Pijl (1979) considered the members of the Sapotaceae with corollas that act as food rewards to be bat-pollinated. The green, scentless flowers of the hemi-epiphytic shrub *Blakea chlorantha* (Melastomataceae) found in montane Costa Rica attracted two species of rat that are believed to pollinate the plant (Lumer 1980).

### *Birds*

POLLINATOR

There are several families of nectarivorous bird found in the tropics that are important pollinators. According to the ranking of Stiles (1981), the hummingbirds (Trochilidae) are the family most uniformly specialised for exploiting flowers. Hummingbirds are confined to the New World, with many of the more than 300 species found in humid tropical forests. The next most specialised are the sunbirds and spiderhunters (Nectariniidae), which are an Old World group. Honeyeaters (Meliphagidae) and Hawaiian honey-creepers (Drepaniidae) are somewhat less specialised. The lorikeets (Psitta-

Figure 4.3 A golden-winged parakeet visiting a *Moronobea coccinea* flower. The parakeet inserts its beak in the flower apex, contacting the anthers and stigmatic lobes. Drawn by Alexandre Kirovsky. From Vicentini & Fischer (1999). Copyrighted 1999 by the Association for Tropical Biology, P.O. Box 1897, Lawrence, KS 66044-8897. Reprinted by permission.

cidae; Loriinae) regularly feed from flowers, but tend to be destructive. Some species of white-eye (Zosteropidae) and honeycreeper (Coerebidae) are also specialised flower visitors. Many small omnivorous birds such as starlings (Sturnidae) and bulbuls (Pycnonotidae) will feed on nectar if it is available and accessible. Hummingbirds are the only family where most foraging is done while in flight by hovering near the flower. Sunbirds are capable of hovering, but perch if a perch is available. This difference in foraging styles is reflected in floral morphology (Westerkamp 1990).

FLOWERS

Bird-flowers are typically diurnal and scentless. Birds, it is widely believed, have limited abilities at olfaction, but they generally have excellent colour vision. Red is a common bird-flower colour, and as insects have a poor capacity to distinguish red it makes this a good colour as a general attractant for birds. Yellow, often in combination with red, is another tint favoured by flowers designed to attract birds. Bird-flowers often have long tubular corollas, sometimes with reflexed lobes at the tube mouth, what Endress has called the 'dogfish flower.' The long corolla tubes match the long beaks of the pollinating birds, a probable example of co-evolution. However, in a study of avian flower visitors in New Guinea there did not appear to be much congruence between floral morphology and beak form (Brown & Hopkins 1995). The flowers are fairly robust, often with inferior ovaries to protect the ovules from the quite large visitors. Another bird-flower form is the brush flower with an array of long stamens as seen in *Calliandra*. Hummingbird-flowers are usually held singly, or in clusters, in a horizontal or pendent position on flexible pedicels. The flowers produce large quantities of nectar, which appears to be the sole reward offered by bird-flowers (Stiles 1981). This generally has low concentrations of sugars and amino acids. Hummingbird-flowers tend to produce sucrose-rich nectar (Baker & Baker 1983; Baker *et al.* 1998), whereas bird-flowers from the palaeotropics are more likely to be rich in glucose and fructose. Many birds are thought to be physiologically intolerant of sucrose.

#### PLANTS

*Erythrina*, *Spathodea* and probably *Hibiscus* are tropical tree genera with bird-pollinated members. The species of *Rhododendron* with brightly coloured flowers found on tropical mountains are mostly bird-pollinated. In the cooler climates of the mountains birds are probably more active than insects.

#### CASE STUDIES

There has been surprisingly little research that focuses on bird-pollination of tropical trees. The classic studies of hummingbirds as pollinators in rain forests (see, for example, Stiles 1975, 1978) largely concern herbaceous and epiphytic species. However, *Symphonia globulifera* is pollinated by hummingbirds in Central Amazonia (Bittrich & Amaral 1996a). Perching birds may also be involved in pollination in the neotropics. Toledo (1977) outlined evidence for a range of small birds being involved in the pollination of two bombacaceous trees, *Bernoullia flammea* and *Ceiba pentandra*, in the rain forests of Southern Mexico. The latter is generally considered chiropterophilous, but birds may forage for nectar during the day. Vicentini & Fischer (1999) have given an account of the pollination of the Central Amazonian canopy tree *Moronobea coccinea* (Guttiferae) by the golden-winged parakeet (*Brotogeris chrysopterus*) (Fig. 4.3). Among the bananas (Musaceae) there are bat- and bird-pollinated species. Itino *et al.* (1991) compared the chiropterophilous *Musa acuminata* ssp. *halabensis* with the ornithophilous *Musa salaccensis* at a site in Sumatra. The bat-pollinated banana had pendent inflorescences with dark purple bracts and gelatinous sugary nectar. Sunbird-pollinated *Musa salaccensis* had erect inflorescences with paler pink-purple bracts and less sweet and runnier nectar.

### *Bees*

POLLINATOR

Current evidence points to bees as the most important pollinating organisms in tropical forests. They can be divided into two pollination guilds according to size: the medium- to large-sized species in the families Andrenidae, Apidae, Anthophoridae, Halictidae and Megachilidae and the small bees in the Apidae (Apini), Halictidae and Megachilidae. Some wasps also fall into the small-bee guild. The two groups are heterogeneous in terms of taxonomy, behaviour and ecology.

Bees are specialist foragers after pollen and nectar and are very efficient at doing so (Fig. 4.4). They are the only group capable of buzz-pollination, with the possible exception of some hoverflies (Syrphidae). That is the stimulation of pollen release from poricidal anthers by vibrating (buzzing) them at the correct frequency (Buchmann 1983). Big bees are more prevalent in the forest canopy, small bees in the understorey. *Xylocopa* (Anthophoridae) is a good example of a big-bee genus, members of which are able long-distance high-fidelity pollinators, but tend to be forest margin species, and their low numbers make them unreliable as pollinators for big trees. The euglossines (Apidae; Euglossini) are a neotropical group that are important long-tongued pollinators. At Lambir Hills National Park in Sarawak, the honeybee *Apis dorsata* was found to be an important pollinator during mass (general) flowering in the rain-forest community (Momose *et al.* 1998d). The number of colonies increased dramatically during this period. This bee is unusual in that it forages for an hour or two before dawn and after dusk. Some tree species at Lambir may open flowers at night to give *Apis dorsata* sole access to the flowers during their crepuscular activity periods.

FLOWERS

Bee-flowers are extremely varied in morphology, from small, simple, open flowers to complex structures that require particular modes of entry to obtain the floral reward. Specialist big-bee-flowers are often brightly coloured, sometimes with nectar guides, markings on the petals directing the bee to the appropriate place to forage for nectar and pick up or deposit pollen. Nectar and/or pollen are the rewards in bee flowers. Bees are often early morning foragers, so bee-flowers open early, sometimes during the night before. Small-bee-flowers are generally small, white or cream and short-tubed.

PLANTS

Leguminosae, Bignoniaceae and Melastomataceae often possess specialised bee-flowers, including poricidal anthers that require buzz-pollination. Burseraceae, Euphorbiaceae, Flacourtiaceae, Lecythidaceae and Sapotaceae are often equipped with less specialised bee-flowers. Tropical trees of the Malpighiaceae and Melastomataceae tribe Memecyleae have been shown to bear

Figure 4.4 Trigonid bee (*Trigona erythrogastra*) on *Melastoma malabathricum* flower in Lambir Hills National Park, Sarawak. Photo: Tamiji Inoue.

flowers offering oils rather than sugary nectar as rewards to bees (Buchmann & Buchmann 1981; Buchmann 1987). Species of *Clusia* and *Dalechampia* (mostly climbers) provide resin as a reward to flower visitors (Armbruster 1984; Bittrich & Amaral 1996b; Lopes & Machado 1998). This is collected by bees to use in building their nests (Armbruster 1984).

CASE STUDIES

Perry & Starrett (1980) found that the emergent *Dipteryx panamensis* was pollinated by at least 13 species of bee. Medium-sized stingless bees (*Trigona*, Apidae) are pollinators of the large dipterocarp *Dryobalanops lanceolata* in Sarawak (Momose *et al.* 1996). Similar bees pollinate *Garcinia hombroniana* in Peninsular Malaysia (Richards 1990b). The palm *Prestoea decurrens* is probably mostly bee-pollinated (Ervik & Bernal 1996). Trigonid bees and drosophilid flies are duped into visiting unrewarding pistillate flowers of the protandrous palm *Geonoma macrostachys* in Ecuador, because of their similarity to the rewarding staminate blooms (Olesen & Balslev 1990). The Mexican understorey shrub *Dalechampia spathulata* is visited by male Euglossine bees that collect aromatic substances from an extra-floral gland in the inflorescences (Armbruster & Webster 1979).

## *Moths*

POLLINATOR

Hawkmoths (Sphingidae) are the commonest moth pollinators, but some species of Noctuidae, Pyralidae and Geometridae are probably also involved. Hawkmoths have the longest tongues of any insect group, are excellent hoverers and are generally nocturnal. They appear to be rare in the mountains, probably because of the low temperatures.

FLOWERS

Moth-pollinated flowers are typically heavily scented (strong and sweet), particularly at night, pale-coloured, and nectariferous with long tubular corollas or spurs, and nocturnal anthesis. An alternative flower form is that of the relatively small corolla with many long exerted stamens as seen in the Capparidaceae. Flowers designed for sphingid visitors can be fairly delicate, even pendulous, as the insect does not need to land. Noctuids have short proboscides and therefore the phalaenophilous flowers are short-tubed. Lepidopteran-pollinated flowers tend to produce sucrose-rich nectar (Baker & Baker 1983).

PLANTS

Tropical tree families that contain moth-pollinated species include the Rubiaceae, Apocynaceae, Meliaceae, Solanaceae (e.g. *Datura*) and Leguminosae (subfamily Mimosoideae).

CASE STUDIES

Prance (1985) provided a comparison of moth and butterfly pollination in two genera of the Chrysobalanaceae of Amazonia. The discovery of moth-pollination in *Gnetum* (Kato & Inoue 1994; Kato *et al.* 1995) (Fig. 4.5) has interesting implications for the evolutionary origins of insect pollination. The Madagascan shrub *Ixora platythyrsa* is visited by noctuid and geometrid moths (Nilsson *et al.* 1990). The markedly protandrous flowers have a secondary pollen-presentation mechanism to facilitate pollination by the moths. The pollen is released onto the unripe stylar heads from where it is picked up by the proboscides of the moths. Linhart & Mendenhall (1977) used dyed pollen to demonstrate pollen movement of more than 350 m by hawkmoths that pollinated the rubiaceous shrub *Lindenia rivalis* in Belize. The plants, which grow along watercourses, were visited regularly by the sphingids, which appeared to have set routes for foraging. Chase *et al.* (1996) used genetic markers to show that the hawkmoths can effectively pollinate *Pithecellobium elegans* over long distances.

### *Butterflies*

POLLINATOR

Butterflies are the other group of the Lepidoptera that is involved in pollination. Butterflies are active during the day, have shorter proboscides than hawkmoths and alight on their chosen flowers before feeding.

FLOWERS

Butterfly flowers usually have long tubular corollas or spurs, and are brightly coloured, often in shades of orange, red or pink, or contrasting combinations of the same, to attract day-flying butterflies to the nectar reward. In large flowers there is usually a landing platform and exserted stamens and style imprecisely deposit, or pick up, pollen from the visiting butterfly. Alternatively, small flowers aggregated in an inflorescence create a landing platform.

PLANTS

Tropical tree genera such as *Delonix*, *Caesalpinia*, *Lantana*, *Cordia* and *Mussaenda* are believed to be butterfly-pollinated.

CASE STUDIES

*Xerospermum intermedium*, a common tree at Pasoh in Peninsular Malaysia,

Figure 4.5 Moth (*Herpetogramma* sp.) on *Gnetum gnemon* strobilus. Photo: Tamiji Inoue.

was visited by trigonid bees and butterflies (Appanah 1982). Kato (1996) reported *Ixora griffithii*, an understorey shrub, to be pollinated by papilionid butterflies, as well as xylocopine bees, at Lambir in Sarawak. *Gardenia actinocarpa* is a rare, dioecious shrub endemic to the rain forests of northern Queensland that is pollinated by butterflies (Osunkoya 1999).

### *Beetles*

#### POLLINATOR

The pollinating coleopterans are a very diverse group, ranging in size from tiny weevils and staphylinids to large scarabs. Beetle pollination was long believed to be the primitive angiosperm condition. However, recent debate over the actual nature of the earliest angiosperms, and the finding that the more primitive extant Magnoliidae are not beetle-pollinated, rather dampens that view. Nevertheless, beetle pollination is very important in tropical forests in both families thought of as primitive and those considered relatively advanced. As Endress remarked, beetles are uncouth floral visitors, performing 'mess-and-soil' pollination. The beetles blunder around the flowers chewing parts, apparently indiscriminately, and becoming covered in pollen.

#### FLOWERS

Beetle-pollinated flowers of tropical trees are quite variable in construction, ranging from small (e.g. palm flowers) to quite large (a few centimetres across in some Annonaceae). They are usually quite open – the beetle traps of the aroids and waterlilies are not encountered among tropical trees – with a strong smell and plentiful pollen. Thermogenicity is sometimes involved in attracting the beetles. Temperatures 6 K above ambient have been recorded in *Annona sericea*. Palms have many small flowers on large inflorescences, and are often beetle-pollinated.

#### PLANTS

Annonaceae, Lauraceae and Myristicaceae are important tropical tree families with many beetle-pollinated species (Schatz 1990; Irvine & Armstrong 1990). Eupomatiaceae, Degeneriaceae (Thien 1980) and Winteraceae are examples of other, relatively primitive, cantharophilous families. Among the monocots, the Palmae and the Cyclanthaceae attract beetles (Henderson 1986; Schatz 1990).

#### CASE STUDIES

The beetle pollination of various nutmeg species has been studied in detail in Australia (Armstrong 1997). Palms found to be beetle-pollinated include *Bactris porschiana* (Beach 1984), *B. gasipaes*, *B. bifida* and *B. monticola* (Listabarth 1996), *Cryosophila albida* (Henderson 1984), *Phytelephas seemannii* (Bernal & Ervik 1996) and *Socratea exorrhiza* (Henderson 1985). Momose *et al.* (1998d) report that chrysomelid beetles are the pollinators of many dipterocarp species during general flowering at Lambir in Sarawak

(Fig. 4.6). In tropical Queensland, beetles have been found to be important pollinators of rain-forest trees. Tree species involved include *Diospyros pentamera* (House 1989), *Flindersia brayleyana*, *Alphitonia petriei* and *Eupomatia laurina* (Irvine & Armstrong 1990). *Eupomatia* is dependent on a tiny weevil for pollination. This species (*Elleschodes hamiltoni*) breeds in the synandria of the flowers and hence is dependent on the tree also.

### *Various small flying insects*

POLLINATOR

This is a catch-all category for relatively unspecialised flowers thought to be visited by small insects. These will include thripses[2] (Thysanoptera).

FLOWERS

Small and unspecialised flowers characterise this group. Thripses appear to be attracted to white, perfumed flowers.

PLANTS

Families such as Anacardiaceae, Euphorbiaceae and Guttiferae often have small, unspecialised flowers.

CASE STUDIES

Sympatric species of *Shorea* section *Mutica* have been shown to be thrips-pollinated in Peninsular Malaysia (Chan & Appanah 1980). The tiny insects reproduce inside the flowers and rapidly increase in population numbers, resulting in a high chance of pollination during a general flowering event. In other regions, endemic dipterocarps have been reported to use different pollinators: bees in Sri Lanka (Dayanandan *et al.* 1990) and chrysomelid beetles in Sarawak (Momose *et al.* 1998d). Sakai *et al.* (1999) found that *Shorea parvifolia*, which had been reported as thrips-pollinated in the Malay Peninsula, was probably more effectively pollinated by chrysomelid beetles at Lambir Hills National Park in Sarawak. Several tropical Annonaceae are apparently thrips-pollinated. In Sarawak *Popowia pisocarpa* is visited by thysanopterans (Momose *et al.* 1998c), as are *Bocageopsis multiflora* and *Oxandra euneura* in Amazonia (Webber & Gottsberger 1995).

### *Flies*

POLLINATOR

Flies (Diptera) are possibly the second most important group of insect pollinators in the rain forest, although myiophily has been very little studied. Relatively few species of fly appear to be dependent on flowers for food. They can be attracted by using a variety of means, including odours and nectar. Saprophilous (carrion and dung) flies can be employed as pollinators

[2] The English name for these insects is derived from the generic name *Thrips*, leading to the apparently infantile plural.

Figure 4.6 Flowers of *Shorea parvifolia* being visited by chrysomelid beetles (*Monolepta* sp.). Photo: Tamiji Inoue.

through floral form and odour imitating the preferred food of these flies.

FLOWERS

Small, pale flowers with nectar attract flies, particularly hoverflies (Syrphidae). Sapromyiophily is not common in tropical trees.

PLANTS

Sterculiaceae have been reported as fly-pollinated, although Momose *et al.* (1998d) state that the species at Lambir are mostly beetle-pollinated.

CASE STUDIES

Cocoa (*Theobroma cacao*) is principally pollinated by tiny midges of the family Ceratopogonidae (Young 1982). House (1989) found that dipterans were the most abundant flower visitors for two lauraceous trees in the Queensland rain forest, *Neolitsea dealbata* and *Litsea leefeana*, and that they were probably the most effective pollinators. Thien (1980) reported fly pollination of *Drimys piperata* (Winteraceae) in the montane forests of New Guinea. Schmid (1970) described syrphid fly pollination of the palm *Asterogyne martiana*. A hoverfly (*Copestylum* sp.) was probably the most efficient pollinator among many insect flower visitors of another understorey palm, *Prestoea schultzeana*, in Amazonian Ecuador (Ervik & Feil 1997).

*Wind*

POLLINATOR

The wind is not a very reliable pollinator in the tropics, as outlined above.

FLOWERS

Small, often with many flowers in an inflorescence. The perianth is often reduced and there is no nectar or perfume. Pollen is produced in large amounts and the stigmas are often feathery to improve pollen capture from the air.

PLANTS

Tropical conifers are probably wind-pollinated (Richards 1996). Some palm and pandan (screw-pine) species may be also.

CASE STUDIES

Wind pollination is not entirely absent from the rain-forest understorey. There is some evidence that it may occur quite widely in the understorey Moraceae. The small Central American tree *Trophis involucrata* has explosive anthesis (Bawa & Crisp 1980), as does *Streblus brunonianus* from Australia (Williams & Adam 1993).

## Tropical flowers

Endress (1994) notes a number of special features of flowers in the tropics, which are relevant to a discussion of tropical rain-forest trees. The rarity of wind pollination has already been mentioned. The presence of some very large flowers is notable. With large pollinators, such as birds and bats and even substantial bees and beetles, lone individual flowers, if they are to attract pollinators and survive their visitation, need to be big. There are many small-flowered species in the rain forest, but the range of flower size is very wide. There are probably also more short-lived flowers in the tropics than elsewhere. Such flowers are usually only receptive for a single day. This probably reflects an abundance of ‘trap-lining’ pollinators. Trap-liners are species that visit individual plants on a regular basis to exploit their floral resources, as a trapper visits a set of traps in repeated, routine fashion to collect what has been caught and re-set the trap. Trap-lining pollinaters include species of bat, bird, bee, butterfly and moth. Trap-liners are advantageous as pollinators because they often move long distances between flower visits and frequently specialise on relatively few species. The propensities of trap-liners tend to improve the chances of pollen being transferred between distant individuals of the same species. Plants can encourage trap-liners by the regular production of few flowers over a long flowering season. The pollinators then visit the plant on a consistent basis. The relatively small floral

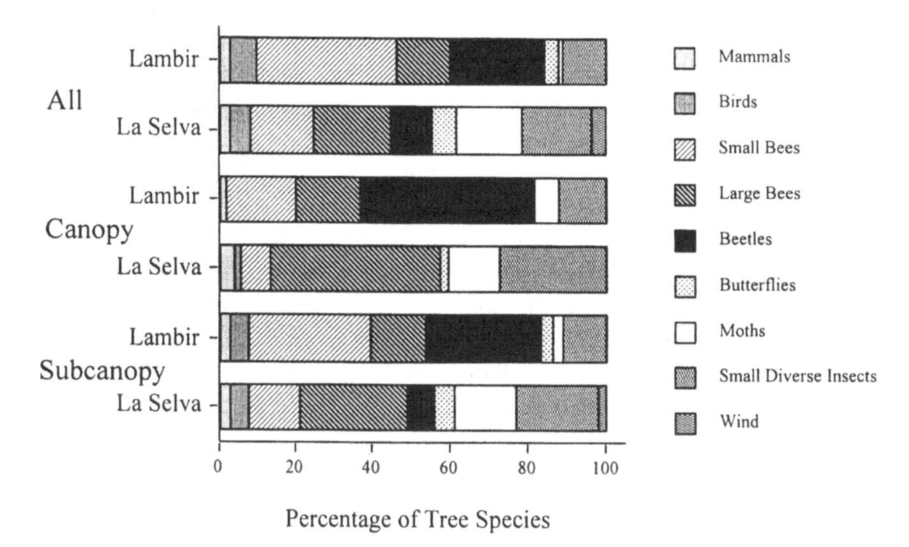


Figure 4.7 Relative frequencies of pollinator type among tree species at La Selva, Costa Rica, and Lambir, Malaysia. Comparisons are given for all trees and for canopy and subcanopy components. Data from Kress & Beach (1994) and Momose *et al.* (1998d).

display does not attract many opportunist exploiters of flowers, but is sufficient to maintain daily attendance by the reliable pollinators. Hence the presence of one-day flowers in tropical forests.

A great variety of positions on the tree for the production of flowers is another feature of the tropical forest. Flowers are generally borne at the ends of twigs, but in some species flowers or inflorescences can arise from the branches, the trunk or even the roots. A few species also exhibit epiphyllous flowers. Ramiflory, cauliflory and rhizoflory are probably adaptations to exploit pollinators and/or seed dispersers active in the forest understorey.

## Relative importance of different pollinators

There are only two tropical rain-forest sites where detailed studies of pollination of a large number of tree species have been conducted. These are La Selva in Costa Rica (Bawa *et al.* 1985a; Kress & Beach 1994) and Lambir Hills National Park in Sarawak (Momose *et al.* 1998d). At La Selva, bees are the most important pollinators, particularly in the upper layers of the forest. Small diverse insects are second and moths third in importance (Fig. 4.7). Vertebrate pollination is relatively rare in the canopy, being commoner among the large herbs (*Heliconia* spp. and other Zingiberales). The understorey generally exhibits a greater diversity of pollination syndromes than the canopy. The dipterocarp forest at Lambir differs from La Selva in having fewer trap-lining pollinator species. This may be because of the mass-flowering pattern of the community with fewer species flowering regularly. The use of migratory bees to allow adequate pollination during the mass flowering means that there is less selection pressure in favour of maintaining pollinator numbers all the time. The report that possibly herbivorous chrysomelid beetles are important pollinators of dipterocarp species during the general-flowering events provokes the fascinating speculation that dipterocarps may maintain their pollinators by providing foliage as food. The forest community at Lambir shows greater reliance on beetles, particularly the canopy species, and small bees than that at La Selva, but contains fewer species pollinated by moths and the small-diverse-insect group.

Momose *et al.* (1998b) have proposed that much of the vertical variation in pollinator type (specialist versus generalist) and interval between reproductive events in the tropical rain forest can be explained in terms of display effects. They have developed a mathematical model based on the premise that large plants can produce large displays that attract more generalist pollinators and increase pollination success. Canopy trees with large stature and low rates of mortality can maximise pollination success by producing huge displays at long intervals. Small, understorey plants are unable to compete in terms of attracting generalist pollinators with floral displays and therefore employ specialist pollinators. The greater risk of mortality in the lower parts of the forest also makes it risky to delay reproduction for long intervals, therefore flowering is predicted to be more frequent in the understorey. These general predictions, greater use of generalist pollinators and longer reproductive intervals in the canopy compared to the understorey, appear to fit with observations at Lambir Hills National Park (Momose *et al.*1998d).

Grubb & Stevens (1985) found an increasing frequency of ornithophilous and anemophilous species with altitude in forests on the mountains of New Guinea.

## Figs

Figs are members of the genus *Ficus* in the family Moraceae. The many species share the syconium as the form of inflorescence and the use of tiny wasps as pollinators. Otherwise, the figs show a huge diversity of form from small shrubs and climbers to huge trees and stranglers. Figs are unique in the tropical rain forest in that each species has its own obligate pollinating species. Despite being tiny and unable to survive for long, recent studies employing genetic markers have shown that fig wasps are very effective long-distance pollinators. The tiny fig wasps are probably dispersed on the wind, but once near a fig tree of the correct species at the right stage of syconium development they can fly to the ostiole of an inflorescence and make their way inside. Concomitant to the necessity of maintaining populations of the fig wasps that have relatively short life-cycle times and short adult lives is the virtually continuous reproductive activity of a fig population. This results in a regular fruit supply. Fig fruits (mostly made up of the fleshy receptacle of the syconium) are palatable to many frugivorous animals and because of their regular availability they can be relied on in periods of relative fruit scarcity. For this reason, figs have been termed keystone species in tropical rain forests because of their role in seeing frugivore communities through hard times (Terborgh 1986). Fig fruits are also relatively high in calcium concentration compared with other fruits (O'Brien *et al.* 1998) which will increase their nutritional importance to frugivores. However, in some forests figs are too rare to represent keystone resources (Gautier-Hion & Michaloud 1989).

## Dispersal mechanisms

After successful pollination, fertilised ovules will develop into seeds and the ovary will become a fruit. The seeds, the offspring of the tree, will then require to leave their mother plant in order to establish as new individuals. There are four main agents of seed dispersal: gravity, wind, water and animals. Their relative importance in the dispersal of a species allows its classification as autochorous, anemochorous, hydrochorous or zoochorous. In addition to the primary dispersal of the seed from the tree, there may also be secondary dispersal by another agent from the first landing site. For instance, there is growing evidence of an important role played by ants in removing tiny seeds from animal droppings and dispersing them further (Roberts & Heithaus 1986; Kaufman *et al.* 1991; Levey & Byrne 1993).

### Autochory

Autochorous species include those with explosive dehiscence (ballistochores) and those with apparently no other means of dispersal than falling off the parent tree. The biggest seed of all, the double coconut or coco-de-mer *Lodoicea maldivica*, is an example of the latter. Unlike the true coconut, it cannot survive long immersed in water and is therefore not hydrochorous. In the rain forest, the Euphorbiaceae (the rubber tree *Hevea brasiliensis* is a good example), Leguminosae and Bombacaceae are the families most likely to be present as autochorous tree species. Nearly 60% of seeds fall less than 10 m away from the parent plant in ballistochorous *Eperua falcata* in French Guiana (Forget 1989), with maximal dispersal distance of 30 m. An 11 m tree of *Hura crepitans* growing in an open site in Ghana did somewhat better, sending seeds up to 45 m, with a modal dispersal distance of 30 m (Swaine & Beer 1977).

**Anemochory**

Wind-dispersal involves modification in the form of the disseminule to increase its time of descent. This is achieved by either the use of wings, hairs or plumes or a reduction to very small size. Augspurger (1986) introduced a functional terminology for anemochores (Table 4.4). She found that the winged disseminules fell significantly more slowly than the floaters. Wind direction and force at the time of fruit maturity will be highly influential on the spatial distribution of dispersed seeds of anemochorous species. The density patterns for seeds dispersed from two parent trees of *Tachigali versicolor* showed a strong skew in the direction of the prevailing wind (Fig. 4.8).

The taller a tree, the greater the release height for anemochores and the longer the period of descent. It has often been argued therefore that anemochory should be favoured in tall species. Short trees in the forest understorey are rarely exposed to high winds that facilitate anemochore dispersal and the low release height does not allow prolonged descent. In a study of the rolling autogyro disseminules of *Lophopetalum wightianum*, Sinha & Davidar (1992) did indeed find that tall trees had a significant dispersal advantage (Fig. 4.9). Average and maximum dispersal distances increased with tree height. Most surveys of dispersal type among tropical forest tree species have found an increase in abundance of anemochory (or at least possession of dry rather than fleshy fruits) in taller trees (see Table 4.5).

Wind dispersal also seems to become commoner at higher elevations (Table 4.6; Grubb & Stevens 1985), and in drier forests (Gentry 1982). Suzuki & Ashton (1996) report that among members of the largely anemochorous Dipterocarpaceae wingless fruits tend to occur in understorey, riverine or very large-seeded species. The riverine species are hydrochorous and the investment needed in wings for very large seeds is probably too big to be worth while.

Wind dispersal tends to be poorly effective; for instance, in the tall Indonesian rain-forest tree *Swintonia schwenkii* few seeds landed more than 50 m from the parent (Fig. 4.10). Important families of tropical trees with many anemochorous species include the Bignoniaceae, Bombacaceae, Leguminosae, Vochysiaceae, Dipterocarpaceae and Apocynaceae. Anemochory appears to be restricted to advanced subclasses of flowering plants in French

Table 4.4. *A functional classification of anemochores.*

| Type | Description | Examples |
|---|---|---|
| Dust diaspores | microscopic seeds so small they float on the lightest breeze | probably only species of *Eucalyptus*, *Rhododendron* and *Clethra* among tropical trees have dry seeds that approach the dust size range |
| Floaters | plumed fruits or seeds where feathery hairs provide buoyancy reducing the rate of descent | many species of Bombacaceae and Apocynaceae have disseminules of this type |
| Undulators | flat, thin disseminules that glide and undulate downwards, with frequent stalls, like falling sheets of paper | *Pterocarpus* |
| True gliders | similar to undulators, but better aerodynamic performance allows them to glide forwards in still air | not as yet reported in tropical tree species, but likely to exist; some rain-forest lianas probably produce such seeds |
| Tumblers | rotate about an axis parallel to the ground as they fall; they often resemble tiny paddle wheels | Combretaceae often have this fruit type |
| Helicopters | spin about their central axis as they fall, with wings above to provide lift | many members of the Dipterocarpaceae and some Anacardiaceae have helicopter fruits |
| Autogyros | winged disseminules that spiral around one end (usually where the seed is placed) of their length as they fall | a relatively common disseminule type among anemochores; many legumes produce autogyros |
| Rolling autogyros | a form of autogyro where the disseminule rotates about its long axis while spiralling around one end; the seed tends to be more centrally placed on the wing than in the simple autogyro | *Lophopetalum wightianum* |

Guiana (Mori & Brown 1994), but it remains unclear whether this observation can be extrapolated to all tropical areas.

### Hydrochory

The use of water as a dispersal agent is common in swamp, riverine and coastal forest species. Hydrochores generally float and often have corky tissues to provide extra buoyancy. They remain viable despite prolonged periods of immersion. Seeds of the riverine Amazonian tree *Swartzia polyphylla* exhibit a range of specific gravity about unity (Williamson *et al.* 1999).

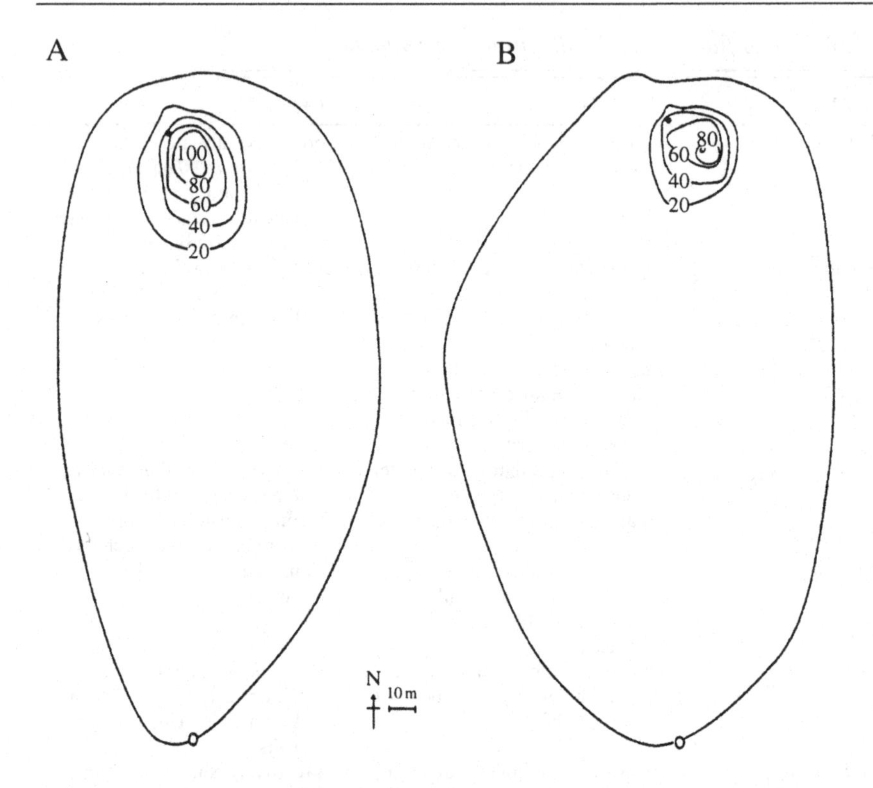


Figure 4.8 Isoclinic lines of density (number per square metre) of dispersed seeds around two *Tachigali versicolor* trees (A and B). Black dots represent the positions of the parent trees. Line 0 represents the distance of the last seed beyond which no seed was found within 20 m. After Kitajima & Augspurger (1989).

Therefore some seeds float in water and others sink. The floaters are dispersed greater distances than the sinkers.

### Zoochory

Zoochores are divided into two groups: epizoochores that travel on the animal's body, usually as accidentally attached burrs or sticky fruits, and endozoochores that are transferred in the animal's mouth or gut. Epizoochory is rare among tropical trees. In a survey of Central French Guiana, Mori & Brown (1998) only found one tree species, *Trymatococcus amazonicus*, with morphological adaptations to epizoochory among a flora of 1861 species. The very sticky fruits of *Pisonia grandis* are believed to be dispersed by adhering to birds (van der Pijl 1982).

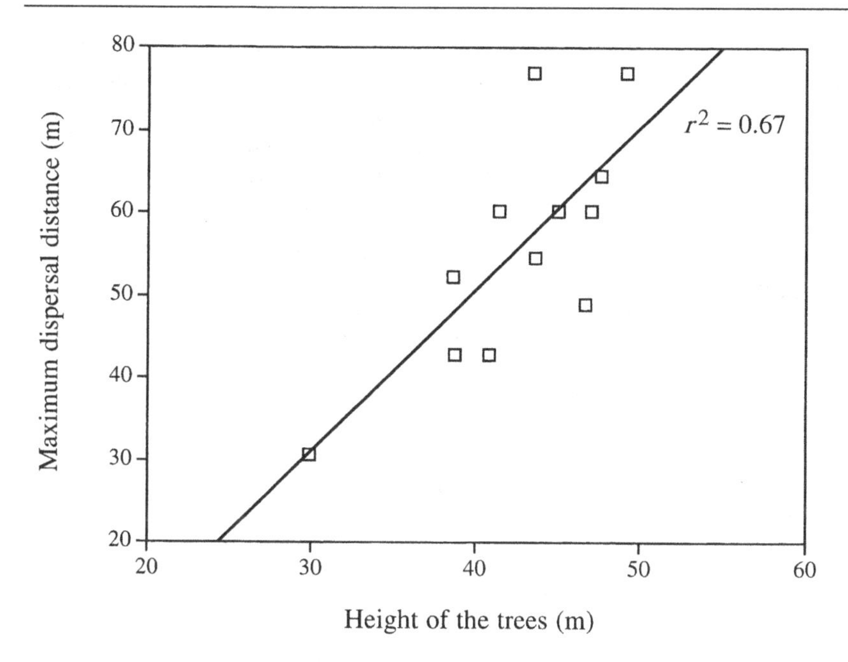


Figure 4.9 Maximum dispersal distance with relation to the height of the tree for *Lophopetalum wightianum* in the Western Ghats, India. After Sinha & Davidar (1992).

Animals generally disperse seeds because they are given some reward to attract them to the fruit. This is usually a fleshy expendable part of the fruit or seed. However, hard dry fruits can also be endozoochores. In such fruits, the seed itself is the reward. Dispersal is facilitated by hoarding seed-eaters that hide away excess food and forget about some, allowing germination to take place.

Strict frugivory is relatively rare in animals. Only 17 bird families (15.6% of total) are strictly frugivorous, but many birds are opportunistically frugivorous. Among the mammals, primates are the only group where strict frugivory is common. Bats also have quite a lot of frugivorous species and some forest ruminants, like the brocket deer and African cephalophanes (*Cephalophus* spp.) (Dubost 1984) can be included. Other partly frugivorous mammals include marsupials, tree-shrews, bears, foxes, elephants and civets. Reptiles, fish, and ants are the only other groups to be involved in seed dispersal in tropical rain forests in a major way.

Given the abundance of frugivorous birds and mammals in tropical forests it is not surprising that a high proportion of tree species in most rain forests have fleshy fruits (Table 4.7); generally at least 60%, and often much higher. Fruits offer food rewards to frugivores. The nutritional composition of fruit flesh varies considerably. In an analysis of published data on fruit-flesh composition, Jordano (1995) found the main variation in fruit pulp is among three nodes. These are: high lipid, low fibre, low non-structural carbohydrate

Table 4.5. *Stratification of fruit types with height in tropical forests*

Values are the percentage of species in the tree community of a particular height class.

| Location | Height class | Dispersal type: fleshy | dry/wind | other |
|---|---|---|---|---|
| Venezuela[1] | 10–30 m | 56 | 43 | — |
| | >30 m | 42 | 58 | — |
| Ghana[2] | <30 m | 78 | 6 | 16 |
| | ≥ 30 m | 60 | 29 | 11 |
| Ivory Coast[3] | > 30 m | 73 | 27 | — |
| | 20–30 m | 81 | 6 | 14 |
| Santa Rosa, Costa Rica[4] | < 10 m | 77 | 8 | 15 |
| | ≥ 10 m | 64 | 29 | 7 |
| La Selva, Costa Rica[4] | < 10 m | 98 | 2 | — |
| | ≥ 10 m | 85 | 13 | 3 |
| Barro Colorado Island[5] | < 10 m | 87 | 5 | 8 |
| | ≥ 10 m | 78 | 16 | 5 |
| Rio Palenque, Ecuador[4] | < 10 m | 91 | 6 | 4 |
| | ≥ 10 m | 93 | 4 | 3 |

References: [1]Roth (1987); [2]Hall & Swaine (1981); [3]Alexandre (1980); [4]Gentry (1982); [5]Howe & Smallwood (1982).

Table 4.6. *Percentage of species (trees & shrubs) in different dispersal/fruit classes in forests in New Guinea*

| Fruit/ dispersal type | Lower montane | Upper montane | Subalpine |
|---|---|---|---|
| fleshy | 85 | 81 | 75 |
| wind | 7 | 13 | 20 |
| none obvious | 9 | 5 | 5 |

Data from Grubb & Stevens (1985).

Figure 4.10 (*opposite*) Spatial distribution of wind-dispersed fruits of *Swintonia schwenkii* in Sumatra. Upper panel: (a) Positions of mature fruits on the forest floor in the plot. Crowns and root collars of emergent trees are also shown. Three areas marked A, B and C are defined for analysis of dispersal distance (cf. lower panel). (b) Topography of the plot. Lower panel: The relation between distance from the tree no. 381 and density of mature fruits in three domains, A, B and C. After Suzuki & Kohyama (1991).

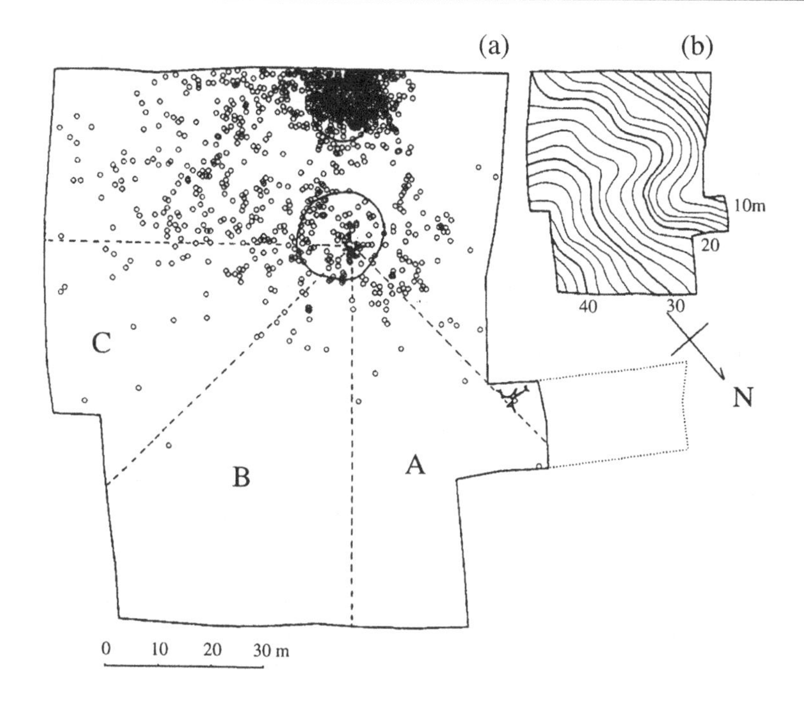
(a)
(b)
C
B
A
0 10 20 30 m
10m
20
40
30
N

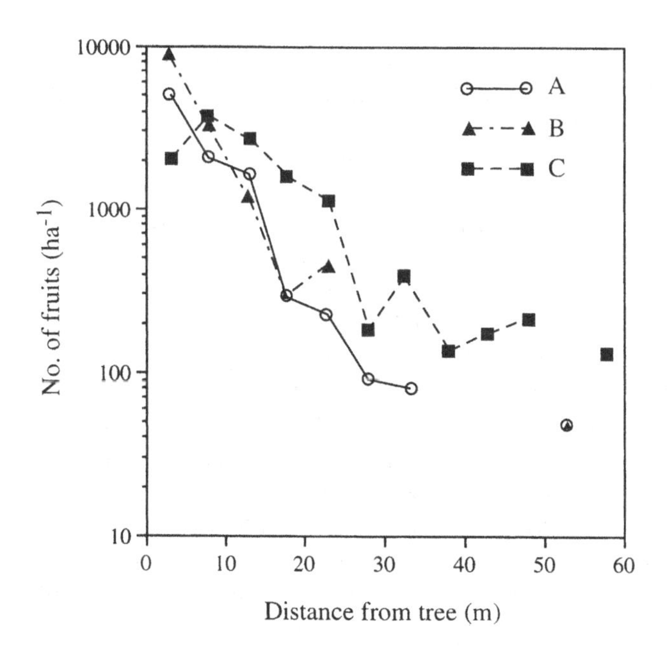
A
B
C
No. of fruits (ha$^{-1}$)
Distance from tree (m)
10000
1000
100
10
0
10
20
30
40
50
60

Table 4.7. *Percentage of tree and shrub species among different disseminule categories*

| | Dispersal/fruit type | | |
|---|---|---|---|
| Location | fleshy | dry/wind | other |
| French Guiana[1] | 96 | 4 | — |
| La Selva, Costa Rica[2] | 95 | 5 | — |
| Rio Palenque, Ecuador[2] | 93 | 5 | 2 |
| Malay Peninsula[3] | 92 | 8 | — |
| Queensland[4] | 84 | 16 | — |
| Barro Colorado Island[5] | 82 | 13 | 5 |
| Hong Kong[6] | 76 | 24 | — |
| Santa Rosa, Costa Rica[2] | 68 | 23 | 9 |
| Venezuela[7] | 68 | 13 | 18 |
| Ivory Coast[8] | 66 | 16 | 5 |

[1]Charles-Dominique *et al.* (1981); [2]Gentry (1982); [3]Putz (1979); [4]Willson *et al.* (1989); [5]Howe & Smallwood (1982); [6]Corlett (1996); [7]Roth (1987); [8]Alexandre (1980).

(NSC); low lipid, high NSC, low fibre; and medium lipid, medium NSC, high fibre. Protein and water contents vary independently of this. In general, fruits tend to contain low concentrations of protein. Leaves are richer in protein than most fruits. However, because of the relatively high lipid or carbohydrate concentration in the flesh, fruits are valuable sources of energy for animals. Myristicaceae, Lauraceae, Meliaceae and Palmae are families that often have large, oily fruits. Myrtaceae, Moraceae, Rutaceae and Rubiaceae tend to have sugary but watery fruits. For species dispersed by birds and bats, the sugar profile of the fruit juice was found to be similar to that of nectars of flowers pollinated by the same animal group (Baker *et al.* 1998). Thus passerine and microchiropteran fruits had relatively high concentrations of hexoses, whereas megachiropteran fruits were more rich in sucrose.

Frugivores process fruits in their mouths. Fruit flesh can be dealt with in such a way that seeds are spat out, crushed or swallowed whole. Swallowed seeds may be regurgitated, or may pass right through the digestive tract and be voided through defaecation. Frugivores that habitually destroy seeds by mashing or chewing in the mouth or through digestion in the gut are of no benefit to the plant. These are usually discouraged by protecting the seeds in hard, smooth coats that are difficult to rupture and by defending the seed contents with noxious chemicals. Small seeds may be more likely to escape damage in the mouth than large ones. Frugivores that spit or rapidly regurgitate seeds are not going to be very effective seed dispersers because they will probably perform these actions while still in the fruiting tree's crown, resulting in very short dispersal distances. Dispersal via defaecation is likely to result in longer residence times in the disperser and hence greater dispersal distances. Frugivores, particularly birds and bats, have very rapid gut passage times. Presumably, this is because they are eating readily digestible food and, as flying animals, deadweight in the form of indigestible fibre and seeds is energetically expensive to carry around and best voided as soon as possible. There has been speculation that fruit flesh may include laxative chemicals to speed up gut action. We all know of the tendency of a high-fruit diet to promote bowel action. The advantage of this to the fruiting tree is that shorter time in the gut may reduce seed damage from digestive enzymes. Experimental evidence in favour of this 'fruit laxative' hypothesis was provided by Murray *et al.* (1994) for dispersal of *Witheringia solanacea* by *Myadestes melanops*. However, Witmer (1996) produced an alternative explanation for these results based on the influence of sugar concentrations in fruit pulp on frugivore gut action. The role of secondary chemicals in ripe fruit flesh remains controversial (Cipollini & Levey 1998; Eriksson & Ehrlén 1998).

The removal of fruits from the parent tree before the seeds are mature is likely to cause the death of the embryos because they require further resources from the maternal plant to complete development. Therefore, fruits usually only develop attractions for frugivores to coincide with seed maturity. The process of ripening involves a change from an unattractive to an attractive state. Unripe fruits are often hard and dry, bitter or sour to taste (Ungar 1995), with an inconspicuous coloration.

Big fruits tend to attract only large frugivores, probably because of the influence of size on efficiency of foraging. Small frugivores waste too much time and effort in trying to deal with fruits larger than they can comfortably manage. Mack (1993) speculated that fruit size maxima might reflect frugivore size. He tested this theory by comparing fruit and frugivore sizes between the neotropics and palaeotropics. Using fruit length for species from eight important, tree-dominated, plant families, he found that there was a tendency for the average of the mean generic fruit lengths to be greater in the Old World. This probably reflects the availability of larger frugivores in the palaeotropics. In the Old World, 12 families of frugivorous mammals have species with adult mass of more than 20 kg, whereas only three families do in the New World. Among frugivorous birds only one family has species with individuals of more than 1.5 kg in the Americas, the Old World has four families, including the cassowaries, which are an order of magnitude larger than any New World frugivorous bird. Old World primates are generally bigger than New World ones also (Lambert & Garber 1998).

## Dispersal syndromes

In a similar manner to flower form for pollination, fruit form can be assigned to a number of syndromes associated with each type of dispersal agent. These are described (based on van der Pijl (1982), Corlett (1998) and others) below.

### *Birds*

#### DISPERSERS

Birds are probably the most important frugivores in tropical forests. Major New World frugivores include the cotingas (Cotingidae), toucans (Ramphastidae) and the trogons (Trogonidae). In the Old World, hornbills (Bucerotidae) and fruit pigeons (various genera) are important. New Guinea is unique in the domination of the frugivore community by birds-of-paradise (Paradisaeidae) and cassowaries (Casuariidae), the latter also found in tropical Australia. Small frugivore/insectivore species, such as bulbuls (Pycnonotidae) and starlings (Sturnidae), are also very significant tropical rainforest seed dispersers. Vultures, and other carrion birds, may eat some oleaginous fruits such as those of the oil palm (*Elaeis guineensis*) (van der Pijl 1982).

Snow (1981) introduced the concept of specialist and generalist avian frugivores. Specialist frugivores are almost entirely frugivoruous, whereas generalists include other items in their diet, notably insects. He also argued that specialist frugivores were attracted to a different type of fruit, more lipid-rich and less sugary, than generalist frugivores. The specialist–generalist concept has rather fallen out of favour because it tends to rely on arbitrary definitions of how large a proportion of the diet is made up of fruit. In addition, it has been found that so-called specialist-frugivorous-bird-fruits are often also exploited by the generalist category, and vice versa.

#### FRUITS

Bird-fruits are generally characterised by easy access to the fruit on the tree by the bird. This frequently means a terminal placement on the branches close to the foliage, with firm attachment to the tree. Birds generally forage on the tree, although terrestrial birds such as cassowaries exploit fallen fruit, and species adapted for dispersal by such birds must have fruits that fall from the crown. Bird-fruits are usually fleshy (aril, sarcotesta, flesh of baccate or drupaceous fruits) with relatively little protection. Birds, with the possible exception of birds-of-paradise (Pratt & Stiles 1985), are not good at manipulating fruits before ingestion, so thick peels are not a feature of the fruits they exploit. Woody or leathery capsules generally dehisce to reveal the fleshy reward, before birds visit. Relatively small, spherical berries or drupes are more typical. The seeds may be protected (hard coat, bitter taste, toxic substances) to allow them to pass through the frugivores undamaged. Large seeds are usually elongate as ease of swallowing is related to the second dimension of an object; cf. pharmaceutical capsules.

There have been a number of surveys of the colours of bird-fruits in tropical forests (Wheelwright & Janson 1985; Willson *et al.* 1989; Gorchov *et al.* 1995). These have found black to be the commonest bird-fruit colour followed by red. Interestingly, in Gabon, Gautier-Hion *et al.* (1985) did not record any black fruits eaten by birds. Dowsett-Lemaire (1988) reported purple/black as a common colour in fruits eaten by birds in montane forest in Malawi, so black bird-fruits do occur in tropical Africa. Green is not a common ripe-fruit colour for bird-dispersed species in the tropics. Blue also seems quite rare in the neotropics, but less so in the palaeotropics where genera such as *Elaeocarpus* and *Lasianthus* often have bright blue fruits. Birds generally find fruit by using visual cues, so contrasting black or red against the green foliage is a clear signal by the tree. Birds have good colour vision, probably over a wider spectral range than our own, making colour surveys of bird-fruits based on human observation a poor, and possibly inaccurate, approximation. In addition, there is often a tendency to record unripe fruit colours, as truly ripe fruits disappear quickly from the tree. The avian sense of smell is poor, so bird-fruits rarely have a strong odour. The oilbird (*Steatornis caripensis*) is an exception. It uses olfaction to find ripe fruit, mostly Lauraceae and Palmae, at night.

Nutritionally, bird-fruit flesh most commonly offers a sugary reward, being low in lipid and protein. However, more lipid is offered by some trees.

#### PLANTS

All tropical trees that produce small, brightly coloured berries are primarily bird-dispersed, although the tamarins and marmosets of the New World tend to be 'bird-fruit' eaters also (Levey *et al.* 1994). Families with many bird-dispersed species include the Rubiaceae, Melastomataceae and Lauraceae. Kubitzki (1985) reported Amazonian Lauraceae to be invariably bird-dispersed.

The large genus of tropical trees *Aglaia* (Meliaceae) produces arillate seeds dispersed by birds and primates. The bird-dispersed species tend to have lipid-rich arils whereas the primate-dispersed ones are more sugar-rich (Pannell & Kozioł 1987).

#### CASE STUDIES

Perhaps the best-studied tropical tree with respect to its coterie of dispersers is the central American nutmeg *Virola nobilis* (mis-identified as *Virola surinamensis* in the early publications) (Howe 1993). About 50% of fruits are removed from the parent plant, mostly by 8 of the 68 species of frugivore recorded from the tree over many hours of observation. The most important frugivore for the tree is the chestnut-mandibled toucan *Ramphastos swainsoni*. Toucans and guans drop more then half of the seed carried 40 m or further from the parent tree.

Three shrubs (*Phytolacca rivinoides*, *Witheringia solanacea* and *W. coccoloboides*) in montane forest at Monte Verde in Costa Rica were shown to be dispersed effectively by three species of bird (black-faced solitaire *Myadestes melanops*, black-and-yellow silky flycatcher *Phainoptila melanoxantha* and prong-billed barbet *Semnornis frantzii*) (Murray 1988).

Using estimates of gut passage rates and studies of bird movement, including telemetry, it was calculated that median dispersal distances for the nine bird–plant combinations were 35–60 m with maxima over 500 m in some cases. Howe & Primack (1975) demonstrated that obligately frugivorous toucans achieved longer dispersal distances for the Central American tree *Casearia corymbosa* than opportunistic frugivore/insectivores such as fly-catchers. Cassowaries are important dispersers of large seeds in Queensland (Stocker & Irvine 1983) and New Guinea (Mack 1995).

A feature of certain legume species is the production of hard, bright and shiny red or red-and-black seeds. These are found, for example, in the genera *Adenanthera* and *Ormosia*. The legume splits to reveal these seeds, which have been considered as mimetic of arillate seeds and likely to be dispersed by birds fooled into picking up the seeds believing them to have a rewarding fleshy appendage. Peres & van Roosmalen (1996) have come up with a new theory to explain the supposedly mimetic seeds. They observed *Ormosia lignivalvis* trees over long periods and noted very little frugivore activity in them. Most of the seeds fell to the forest floor where they were picked up by terrestrial birds such as tinamous. Peres & van Roosmalen hypothesised that the very hard seeds were excellent alternatives to grit in the gizzards of these large granivorous birds. The *Ormosia* seeds can eventually pass through the birds unharmed. Observations on *Ormosia* spp. in Mexico showed that although some seeds were removed by frugivorous birds from the canopy more seeds were probably dispersed by being picked up from the forest floor by rodents and granivorous birds (Foster & Delay 1998).

### *Bats*

#### DISPERSERS

The fruit bats of the New World are all members of the family Phyllostomidae. They do not attain the large size achieved by the flying foxes of the Old World tropics, although not all palaeotropical fruit bats are big. Fruit bats forage nocturnally. They eat soft fruit flesh, sometimes sucking the juice and spitting out fibrous matter. Fruit bats often use feeding roosts, sites to which they return during the night to eat the fruit they have collected. Many seeds are discarded beneath the feeding roost. The ability of fruit bats to fly long distances means they have colonised many relatively remote islands where there are few other large frugivores.

#### FRUITS

Bat-fruits are usually held away from the foliage, often dangling beneath the crown like mangoes (*Mangifera* spp.), or being borne on the stem or trunks. The fruits are typically pale in colour, commonly green (Phua & Corlett 1989; Gorchov *et al.* 1995), soft (weakly protected), juicy and often odoriferous (musty or rancid smells predominate). Kubitzki (1985) remarked on a frequent reminiscence of stables when near bat-fruits. He recognised three main bat-fruit types in the Amazon: drupes, fruits with soft flesh and hard, shiny seeds (Sapotaceae and Annonaceae) and fruits with many tiny seeds in soft or firm pulp (*Piper*, *Cecropia*, *Ficus*). New World, microchiropteran-dispersed fruits tend to have lower sucrose contents than Old World, mega-chiropteran-dispersed species (Baker *et al.* 1998).

PLANTS

Bat-fruits occur across a wide range of taxa in the rain forest, as seen in the examples given in the preceding paragraph. Gautier-Hion *et al.* (1985) claimed bats are not important seed dispersers in Africa. This seems surprising.

CASE STUDIES

Fleming (1988) has studied the short-tailed fruit bat *Carollia perspicillata* of Central America in detail. It disperses seeds in a rather clumped way. About two thirds of ingested seeds were defaecated close to the parent tree, mostly under the tree of a nearby night roost, which was usually within 50 m of the fruiting tree, but some seed were moved long distances, probably up to 2 km. The bat had gut passage times of 20–40 min.

### *Primates*

DISPERSERS

Many primates are frugivorous, although most colubines and some cercopithecines are seed destroyers, using foregut fermentation to assist in detoxification of the seeds. Old World primates have colour vision similar to our own, whereas New World monkeys are dichromatic or polymorphic within a species for two or three colour receptors (Jacobs *et al.* 1996; Osorio & Vorobyev 1996).

FRUITS

Primate-fruits are often protected with a peel that the dextrous apes and monkeys can remove with the assistance of their large teeth to gain access to the fruit flesh, although primate-fruits that are dehiscent with arillate seeds also occur. The fruits are generally quite big as befits dispersers of large body size. Different authors have tended to emphasise different colours in referring to primate fruits. Janson (1983) reported that fairly muted colours predominate, with shades of orange, yellow, green and brown being typical; likewise Dew & Wright (1998) found green, brown, purplish and black fruits to be favoured by rain forest lemurs in Madagascar. However, others have found that primates eat fruits that are brightly coloured, particularly in yellow, orange and red (Gautier-Hion *et al.* 1985; Julliot 1996; Lambert & Garber 1988). The fruits are quite often scented. They remain attached to the tree where the arboreal primates can forage for them.

PLANTS

Families with indehiscent arillate fruits such as the Sapindaceae, Meliaceae and Euphorbiaceae are typically primate-dispersed.

CASE STUDIES

*Cola lizae* is a tree from Gabon reliant on lowland gorillas (*Gorilla gorilla gorilla*) for its seed dispersal (Tutin *et al.* 1991). Chimpanzees are effective seed dispersers in Kibale, Uganda (Wrangham *et al.* 1994). Garber (1986) found that 87% of seeds recovered from moustached and saddle-back tamarins travelled at least 100 m from the parent tree in a Peruvian rain forest. Red howler monkeys (*Allouatta seniculus*) have relatively long gut-passage times, leading to dispersed seeds being clumped beneath the howlers' sleeping trees (Julliot 1997). Lowland gorillas also deposit large numbers of seeds near their nest sites, which can provide favourable locations for seedling establishment (Rogers *et al.* 1998; Voysey *et al.* 1999).

### *Scatter-hoarding mammals*

DISPERSERS

Seed-eating mammals may be responsible for the dispersal of some tree species, largely through their habit of hiding seeds excess to immediate nutritional needs. These hoarded seeds in scattered caches may be found again and eaten in times of seed shortage, but some will be forgotten and if buried in the ground have a good chance of germination. The tree is therefore using some of the seeds as the reward to the disperser. The best-studied group of tropical forest scatter-hoarders are the caviomorph rodents – the agoutis, acouchies and their kin – in the neotropics. Forget & Milleron (1991) claimed that these are the only hoarding mammals that actually bury their seed caches rather than just cover them with litter. However, Leighton & Leighton (1983) referred to seed burial by the horse-tailed squirrel (*Sundasciurus hippurus*) in Borneo, Brewer & Rejmánek (1999) reported the spiny pocket mouse (*Heteromys desmarestianus*) to bury seeds in the soil and Vander Wall (1990) cited several other examples of tropical Sciuridae burying seeds. Seed burial may be advantageous to the hoarder if it provides a source of new seedlings in the future, the cotyledons or hypocotyls of which can be exploited for food (Forget 1992b). It is likely that rodents are involved in the scatter-hoarding of seeds throughout the tropics. Many forest-dwelling small mammals may be primarily seed predators, rapidly consuming seeds or carrying them back to their burrows, which may be too deep to favour successful germination (Sánchez-Cordera & Martínez-Gallardo 1998; Adler & Kestell 1998), but a small proportion of seeds may be scatter-hoarded. In the temperate zone birds, such as jays and nutcrackers (Corvidae), are also scatter-hoarders. Tropical jays have been reported to bury seeds (Vander Wall 1990) but little is known of their role as dispersers.

FRUITS

Tree species that employ scatter-hoarders have hard, dry fruits with very well-protected seeds. The mammalian scatter-hoarders are equipped with large sharp teeth and powerful jaws, and can bite through thick endocarps or seedcoats to eat a seed's contents. The protection of the seeds limits their availability to just a few species, making it more likely that there will be a surplus of seeds for hoarding. Large, well-protected seeds suffered low predation rates in Borneo (Blate *et al.* 1998). A poorly protected seed would be at the mercy of many seed predators and few would be cached. The hard work involved in opening the seeds is usually rewarded with a large endosperm rich in energy and protein. Indeed, Grubb (1998b) found that rain-forest tree species in Queensland with well-protected seeds had higher nitrogen concentrations in the embryo and endosperm than poorly protected seeds of similar size. Forget *et al.* (1998) found that agoutis were more likely to cache large seeds than small ones, and tended to move large seeds longer distances.

PLANTS

The acorns of the Fagaceae, the pyxidia of the Lecythidaceae, some tough-podded legumes and nuts of the Proteaceae are some examples of fruits of scatter-hoarded species.

CASE STUDIES

Studies of rodent scatter-hoarding in the neotropics include that of *Vouacapoua americana* by the acouchy (*Myoprocta exilis*) in French Guiana (Forget 1990), and *Dipteryx panamensis* (Forget 1993) and the palm *Astrocaryum standleyanum* (Smythe 1989) by agoutis (*Dasyprocta punctata*) on Barro Colorado Island, Panama. Smythe (1989) found that palm seeds buried by the agoutis had a much higher germination rate than those left on the forest floor (29.2% germination after 11 months, compared with 2.6%).

### *Others*

African elephants (*Loxodonta africana*) are the largest of tropical forest frugivores. Elephants incorporate fruit as quite a large part of the diet in the rain forests of Africa and Asia. Elephant fruits are generally large, with big, well-protected seeds, often with a strong 'yeasty' smell (Yumoto *et al.* 1995; White *et al.* 1993). The pericarp is often thick, to deter other herbivores, and relatively dull in colour. The fruit flesh is generally firm and fibrous and quite sweet. Passage through an elephant greatly improved the germination of the African tree *Balanites wilsoniana* (Chapman *et al.* 1992). Tapirs are the closest ecological equivalents of elephants in the neotropical forests. Giant mammals that became extinct during the Pleistocene may also have been responsible for dispersal of some neotropical species (Janzen & Martin 1982), although Howe (1985) has questioned the existence of a 'giant-mammal syndrome' in the Central American flora.

Tree-shrews (Tupaiidae) were believed to be largely insectivorous, but Emmons (1991) found four species studied in Sabah to be highly frugivorous. The tree-shrews exploited small berries of the 'bird-fruit' type. These animals have a simple gut anatomy reminiscent of fruit bats and similar rapid gut passage times.

Fish may sound unlikely seed dispersers in tropical forests, but in seasonally flooded forest they may pay an important role. This has been demonstrated for the extensive varzea (whitewater) and igapo (blackwater) forests of the Amazon basin (Goulding 1980; Kubitzki & Ziburski 1994). Many swamp-forest tree species are hydrochorous, but the fruit of some trees sink and require a fish to break open the outer covering to obtain the flesh and seeds. In Amazonian swamps the palm *Astrocaryum jauari* is dispersed by catfish. *Ficus glabrata*, a riverside tree fig in Costa Rica, is dispersed by the characid fish *Brycon guatemalensis* (Horn 1997).

Two herbivorous turtles have been demonstrated to be seed dispersers in tropical forests (Moll & Jansen 1995). Gray's monitor lizard (*Varanus olivaceus*), from the Southern Philippines, is quite strongly frugivorous as an adult and viable seeds are passed in the faeces (Auffenberg 1988). Red crabs (*Gecarcoidea natalis*) are probably dispersers of well-protected seeds on Christmas Island (O'Dowd & Lake 1991).

## Secondary dispersal

Ants remove seeds from the faeces of frugivores. Minuscule litter-dwelling species may be important removers of tiny seeds from bird droppings at La Selva, Costa Rica (Kaspari 1993; Levey & Byrne 1993). *Miconia* seeds were extracted by the foraging ants from frugivore excrement and taken to their nests, which were usually located in hollow twigs on the forest floor. About 94% of the seeds taken were eaten by the ants, but the remainder were discarded on the waste piles in the nest (Levey & Byrne 1993). Viable seeds were found in abandoned nests, and these may represent safe sites in which to germinate for some tiny-seeded species, such as those of *Miconia* and other Melastomataceae. Small arillate seeds, either fallen from the parent tree, or dropped by frugivores, attract ants. Pizo & Oliviera (1998) observed a range of ant species to move seeds of *Cabralea canjerana* in the Atlantic forests of Brazil. Individuals of the larger species (ponerines) carried seeds back to their nests; smaller species worked in groups. The ants moved the seeds short distances – very rarely more than 5 m – but they may play an important role through hiding seeds from seed predators. In addition, the removal of the adherent flesh may be beneficial as it is usually rapidly colonised by fungi that may also attack the seed. Oliveira *et al.* (1995) found that removal of the fruit flesh by ants had a positive effect on germination in *Hymenaea courbaril.* Leaf-cutting ants (*Atta colombica*) were observed to remove fallen fruits of *Miconia argentea* from the forest floor on Barro Colorado Island, Panama, and discard large quantities of viable seed on their colony refuse piles. The ants also took fruits from the crowns of the trees, thus acting as primary dispersers also (Dalling & Wirth 1998).

Small mammals, particularly rodents, may also remove seeds from material defaecated or regurgitated by large vertebrate frugivores. Caching by rodents probably improved the survival of dispersed seeds of *Guarea* spp. in Monteverde, Costa Rica (Wenny 1999).

Seeds buried during the activities of dung beetles may also represent a form of secondary dispersal beneficial to trees. Beetles that bury monkey dung allow seeds to escape predation by rodents that extract the seeds from the faeces (Estrada & Coates-Estrada 1991; Shepherd & Chapman 1998; Andresen 1999). Dung beetles generally bury seeds at depths of 1–3 cm, which are about optimal for seed survival and germination. Small seeds are more likely to be buried by the beetles than big ones (Andresen 1999).

## Efficacy of seed dispersal

How good at dispersing their seeds are trees in the rain forest? Observations on wind-dispersed species make it seem unlikely that any but a tiny proportion of seeds reach beyond 100 m (Table 4.8). Augspurger (1986) estimated mean dispersal distances from measuring propagule rate of descent and assuming a wind speed of 1.75 m $s^{-1}$ to be in the range 22–194 m for 34 species on Barro Colorado Island. Canopy filtering and low wind speeds may make actual dispersal distances lower than these estimates. Some vertebrate frugivores, including bats, birds, primates and terrestrial mammals, have been observed or inferred to transport seeds hundreds of metres (Table 4.9), but these studies generally do not report the fate of the entire seed crop. Many fruits and seeds fall directly beneath the parent, or are dispersed only short distances by inefficient dispersers. The few studies of seed dispersal patterns in zoochorous tropical trees (Table 4.9) demonstrate at least half of any seed crop falling beneath the parent crown. However, the tail of the distribution is generally long and some seeds arrive relatively far from their origin. The successful colonisation by species of distant sites (e.g. tree invasion of the Krakatau Islands, Indonesia, after sterilisation through volcanic eruption (Whittaker *et al.* 1989)) indicates that long-distance dispersal events do occasionally occur.

However, a community-level study of seed rain on Barro Colorado Island, Panama (Hubbell *et al.* 1999) gave a rather different impression of the effectiveness of seed dispersal. Over a period of 10 years, a network of 200 seed traps in the 50 ha plot caught seed of 260 of the 314 tree species present. Only 7 species dispersed seed to more than 75% of the trap sites. Fecundity and/or dispersal limitation apparently preclude recruitment of many of the species at any particular site in the forest.

Table 4.8. *Studies of seed dispersal distance in tropical rain-forest tree species where the spatial pattern of seed fall around parent trees was monitiored*

| Species | Ref. | Dispersal agent | Fruit/seed size | Measure of distance | *n* | Site | Notes |
|---|---|---|---|---|---|---|---|
| *Chlorocardium rodiei* | 1 | gravity | seed *ca.* 40 g dry mass | no seeds more than 14 m from adult tree | ? | Guyana | secondary dispersal by rodents may also be involved |
| *Eperua falcata* | 2 | autochory | seed 7.4 ± 2.2 g fresh mass | 60% of trees within 10 m; max. dispersal 30 m | 2 | French Guiana | |
| *Lonchocarpus pentaphyllus* | 3 | wind | fruit *ca.* 150 mg fresh mass, seed *ca.* 50 mg dry mass | 40 % of fruits fall under crown; max. dispersal 70 m | 1 | Barro Colorado Island, Panama | tree 32 m tall |
| *Lophopetalum wightianum* | 4 | wind | seed 45–311 mg fresh mass | median dispersal distance 15–43 m, max. 30–80 m | 12 | Western Ghats, India | trees 29.8–49 m tall |
| *Platypodium elegans* | 5 | wind | fruit 2 g dry mass | median dispersal distance 10–23 m; max. dispersal distance 75–105 m | 4 | Barro Colorado Island, Panama | dispersal distances estimated along a transect in direction of prevailing wind |
| *Scaphium macropodum* | 6 | wind | seed 2.21g fresh mass | no fruits beyond 48 m; seed density peaked directly beneath parent tree | 1 | West Kalimantan | |
| *Swintonia schwenkii* | 7 | wind | fruit 2 g fresh mass | few fruits beyond 50 m | 1 | Gunung Gadut, Sumatra | tree 61.7 m tall |
| *Tachigali versicolor* | 8 | wind | seed 500–600 mg dry mass | 15% seeds fell under parent tree, 95% of seeds within 100 m of parent | 2 | Barro Colorado Island, Panama | trees 30 and 37 m tall |

| | | | | | | | |
|---|---|---|---|---|---|---|---|
| *Brosimum alicastrum* | 9 | ?small mammals | seed 10–14 mm diam | few seeds beyond 20 m; seed numbers peak at 10 m from tree trunk (?crown edge) | 14 | Los Tuxtlas, Mexico | tree size? what disperses fruits away from tree |
| *Ficus stupenda* | 10 | vertebrate | no data | more than 50% of seeds fall beneath crown; up to 45% of seeds dispersed more than 60 m from parent | 4 | Kalimantan, Indonesia | strangler fig; tail of seed dispersal-distance distribution estimated by curve fitting |
| *Ficus subtecta* | 10 | vertebrate | no data | more than 50% of seeds fall beneath crown; up to 45% of seeds dispersed more than 60 m from parent tree | 3 | Kalimantan, Indonesia | same as previous entry |

1, Zagt & Werger (1997); 2, Forget (1989); 3, Augspurger & Hogan (1983); 4, Sinha & Davidar (1992); 5, Augspurger (1983b); 6, Yamada & Suzuki (1997); 7, Suzuki & Kohyama (1991); 8, Kitajima & Augspurger (1989); 9, Burkey (1994); 10, Laman (1996).

Table 4.9. *Studies of seed dispersal in tropical rain-forest trees where seed dispersal by particular animal species was studied*

| Disperser species | Ref. | Seed dispersed | Measure of dispersal distance achieved | Methods | Site |
|---|---|---|---|---|---|
| **bat** | | | | | |
| *Carollia perspicillata* | 1 | *Piper amalago* | 90% of seeds dispersed at least 50 m; few seeds moved more than 300 m | estimated from foraging studies | Costa Rica |
| **bird** | | | | | |
| black guan (*Chamaepestes unicolor*), resplendent quetzal (*Pharomachrus mocinno*), emerald toucanet (*Aulacorhynchus prasinus*), mountain robin (*Turdus plebejus*) | 2 | *Ocotea endresiana* | most seed within 20 m of the parent | foraging birds followed and seed dispersal observed | Monteverde, Costa Rica |
| three-wattled bellbird (*Procnias tricarunculata*) | 2 | *Ocotea endresiana* | 59% of seeds more than 40 m from parent tree | see above | Monteverde, Costa Rica |
| three species of turaco (Musophagidae) | 3 | various | mean and median 100–300 m | estimated from gut passage times and foraging studies | montane forest, Rwanda |
| *Myadestes melanops*, *Phainoptila melanoxantha* & *Semnornis frantzii* | 4 | *Witheringia solanacea*, *Witheringia coccoloboides* & *Phytolacca rivinioides* | 20–36 % seed dispersed within 30 m from the parent plant; max. for the 3 bird spp. range 220–510 m | estimated from gut passage times and foraging studies | Costa Rica |

| | | | | | |
|---|---|---|---|---|---|
| dwarf cassowary (*Casuarius bennetti*) | 5 | *Aglaia mackiana* | mean 388 m; max. nearly 1 km | tagged seeds recovered from faeces | Papua New Guinea |
| Salvin's curassow (*Mitu salvini*) | 6 | *Ficus stenophylla* | mean 329 m; max. 451 m | seeds recovered from faeces | La Macarena, Colombia |
| **non-flying mammal** | | | | | |
| tapir (*Tapirus terrestris*) | 7 | *Maximiliana maripa* | seeds transported up to 2 km from parent | seeds identified in tapir latrines | Maracá Island, Brazil |
| agouti (*Dasyprocta punctata*) | 8 | *Gustavia superba* | nearly half of seeds scatter-hoarded within 10 m | tagged seeds recovered from caches | Barro Colorado Island, Panama |
| red howler monkeys (*Alouatta seniculus*) | 9 | various | mean 76–440 m; max. 288–575 m | seeds recovered from faeces | La Macarena, Colombia |
| Humboldt's woolly monkey (*Lagothrix lagotricha*) | 9 | various | mean 126–454 m; max. 126–1106 m | seeds recovered from faeces | La Macarena, Colombia |

1, Fleming & Heithaus (1981); 2, Wenny & Levey (1998); 3, Sun *et al.* (1997); 4, Murray (1988); 5, Mack (1995); 6, Yumoto (1999); 7, Fragoso (1997); 8, Forget (1992b); 9, Yumoto *et al.* (1999).

## Clumped or scattered seed distributions

Dispersers differ in the relative amount of clumping found in the pattern of seed distribution they produce. Disperser body size seems to be a correlate of degree of aggregation in the seed dispersion pattern. Tiny frugivores, such as small birds, may void seeds singly, or in groups of a few each time. Large herbivores like elephants or rhinoceroses may produce huge dung piles containing thousands of seeds. Howe (1989) hypothesised that tree species should, to some extent, be adaptated to cope with the typical degree of aggregation expected in their seed dispersion pattern. Species that are typically clumped after dispersal should be better able to survive as seeds and seedlings in those conditions than species that are normally more evenly distributed. They could do this by having greater physical and chemical defences against pathogens and predators. The large-seeded species dispersed by the large frugivores that produce the clumped distributions also tend to have more skewed distributions, with many fruits not being dispersed and falling beneath the parent tree. This is because there are more species of small-sized frugivore in the forest to disperse seeds from small fruits. Howe argued the same point for survival under the parent tree. The species that characteristically drop many seeds directly beneath the parent tree should have better survival there than those species that are more successful at dispersing the seeds away from the parent tree. This hypothesis was tested by Chapman & Chapman (1996) using observations on six species in Kibale Forest, Uganda. They found that the two species with low dispersal (< 1% of fruits removed from the crown zone of the tree) did show low seed and seedling mortality near the parent compared with the four species that had higher fruit removal rates.

## Seed survival

Seeds are susceptible to attack by a wide range of organisms. Janzen (1971) listed a large number of both vertebrate and invertebrate groups that are seed predators. In addition, bacteria and fungi will destroy seeds if given the chance. Seeds can be defended through hard or tough endocarps or seed coats (Blate *et al.* 1998; Grubb *et al.* 1998) and by chemicals toxic to predators and pathogens. However, the incentive of the high-quality food contained in a seed has led to strong selection pressure among granivores to overcome the plant defences. Many studies have discovered very high rates of predation on newly dispersed seed of rain-forest trees. In 30 days, seeds of 40 species scattered individually showed an average mortality of more than 50% in West Kalimantan, Indonesia (Blate *et al.* 1998). All of a sample of 1800 seeds of

*Gilbertiodendron dewevrei* monitored in a rain forest in the Congo were destroyed by beetles and mammals within four weeks (Hart 1995).

Animals foraging for food on the forest floor are likely to find clumps of seeds more easily than those scattered widely, and hence act as agents of density-dependent mortality on the seeds. A seed-eater might also look for fruiting parent trees and search underneath for fallen seeds and hence act as a predatory force inversely proportional to the distance from the parent tree. Density- and distance-dependent mortality effects will interact with the seed dispersion pattern to produce the spatial distribution of established seedlings with reference to the parent tree. Seedlings are also susceptible to predators and disease and similar density- and distance-dependent mortality factors can act further to alter the spatial distribution of later juvenile size classes. Seed dispersal-distance frequency distributions are nearly always strongly skewed toward the origin. In other words, most seeds come to rest very near to the parent tree. In addition, because a tree usually acts as a point source sending seeds outwards into progressively larger annuli at successive distance increments there will be a strong tendency for seed density on the forest floor to drop quickly with distance from the parent tree unless dispersal is extremely efficient. Portnoy & Willson (1993) found that an algebraic function best fitted most available data for seed dispersal such that a linear decrease is observed when distance and seed density are both plotted on logarithmically scaled axes.

Density- and distance-dependent mortality has been put forward as a mechanism to account for the co-existence of many tree species in the tropical rain forest. It is often referred to as the Janzen–Connell model of community structure, taking its name from the two workers who independently proposed the hypothesis with emphasis on different life history stages (Janzen (1970) seeds; Connell (1971) seedlings). They both suggested that density- and distance-dependent mortality could be strong enough to act as a compensatory mechanism preventing common species becoming too common and hence causing the local extinction of rare species. This could be achieved via the probabilistic action of mortality across the community, or more directly by the prevention of regeneration of juveniles beneath adults of the same species for common trees. Howe & Smallwood (1982) referred to the Janzen–Connell model as the 'escape hypothesis' and used it to explain some of the advantages of seed dispersal to plants. This is based on the Janzen–Connell hypothesis of density- and distance-dependent mortality, and contends that owing to this phenomenon it is advantageous for a plant to disperse its seeds so that more escape the high mortality near the parent. I prefer to consider the Janzen–Connell model as a special case of the escape hypothesis, rather than as the same concept. The escape hypothesis conjectures an advantage in offspring survival to increased dispersal distance. The Janzen–Connell hypothesis predicts that offspring mortality is so high for common species that it effectively limits their population size. The escape effect might be operating in a forest, but not at the level needed to cause sufficient compensatory mortality to influence species co-existence.

There has been considerable confusion in the literature about testing the Janzen–Connell hypothesis. Many purported tests have been searches for density-dependent mortality within a species, not tests of species coexistence in the forest community. Evidence for the presence of the required mechanism is not proof of the predicted outcome. However, such evidence is a test of the more general escape hypothesis, in the sense that is used here.

It would actually be very surprising if the escape hypothesis were not found to operate because of the very skewed dispersal-distance frequency distributions exhibited by most tree species. The density of seeds beneath the parent is generally so high that, if nothing else, self-thinning of this dense monospecific stand must lead to high mortality of recruits. Mortality in crowded seedling stands in the shade is often brought into effect by pathogens, making it difficult to distinguish thinning from seedling predation. However, the strong skew in the seed distribution also means that the mortality must be extremely high to be large enough to result in the hyper-dispersed pattern of recruits with respect to adults predicted by the Janzen–Connell model. Janzen (1969, 1971, 1974), in fact, proposed the existence of another mechanism that might act to counter the expected mortality. This is predator satiation. If there are many seeds available in a site at a particular time the predators may be unable to eat them all. The predator satiation hypothesis was put forward to explain the mass flowering and fruiting observed in some plant species. Notably in the lowland forests of West Malesia, particularly of the Malay Peninsula, Sumatra and Borneo, there is suprannual and irregular periodicity to reproduction (Appanah 1985; Ashton *et al.* 1988), with many species, not just Dipterocarpaceae, participating in mast years. It is notable that fruiting is more synchronised than flowering, at least in dipterocarps. The surfeit of seeds produced in the mast years is more than can be eaten by the population of seed predators, and hence the mortality of seeds is relatively low. In addition, the irregular production of seeds may make it difficult for species of seed-eating animal to maintain large populations in the forest. It is possible also that predator satiation might exist in non-masting forests if seed predators are strongly territorial or relatively sedentary, because then there would only be a few individuals to attack the many seeds falling near the parent tree. Invertebrate seed predators may be less likely to be satiated by large seed crops because their short life cycles allow numbers to build up very quickly to exploit the increase in food availability. The weevil that attacks seeds of *Shorea* species, *Nanophyes shoreae*, appears able to track flowering trees of potential hosts at Pasoh, Malaysia, during a mass flowering, and hence successfully infest a relatively high number of seeds (Toy 1991). However, among five species of *Piper* at Los Tuxtlas, Mexico, predispersal seed loss to insect predators as a proportion of the total was found to be highest in the species that typically produced fewer seeds than those that produced many (Grieg 1993). Augspurger (1981) found about twice the rate (11% compared with 5%) of seed infestation with microlepidopteran larvae in individuals induced to flower compared with synchronous flowerers of *Hybanthus prunifolius*, although the infestation rate was not particularly high.

## Tests of the escape hypothesis

There have been many studies of the survival of seeds and seedlings of tropical trees with respect to conspecific density or distance to adult trees (Hammond & Brown 1998) (Table 4.10). Some of the research published had poor degrees of replication, studying just the offspring of one tree at one fruiting season. Despite this, the evidence seems to be in favour of the escape hypothesis. Seedlings of *Dipteryx panamensis* at La Selva, Costa Rica, showed a 100% mortality over the 7–20 month old period within 8 m of the parent tree (Clark & Clark 1984). The seeds of *Virola nobilis* are attacked by weevils (Curculionidae) and bark beetles (Nitidulidae) and probably mammals (Howe 1993). The mortality due to these predators was very high near parent trees. There was a 22–44-fold advantage in survival for seeds getting 45 m from the parent tree compared with 5 m away (Fig. 4.11). This appeared to be mostly a density-dependent effect because seeds 5 m from an adult male tree had the same survival rate as those 45 m from a fruiting female tree. Well-dispersed seeds may escape death from insect attack because the pests take longer to find them giving the seeds an opportunity window in which to germinate and establish. *Chlorocardium rodiei* seeds in a Guyanan forest were equally likely to be infested by beetles over an 85 week period at a range of dipseral distances, but those dispersing further were more likely to establish as the beetle attack came later (Hammond *et al.* 1999).

Besides animals, other agents of mortality have been shown to act in a density- and distance-dependent manner. Augspurger (1983a) found that seedling mortality due to fungal pathogens was greater closer to parent trees of *Platypodium elegans*. A canker disease of *Ocotea whitei* that led to sapling mortality on Barro Colorado Island also exhibited density dependence (Gilbert *et al.* 1994). Yamada & Suzuki (1997) identified parental leaf litter as the factor inhibiting seedling establishment beneath seeding trees of *Scaphium*

Table 4.10. *Comparison between vertebrate and insect attack on seeds and seedlings of tropical rain-forest tree species based on their support of the escape hypothesis*

*n*, The number of mature individuals around which seed/seedling survival was studied; h/l indicates a comparison between high- and low-density sites for mature individuals; nd, no data available to author.

| Species | *n* | Attack type | Seeds or seedlings or both? | Supports escape hypothesis? |
|---|---|---|---|---|
| *Aglaia mackiana*[1] | 6 | insect/pathogen | seedlings | no |
| *Astrocaryum macrocalyx*[2] | 2 | insect | seeds | yes |
| *Carapa guianensis*[3] | nd | insect | seeds | yes |
| *Chlorocardium rodiei*[4] | 10 | insect | seeds | yes |
| *Copaifera pubiflora*[5] | 20 | insect | seeds | yes |
| *Gilbertiodendron dewevrei*[6] | h/l | insect | seeds | yes |
| *Julbernardia seretii*[6] | h/l | insect | seeds | yes |
| *Mora gonggrijpii*[3] | nd | insect | seeds | no |
| *Macoubea guianensis*[7] | 4 | insect | seeds | no |
| *Maximiliana maripa*[8] | 6 | insect | seeds | yes |
| *Normanbya normanbyi*[9] | 5 | insect | seeds | yes |
| *Pouteria* sp.[7] | 4 | insect | seeds | no |
| *Scheelea zonensis*[10] | 7 | insect | seeds | yes |
| *Scheelea zonensis*[11] | 14 | insect | seeds | yes[a] |
| *Virola nobilis*[12] | 5 | insect | seeds | yes |
| *Virola michelii*[3] | nd | insect | seeds | no |
| *Astrocaryum macrocalyx*[2] | 2 | vertebrate | seeds | no |
| *Bertholletia excelsa*[2] | 4 | vertebrate | seeds | no |
| *Brosimum alicastrum*[13] | 14 | vertebrate | seeds | no |
| *Carapa guianensis*[3] | nd | vertebrate | seeds | no |
| *Chlorocardium rodiei*[4] | 10 | vertebrate | both | no |
| *Dipterocarpus acutangulus*[14] | 1 | vertebrate | seedlings | no |
| *Dipterocarpus globosus*[14] | 1 | vertebrate | seedlings | no |
| *Dipteryx micrantha*[2] | ? | vertebrate | seeds | no |
| *Dipteryx panamensis*[15] | 6 | vertebrate | seedlings | yes |
| *Dipteryx panamensis*[16] | 19 | vertebrate | seeds | no? |
| *Dipteryx panamensis*[17] | nd | vertebrate | seeds | no |
| *Dryobalanops aromatica*[14] | 1 | vertebrate | both | no |
| *Dryobalanops lanceolata*[14] | 1 | vertebrate | both | no |
| *Eperua grandiflora*[18] | 1 | vertebrate | both | no |
| *Eperua falcata*[19] | 2 | vertebrate | both | no |
| *Gilbertiodendron dewevrei*[6] | h/l | vertebrate | both | no |
| *Gustavia superba*[20] | h/l | vertebrate | seed | no? |
| *Gustavia superba*[21] | h/l | vertebrate | seed | no |
| *Hymenaea courbaril*[2] | 2 | vertebrate | seed | no |
| *Julbernardia seretii*[6] | h/l | vertebrate | both | no |
| *Macoubea guianensis*[7] | 4 | vertebrate | seeds | no |
| *Normanbya normanbyi*[9] | 5 | vertebrate | seeds | no |
| *Pouteria* sp.[7] | 4 | vertebrate | seeds | no |
| *Scheelea zonensis*[22] | 28 | vertebrate | seeds | yes |
| *Tachigali versicolor*[23] | 2 | vertebrate | both | no |

Table 4.10. (*cont.*)

| Species | *n* | Attack type | Seeds or seedlings or both? | Supports escape hypothesis? |
|---|---|---|---|---|
| *Virola michelii*[3] | nd | vertebrate | seeds | no |
| *Virola nobilis*[12] | 5 | vertebrate | seeds | no |

[a]Density, but not distance, effect detected.
After Hammond & Brown (1998). Individual studies: [1]Mack *et al.* (1999); [2]Terborgh *et al.* (1993); [3]Hammond & Brown (1998); [4]Hammond *et al.* (1999); [5]Ramirez & Arroyo (1987); [6]Hart (1995); [7]Notman *et al.* (1996); [8]Fragoso (1997); [9]Lott *et al.* (1995); [10]Wright (1983); [11]Wilson & Janzen (1972); [12]Howe *et al.* (1985); [13]Burkey (1994); [14]Itoh *et al.* (1995); [15]Clark & Clark (1984); [16]De Steven & Putz (1984); [17]Forget (1993); [18]Forget (1992a); [19]Forget (1989); [20]Sork (1987); [21]Forget (1992b); [22]Forget *et al.* (1994); [23]Kitajima & Augspurger (1989).

*macropodum*. The thick layer of large leaves appeared to prevent root penetration to the soil in the germinating seeds.

After reviewing the evidence from the literature (Table 4.8), Hammond & Brown (1998) came to the conclusion that density- and distance-dependent mortality was generally exhibited more strongly where invertebrates were the main agents of destruction rather than vertebrates. This, they argued, reflected the different food choices and foraging strategies of the two groups. Seed-eating insects are frequently highly specialised, attacking a very limited range of species and foraging based on cues such as chemical attractants that are likely to lead them to sites of high seed density. Vertebrates are more catholic in food choice and will forage more opportunistically. They also seem easier to satiate than invertebrate predators that increase rapidly in population size in response to large seed crops.

On Barro Colorado Island, Schupp (1992) found that *Faramea occidentalis* showed clear density-dependent seed mortality around parent trees, but when the forest was considered as patches of low and high adult *Faramea* population densities, seed survival was higher in the areas of forest with more adult *Faramea* trees. This Schupp ascribed to satiation of territorial rodents in the areas of the forest with denser *Faramea* populations. Burkey (1994) found that seed predation rates by small mammals near individuals of *Brosimum alicastrum* at Los Tuxtlas, Mexico, were inversely related to tree seed-crop size (Fig. 4.12). Those trees producing big seed crops were more effective at satiating the rodents.

Augspurger & Kitajima (1992) manipulated the seed distribution of two individuals of *Tachigali versicolor* on Barro Colorado Island, Panama. They created transects of even distribution of seeds, extending the tail of the distribution beyond the normal maximal seed dispersal distance to 100–1800 m. Mortality was largely due to mammalian predation of seeds and early-stage seedlings. Density-dependent mortality was found to act at two spatial scales. At the fine scale there was higher mortality among the dense populations of seeds and seedlings near the parent tree. On a broader scale, high mortality occurred in the extended tail of the seed distribution. Augspurger & Kitajima argued that this reflected predator satiation near the parent. The distribution of established seedlings after mortality was still strongly skewed toward the parent tree. This might not be disadvantageous in this species, because, being monocarpic, the death of the seed parent would leave a gap that could be filled by one of its own offspring that fell beneath it.

In the Ituri Forest of Congo, Hart (1995) studied the seed and seedling survival of two caesalpinoid legume trees with very big seeds. *Gilbertiodendron dewevrei* forms mbau forest groves where it is the single dominant species. *Julbernardia seretii* occurs in the more species-rich matrix forest around the mbau patches, but it also tends to be clumped in adult distribution. Both species possess relatively ineffective ballistochorous seed dispersal, with *Gilbertiodendron* getting few seeds more than 10 m from the parent crown, and *Julbernardia* doing slightly better with maximal dispersal distan-

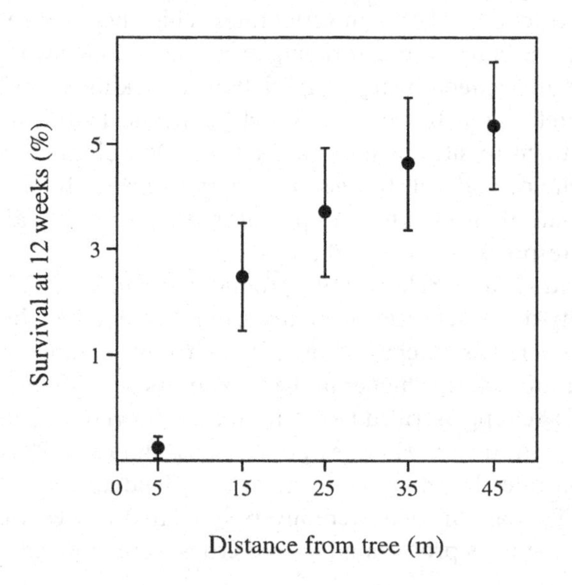


Figure 4.11 Seedling survival to 12 weeks after fruit fall as a function of distance from 17 fruiting *Virola nobilis* trees for 3400 seeds planted in 1982. Twelve weeks marks the end of dependence on parental endosperm, and of vulnerability to insect seed predators. After Howe (1993).

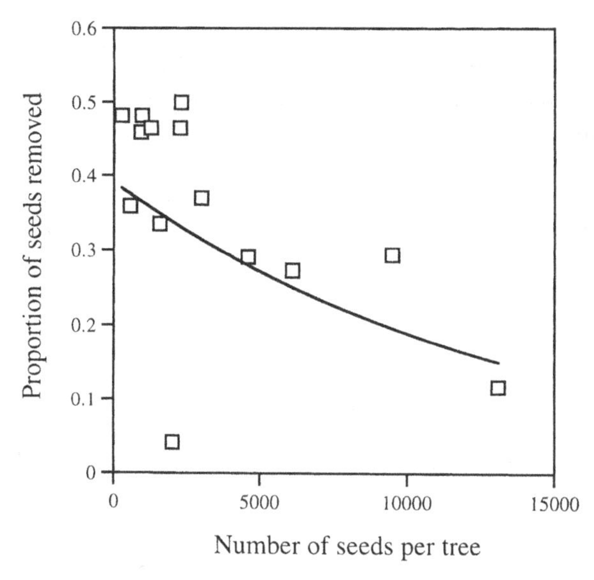


Figure 4.12 Seed predation as a function of seed production in trees of *Brosimum alicastrum* growing at Los Tuxtlas, Mexico. After Burkey (1994).

ces of 30–40 m. Hart found that seed predation of *Gilbertiodendron* was very high, with only 0.3% seed survival. Seed predators included two curculionid beetles that did not appear to exhibit satiation, and mammals (antelopes and rodents) which probably were satiated in the groves. The mammalian predators were efficient destroyers of any seeds that were dispersed out of the mbau grove. Seed survival was best at the periphery of the mbau grove. *Julbernardia* showed a much higher seed survival, about 60%. *Gilbertiodendron* had a high seedling survival, 49% over 10 years, allowing it to maintain its dominance in the mbau groves.

Another grove-forming tropical tree species is the brazil-nut tree (*Bertholletia excelsa*) of Amazonia. Peres *et al.* (1997) demonstrated that predator satiation did not occur in this species, quite the reverse. The rate of seed removal from the large pyxidia produced by the brazil-nut trees was higher in the *Bertholletia* groves than outside them. Agoutis were the main removers of the brazil nuts. Peres *et al.* argued that the agoutis outside the brazil-nut groves did not have the search image for the brazil nuts and so ignored them. Brazil-nut removal increased in the groves at the period of maximum pyxidia production. Thus there was no evidence of predator satiation in this system, which is beneficial to the brazil-nut trees because the agoutis are the main dispersers of the species through scatter-hoarding.

Many studies equate seed removal by mammals with predation, but if scatter-hoarders are responsible a proportion of the removed seeds may survive (Forget *et al.* 1998). Burial of seeds by caching mammals may be important in some species because this allows the seeds to escape insect attack (Terborgh *et al.* 1993).

In conclusion, there is evidence that the escape hypothesis is correct and there is an advantage in terms of survival to dispersal to greater distances from the parent tree, particularly where insects are the main predators. Vertebrate predation has rarely been documented to operate on tropical trees in a manner that makes greater dispersal distance strongly advantageous in terms of survival. Satiation of vertebrate predators may blur some of the expected patterns of offspring survival. Invertebrates seem less liable to satiation.

Many studies of seed and seedling survival have been conducted for large-seeded species because of the logistic advantages of working with readily visible objects in relatively low numbers. The fates of the tiny seeds of many species of tropical tree have been much less well studied. It seems less likely that predator satiation will operate with very small seeds. Another factor that is likely to disrupt the recruitment pattern predicted from knowledge of the effectiveness of seed dispersal and predation is the presence of superior sites for establishment in the forest. This is dealt with in the next section.

### Other advantages of dispersal

Howe & Smallwood (1982) proposed two hypotheses, besides the escape hypothesis, that predict advantages to plant offspring that are dispersed away from their seed parent. These are the colonisation hypothesis and the directed-disperal hypothesis. The colonisation hypothesis proposes the existence of above-average sites for juvenile establishment in the landscape where the species of interest occurs. The more widely an individual tree disperses its seeds, the greater the number of such superior sites liable to receive seeds from the tree. At least, this is likely to be the case when comparing a very limited dispersal distance with one larger. When dispersal distances become very large it becomes probable that the low densities inside the region receiving seeds mean that some superior sites within range will be missed. Rather like tests of the Janzen–Connell hypothesis, tests of the colonisation hypothesis have generally demonstrated the existence of the mechanism of the proposed effect, not its predicted outcome. As far as the colonisation hypothesis is concerned, attention has largely been centred on gaps as superior sites. A number of studies have demonstrated a definite advantage of dispersal to gaps (Augspurger 1983a; De Steven 1988; Osunkoya *et al.* 1992; Itoh *et al.* 1995; Cintra & Horna 1997), but these alone are not demonstrations of the validity of the colonisation hypothesis. What is needed for this is evidence that individuals with larger dispersal ranges colonise more gaps (or other superior sites) and hence have a greater chance of descendant individuals reaching the next generation.

The directed-dispersal hypothesis points to the advantage of dispersal modes that result in seeds being taken directly to superior sites. The only well-documented case for directed dispersal involving a tropical tree is that of *Ocotea endresiana* by male three-wattled bellbirds (*Procnias tricarunculata*) in cloud forest in Costa Rica (Wenny & Levey 1998). Among the coterie of five avian dispersers of *Ocotea endresiana*, four species produce a seed-dispersal pattern very strongly skewed toward the parent tree whereas the bellbird gives rise to a bimodal distribution because the male bellbirds habitually return to song perches, which are often branches on dead trees in gaps. Wenny & Levey (1998) found that survival for seeds and seedlings was about twice as high beneath such perches as in understorey sites, largely because of a reduced incidence of fungal disease in gaps.

Directed dispersal is often exemplified by reference to seeds with an elaiosome that are harvested by ants. The elaiosome is eaten and the seed discarded on the nutrient-rich waste heap of the ant colony. Few tropical trees employ ants as their primary dispersers. Germinating in dung piles may provide more nutrients to the seedling than those that did not pass through an animal's gut as seeds, but there may also be more seedlings to compete against on the pile. Many studies have shown germination enhanced by passage through animals, but this probably reflects the evolution of mechanisms to prevent damage during the passage through an animal rather than some fundamental positive value of spending time inside one.

## The search for the Janzen–Connell effect

Many of the analyses of the data from the 50 ha forest dynamics plot on Barro Colorado Island have been aimed at trying to detect the Janzen–Connell effect operating within the tree community (Hubbell 1998). The earliest analysis used the initial census data from the plot to compare the spatial distributions of the saplings and adults of common species (Hubbell & Foster 1986) on 1 ha subplots. These showed a generally positive correlation between adult and juvenile densities for species across the large plot. Only the two commonest species, *Trichilia tuberculata* and *Alseis blackiana*, exhibited significant negative relationships indicative of compensatory mortality. Working in Borneo, Webb & Peart (1999) discovered stronger evidence for compensatory mortality by monitoring the dynamics of seedling populations.

A significant negative relation was found between seedling survival and adult basal-area representation across 149 species in the dipterocarp forest. A focal-tree approach for analysis of saplings around adults on Barro Colorado Island showed evidence of the Janzen–Connell effect for a higher proportion of species, but this was still only a minority of the community (Condit *et al.* 1992b). The problem with using focal adult trees in this type of analysis is that many of the focal trees are probably not the parents of the recruiting saplings because of the long period of time taken for individuals to reach the 1 cm dbh size class (Hubbell 1998). Recently, in line with the findings of Schupp (1992) for one species in the same forest, Wills *et al.* (1997) returned to the plot-based approach in the search for the Janzen–Connell effect, but they used per capita measures of survival and recruitment and employed re-randomisation techniques to compensate for an expected negative autocorrelation within the comparisons made. They found evidence of significantly negative density-dependent effects for 67 out of 84 of the commonest species at one or more plot sizes. In comparison, there was little evidence for significant negative effects within all possible combinations of species pairs from among the 84 species. This suggests that the Janzen–Connell effect is a major force within the tree community on Barro Colorado Island, and that it tends to act to produce a complex patchiness in time and space in population dynamics of individual species. The actual mechanism of the Janzen–Connell effect in the Barro Colorado Island forest remains undetermined, but slow-acting fungal pathogens are the theory favoured at present (Wills *et al.* 1997; Hubbell 1998).

## Co-evolution

For a period, the concept of co-evolution between interacting species was very fashionable. The belief was that many plant species had evolved in tandem with their animal partners in pollination and seed-dispersal, and that the tropical rain forest with its relatively clement and stable environment was a particularly favourable to the development of co-evolution. A good example of the 'co-evolution-acceptive' approach to plant–animal interactions was the wide credence given to the seed dispersal of the tambalocoque (*Sideroxylon grandiflorum*, syn. *Calvaria major*). This Mauritian endemic tree was purportedly nearing extinction because of its reliance on the extinct dodo (*Raphus cucullatus*) as a disperser and abrader of its seeds to achieve germination (Temple 1977). However, Witmer & Cheke (1991) have questioned the notion that the tambalocoque and the dodo were obligate mutualists. They argue that unabraded tambalacoque seeds will germinate and that other dispersers, such as giant tortoises, may have cleaned the seeds of flesh sufficiently well to prevent the fungal contamination that kills many seeds in the wild. Certainly the paper by Temple (1977) contains very limited data to support the conclusion that the tambalacoque had an obligate requirement for ingestion by dodos or similar large birds for germination.

Detailed studies of the real nature of plant–pollinator and plant–disperser interactions have shown that the one-to-one relationships needed for co-evolution are very rare. The figs appear to be the exception that proves the rule of generally diffuse evolutionary change engendered by such interactions. The risks of reliance on just one partner generally outweigh the advantages of greater fidelity and performance. This is true for both the plant and the animal involved. What appears to be closer to reality is groups of species interacting, resulting in guilds of pollinators and dispersers and plant pollination and dispersal syndromes.

# 5

# Seeds and seedlings

## Seeds

### Seed size

The size of seeds interests comparative ecologists because it is so variable among species. The dry mass of seeds ranges over at least six orders of magnitude across species of tropical rain-forest tree. The Melastomataceae and Rubiaceae include tropical tree species with seeds of dry mass as little as 20 $\mu$g (Metcalfe & Grubb 1995; Grubb & Metcalfe 1996). At the other extreme, the seeds of a number of trees, notably legumes, approach 100 g dry mass. Within any tropical forest site, most studies have shown ranges of least five orders of magnitude for tree seed mass (Hammond & Brown 1995; Metcalfe & Grubb 1995; Grubb & Coomes 1997; Lord *et al.* 1997). Of course, mass is a volume-dependent property and so will rise with the cube of the linear dimensions involved, which will rapidly exaggerate size differences between species, but a million-fold range in offspring size is still enormous when compared with animal groups.

Seed size might be under allometric control of other characters. There is evidence of correlations with other size variables. The difficulties of small plants producing big seeds and of small fruits containing big seeds will probably always lead to some degree of positive correlation between plant size and seed size (Fig. 5.1). For tropical trees, a number of studies have shown increases in seed size with adult stature within a particular forest (Hilty 1980; Foster & Janson 1985; Metcalfe & Grubb 1995; Hammond & Brown 1995; Kelly 1995; Grubb & Coomes 1997). Dispersal mechanism may also influence seed size (Fig. 5.1), possibly because of disperser selection for fruit size. In three neotropical sites there was a consistent pattern of mammal-dispersed seeds being 12–14 times heavier on average than bird-dispersed seeds (Hammond & Brown 1995). This might be due to mammalian frugivores choosing the big end of the fruit-size range available in the forest, making it appear as though there was some evolutionary relationship. However, Kelly (1995) found that this pattern was often repeated within different genera at a site in Peru, indicating recurrent evolution of large seed size in mammal-dispersed species.

A feature of tropical rain forests is the presence of very large-seeded species

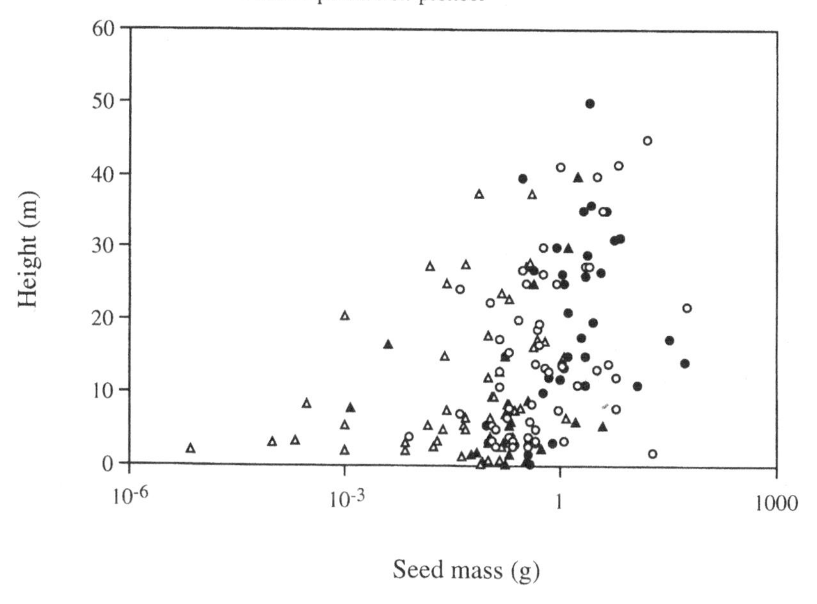


Figure 5.1 Seed mass versus tree height for species from Manu, Peru. Species categorised by dispersal mode (circles, mammal-dispersed; triangles, bird-dispersed) and successional status (filled, non-pioneers; open, pioneers). Data from Foster & Janson (1985).

(Lord *et al.* 1997). Foster (1986), in a review of the adaptive value of large seeds, proposed several advantages. However, critical analysis does not support the acceptance of all of them. Foster put forward the following as likely benefits for bigger seeds.

*Improved seed longevity*

Simplistically, one might argue that bigger seeds will have more reserves and therefore will be expected to survive longer. Foster (1986) pointed out a significant positive correlation between time to germination and seed size in Ng's large data set for Malaysian tree species (Ng 1980). However, Hopkins & Graham (1987) found that it was mostly small seeds that survived burial of up to 2 years well in Queensland (Fig. 5.2). Kanzaki *et al.* (1997) also found a significant negative correlation between seed size and survival time in tropical forest soil for tree species from Malaysia. Respiration rate measurements on some tropical seeds shed light on this paradox. Garwood & Lighton (1990) discovered that seed water content was of greater influence on oxygen consumption rate than seed size. Dormant seeds have low water contents and, hence, low respiration rates. Large seeds tend to have high water contents and therefore respire rapidly. For seeds of similar water content, large seeds have higher absolute respiration rates, but frequently lower rates per unit dry mass. In theory, large seeds that could dry out to enter dormancy should be able to survive for long periods, but this appears to be a relatively rare strategy. Large dormant seeds might be particularly susceptible to vertebrate seed predators in the forest.

*Greater room for secondary compounds*

A large seed might be able to store more chemical defences. These could possibly be mobilised to defend the seed against attack by pathogens or small invertebrate attackers. However, for large seed-eaters the total concentration of chemical defences is more likely to be influential on food choice. It is possible, though, that very large seeds might contain more than the safe

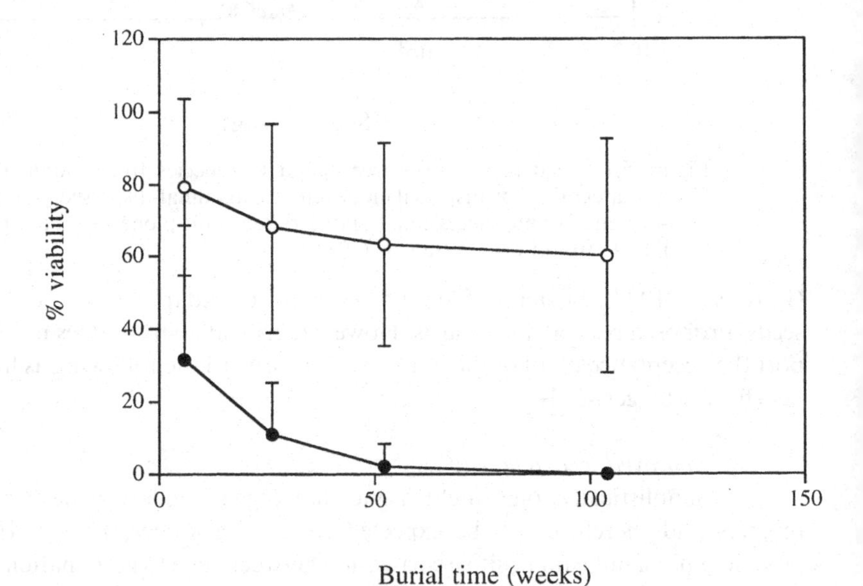


Figure 5.2 Viability of seeds against time buried in soil for species from Queensland, Australia. Solid symbols indicate primary forest species (mostly large seeds), hollow symbols, secondary forest species (mostly small seeds). Data from Hopkins & Graham (1987).

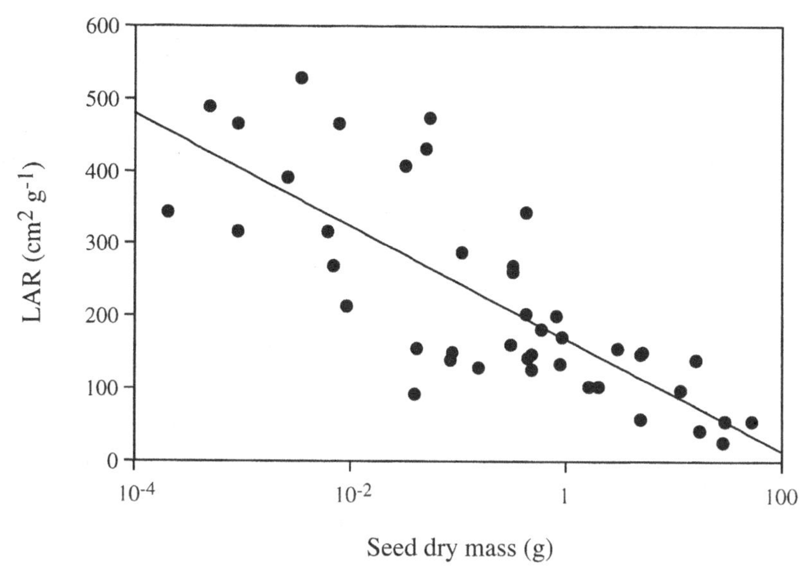


Figure 5.3 LAR as a function of seed dry mass in small seedlings of 43 tropical tree species. Data from ter Steege (1994a), Osunkoya *et al.* (1994) and Kitajima (1994).

dose of a toxin and only part of the seed would be eaten by any one animal at one time, improving the chances of the embryo surviving.

*Increased photosynthetic to non-photosynthetic tissue ratio*

Large seed reserves might allow faster development of greater photosynthetic areas in newly germinated seedlings and hence a greater chance of survival in the shade. This does not seem to be the case. Kitajima (1992) found that there was a strong negative correlation between seed mass and relative photosynthetic ability of newly germinated seedlings of tropical tree species. Small seeds tend to produce seedlings with thin cotyledons of relatively large area (Fig. 5.3), allowing small seedlings to achieve substantially greater relative growth rates. Tiny tropical tree seedlings are capable of persisting in very deep shade (Ellison *et al.* 1993; Metcalfe & Grubb 1997).

*Growth into higher light or deeper soil*

Bigger seeds produce bigger seedlings (Janos 1980; Howe & Richter 1982), and there can be an advantage in greater seedling size. There may be strong gradients in light availability above the forest floor and in water availability beneath it. Large seeds can produce taller seedlings that get more light, or deeper-rooted seedlings that get more water. The seedlings from big seeds may germinate successfully from under greater depths of soil or litter, and can probably compete with neighbours more vigorously and develop mycorrhizas more successfully (Janos 1980). Greater reserves will allow greater chances of survival during unfavourable intervals e.g. when shaded by a fallen leaf, or during a cloudy period. Boot (1996) found a positive correlation between survival of germinated seedlings in the dark over one year and seed size for six tree species from Guyana. Germinants of large-seeded species (e.g. *Castanospermum australe*) in Queensland, Australia, could survive burial for two years (Hopkins & Graham 1987).

*Tolerance of damage or tissue loss*

Damage due to pathogens, herbivores and falling debris is likely to affect seed and seedling survival in the forest. The greater reserves of large seeds can help cope with these vicissitudes. Two very large-seeded species (more than 100 g fresh mass) from New Guinea were more tolerant, in terms of seedling survival and height growth relative to control plants, of removal of more than 50% of non-embryo tissue from the seeds than two other species of seed size more than an order of magnitude smaller (Mack 1998). Fragments of large seeds may also be capable of developing into plantlets as has been shown for *Gustavia superba* (Harms *et al.* 1997). Harms & Dalling (1997) measured the ability of seedlings to recover from decapitation for 13 species with seed fresh mass in the range 0.2–107.6 g. Only the five species with seed mass of 5 g or greater survived decapitation and resprouted. They all had hypogeal germination.

The evidence points to large seed size being advantageous because it produces larger seedlings with reserves sufficient to meet the requirements of physical and chemical defence, periods of resource shortage and repair of damage (Kitajima 1996). The abundance of tree species with very large seeds in tropical rain forests may reflect the deep shade of the forest understorey with fierce competition and ever-present threats from pests and diseases. The presence of potential seed dispersers of large body size may also favour large seeds in the tropical rain forest.

However, many species still have small seeds, probably because of the basic evolutionary tradeoff between seed size and seed number. If a tree has a particular amount of resources it can devote to reproduction then to produce more seeds it will have to reduce the amount of resources given to each seed. More seeds will mean more attempts to produce viable descendants in the next generation. But generally reduced size means increased risk of mortality. So natural selection balances the increasing number of attempts against the reducing chance of survival and comes up with a seed size appropriate to the species. In essence, small seeds are riskier but more can be produced per unit investment.

The tradeoff of seed size against seed number may not be as direct as the foregoing paragraph made it seem, at least as far as fleshy fruits are concerned. Fruits with small seeds tend to have a higher ratio of flesh dry mass to seed dry mass (Grubb 1998b). This is probably because small fruits require to offer proportionally larger rewards than big fruits to attract dispersers.

There may be situations where seed size is less relevant to survival, or where small seeds gain special advantages. It has been argued that in the relatively resource-rich environment of gaps seed size may be less influential on seedling survival and that seedlings from small seeds may rapidly catch up with those from larger ones because of their higher relative growth rates. The idea that strongly light-demanding species have smaller seeds than shade-tolerant ones is quite firmly entrenched in the ecological literature (Swaine & Whitmore 1988), but it has been challenged of late (Kelly & Purvis 1993; Grubb & Metcalfe 1996). Comparisons of seed size between species believed to be gap-demanding for regeneration and those that can persist as juveniles in the forest understorey have generally shown greater mean seed mass (or other size measure) for shade-tolerants in tropical rain forests (Foster & Janson 1985; Hammond & Brown 1995), but this has always involved a very large range of seed size in both species groups. However, statistical analyses that correct for possible phylogenetic influences have shown much less clear evidence for larger seed size in shade-tolerants (Kelly & Purvis 1993; Grubb & Metcalfe 1996). Grubb & Metcalfe (1996) argued that these analyses, which imply a strong phylogenetic influence on seed size, indicate that this character exhibits evolutionary inertia. Within a particular clade, seed size changes slowly in evolutionary terms, and it may not be particularly influential as to whether a species can or cannot be successful as a shade-tolerant or gap-demanding tree. Large-seeded light-demanders include *Aleurites moluccana* (7.8 g dry mass) (Grubb & Metcalfe 1996) and *Ricinodendron heudelotii* (1.4 g dry mass) (Kyereh *et al.* 1999). Various authors (Metcalfe & Grubb 1995; Grubb 1996; Grubb & Metcalfe 1996; Hammond & Brown 1995) have highlighted the presence of very small-seeded, strongly shade-tolerant species among rain-forest tree floras. Hammond & Brown (1995) hypothesised that small adult stature, low resource availability in the forest understorey and use of small-bodied dispersers compound to favour small seed size. These are likely to be selection pressures acting in favour of a trend of decreasing seed size with distance below the top of the canopy, but they are probably still insufficient to explain the very tiny seeds of some species, notably among the Rubiaceae and Melastomataceae. These minuscule seeds are possibly specialist exploiters of certain regeneration sites in the forest. Given the small size of the seeds and their subsequent seedlings, regeneration could not be successful if litter were present, so litter-free microsites on the forest floor are required. Sites on steep banks and slopes are most suitable, where large seeds would fall down, but the minute seeds are caught by tiny irregularities in the soil surface (Grubb & Metcalfe 1996; Metcalfe *et al.* 1998). Grubb & Metcalfe (1996) point to the possibility that these species have evolved from light-demanding ancestors.

Tall trees in the caatinga of Venezuela were shown to have smaller seeds than species of similar stature from forests on more fertile soils nearby (Grubb & Coomes 1997). This was interpreted as the caatinga species maintaining seed number in an environment where soil nutrient poverty had reduced the resources available for reproduction. The same authors argued that root competition was more intense on such low fertility sites (Coomes & Grubb 1998a) which one might predict would favour larger-seeded species, but the possible benefits of increased seed size rise too slowly to outweigh the disadvantages of fewer seeds produced.

### Seed rain and the soil seed bank

Through the activities of dispersers, there is a rain of seeds to most parts of the tropical rain-forest floor. There have been surprisingly few studies of the composition of the seed rain, and of its spatial and temporal variation. At La Selva, Costa Rica, the seed rain was compared between five paired understorey and gap sites by using sterilised soil flats (Loiselle *et al.* 1996). The seed rain was dominated by zoochorous species with the gap sites having a greater representation of wind-dispersed seeds. This could be due to two factors: the greater penetration of wind into gaps bringing seeds from the canopy, and the greater availability of perching places for animal dispersers over understorey sites. These resulted in the seed rain of gap sites being more similar to each other than to the nearby understorey sites. Augspurger & Franson (1988) also found the seed rain of wind-dispersed species to be larger in gap than understorey sites on Barro Colorado Island. At La Selva, only 35% of species and 19–55% of seeds were from plants fruiting within 50 m of the four gaps studied (Denslow & Gomez Dias 1990). The seed rain for highly fecund species can be very large: 40 000 and 66 000 seeds $m^{-2}$ were estimated to fall per year beneath the crowns of *Miconia argentea* and *Cecropia insignis* trees, respectively, on Barro Colorado Island (Dalling *et al.* 1998b). *Cecropia obtusifolia* was found to have a high, but variable, seed rain (184–1925 $m^{-2}$ $yr^{-1}$) at Los Tuxtlas (Alvarez-Buylla & Martínez-Ramos 1990). Distance to nearest seed source explained 60% of this variability.

Many of the seeds that arrive on the forest floor become incorporated in the soil, presumably by the action of animals and drainage water, and by seeds falling down holes and cracks. A proportion of these seeds remains viable for weeks or months and forms a bank of ungerminated seeds. Garwood (1989) has written an extensive review of the soil seed banks of tropical forests. It has been known for many years that if surface soil is taken from inside the forest and spread on trays in a site receiving direct sunlight for at least part of the day and kept adequately moist then many seeds will germinate, even if netting is used to prevent seed rain contaminating the trays. The species that generally dominate such soil-generated regeneration are fast-growing light-demanders because they are present in large numbers and grow rapidly. However, quantitative studies have often also shown an abundance of shade-tolerant herbs and shrubs in the soil seed bank (Putz & Appanah 1987; Kennedy & Swaine 1992; Metcalfe & Turner 1998). Garwood (1989) concluded that 'primary' species contributed a relatively small proportion (0–16%) of seeds in tropical forest soils, but a larger component of the species richness of the seed bank. The abundance of species in the soil seed bank at Barro Colorado Island was strongly negatively correlated to seed size (Dalling *et al.* 1998b): there were more small seeds present than big ones.

Comparisons between the magnitude of the seed rain, and the size of the seed bank allow rough estimates of the potential residence time of seeds in the soil. At Pasoh, Peninsular Malaysia, the rate of germination from the soil seed bank in new gaps was seven times greater than germination from recently rained-in seeds (Putz & Appanah 1987). The stimulation for germination from the soil seed bank provided by gap formation resulted in a significant depletion of the soil seed bank in the subsequent 2–5 years. The *Cecropia obtusifolia* soil seed bank turned over rapidly at Los Tuxtlas (Alvarez-Buylla & Martínez-Ramos 1990), with estimated residence times of 1.02–1.07 years. The seed rain contained a much higher proportion of viable seeds (48%) than the soil (5–17%). *Cecropia* seed samples could be stored in the laboratory for five years or more with no great loss of viability, indicating that seed predation or pathogenesis must be high in the forest soil. The intense rain of seeds kept topping up the seed bank that was rapidly depleted by attacks of soil invertebrates and fungi. More than 90% of *Cecropia* seedlings came from seed less than one year old.

In a detailed study of two species, *Miconia argentea* and *Cecropia insignis*, on Barro Colorado Island, Dalling *et al.* (1998b) found that the seed bank in the top 3 cm of soil beneath crowns of the trees represented only 23% and 2%, respectively, of the annual seed rain. For *Miconia argentea*, once more than 5 m from the crown, the soil seed bank was larger than the annual seed input because of the rapid reduction in seed rain intensity with distance. A decline in the rate of seed loss with distance also contributed to the more efficient incorporation of seeds into the seed bank further from the parent tree. *Cecropia insignis* had high rates (> 90% per year) of seed mortality at all distances from the parent. Treatment of the soil with fungicide greatly enhanced seed survival in both species, indicating that fungi were the main agents of mortality to seeds in the soil.

### Seed germination

The seeds of tropical rain-forest trees show a large interspecific range of time taken to germinate. For a sample of 330 species from the forests of Malaysia, 65% of species showed germination within 20 weeks of sowing fresh seed in a lightly shaded nursery (Ng 1980). The 35% of species that took longer than 20 weeks to germinate often had hard, thick seed coats or endocarps around the seed. Studies of the physiology of germination have shown that a number of factors can cause delayed germination in tropical rain-forest tree seed (Vázquez-Yanes & Orozco-Segovia 1993). These include low water content of seed at maturity, presence of a hard seed coat, small size and early stage of development of the embryo and the presence of chemical germination inhibitors. Rapid germinators tend to have high seed water content at maturity and soft seed coats. Many such species are difficult to store. They lose viability if dried and germinate if stored at high water content. This has led seed physiologists to develop a classification system based on the potential of seeds to be stored. Orthodox seeds (Roberts 1973) are those that can be stored in a dormant state for a relatively long period (many months), usually at relatively low seed water content. Recalcitrant seeds are the rapid germinators that cannot withstand drying. Many tropical tree species produce large, soft-coated seeds of high water content that are recalcitrant.

The seed dormancy of more orthodox species can be brought about in a number of ways. Hard and impermeable seed coats can physically prevent germination, the embryo may need to develop before germination can take place, or environmental conditions may induce dormancy. Seeds with hard coats, or persistent endocarps, may need scarification from animals or environmental factors before germination will take place. Heat may be another factor involved. The tiny seeds of balsa, *Ochroma pyramidale*, remain dormant because of an impermeable testa. The heat of superficial fires or the temperature fluctuations associated with the microclimate of large gaps can cause rupture of the testa, allowing germination to proceed. Shade appears to be the most important inducer of dormancy in the seeds of tropical trees. Photoblastic seeds are responsive to the spectral composition of the impinging radiation. It is generally accepted that phytochrome is the chemical that mediates photoblasty in seeds (Vázquez-Yanes & Orozco-Segovia 1996a).

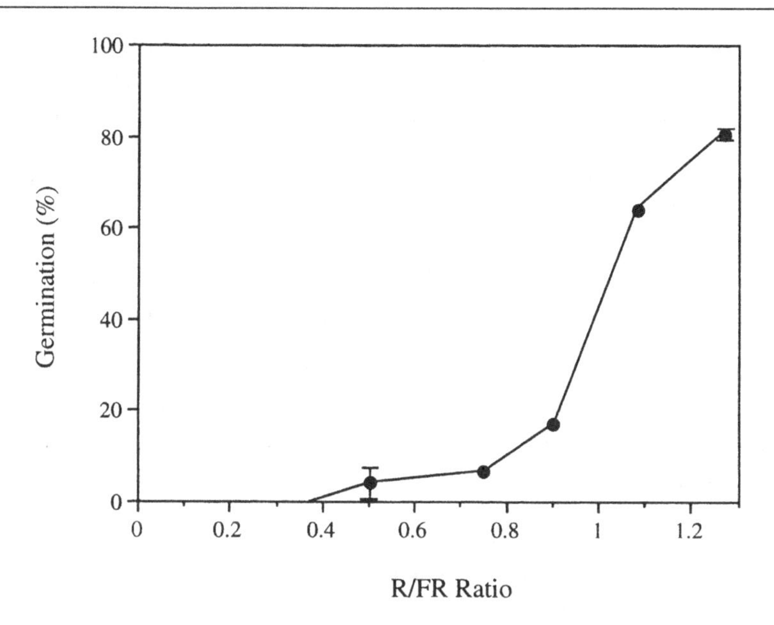


Figure 5.4 Germination of *Cecropia obtusifolia* along a red:far red (R/FR) ratio gradient. Bars represent ±1 SD. After Vazquez-Yanes *et al.* (1990).

Red light of wavelengths around 660 nm stimulates the formation of a physiologically active form of phytochrome that triggers germination in seeds. Far-red light around the 730 nm waveband initiates the conversion of the active to the inactive form of phytochrome, and causes dormancy. The relative fluence rate of the red and far-red wavebands (R : FR) is the critical spectral quality trigger for dormancy or germination in photoblastic seeds. The germination response of *Cecropia obtusifolia* seeds to variation in R : FR is shown in Fig. 5.4. Different microsites in the forest can differ substantially in their R : FR. Canopy leaves absorb red light, leaving the forest understorey relatively enriched in far-red wavelengths. Thus forest shade induces dormancy in photoblastic seeds, but canopy gaps can trigger germination. Leaf litter and soil also have a selective spectral absorptance that tends to reduce the R : FR of light transmitted through them. Sunflecks, brief periods of direct illumination by sunlight, occur at irregular intervals at most sites on the forest floor. These may be sufficient to trigger germination because of their high R : FR, but photoreversion due to the re-establishment of the low R : FR of the shade light in the forest understorey will probably re-implement dormancy before germination takes place. Tropical trees with photoblastic seeds show considerable variation between species, and sometimes within species, in the germination responses to the natural range of R : FR occurring in the forest (Vázquez-Yanes & Orozco-Segovia 1996a; Metcalfe 1996). Light may not be the sole trigger to germination in dormant seeds. There is evidence of an interaction of light with temperature in some species. For instance, Orozco-Segovia *et al.* (1987) found that seeds of *Urera caracasana* required exposure to 4 h of white light at 25 °C for high rates of germination, but at 35 °C 30 min of light was sufficient to achieve comparable germination percentages.

Potentially, environmentally induced dormancy can allow seeds to remain viable in the rain-forest soil for long periods. Fully imbibed seeds of several 'pioneer' species from Mexico retained high viability for more than 5 years when stored in darkness in the laboratory (Fig. 5.5). Seeds of *Mallotus paniculatus* showed no loss of viability after 3 years buried in mesh bags in a Malaysian rain forest (Kanzaki *et al.* 1997). There is some evidence that seeds may change in their environmental requirements for germination with time

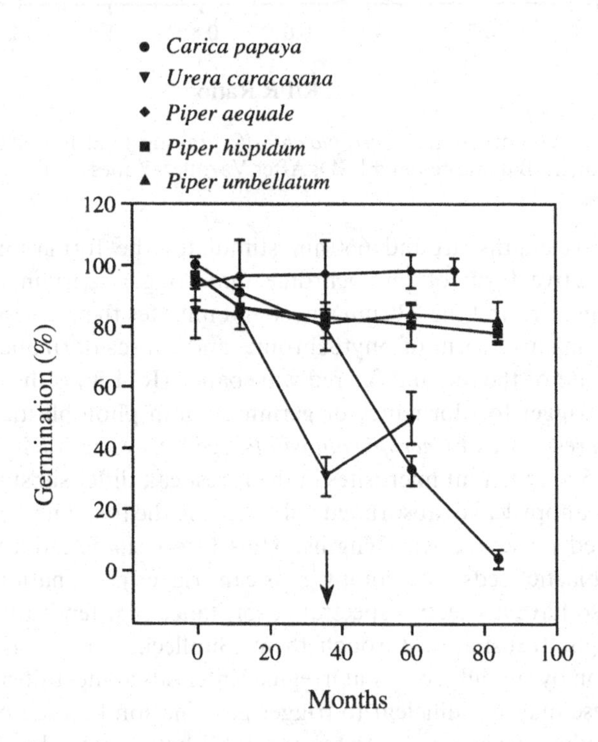


Figure 5.5 Germination after different storage periods of fully imbibed seeds in darkness. After 3 years the germinability of dry seeds of the same species was near zero in all of them. The arrow indicates when the seeds of all the species in dry storage at room temperature lost all germinability. SD indicated by vertical bars. After Vazquez-Yanes & Orozco-Segovia (1996b).

buried in the soil (Vázquez-Yanes & Orozco-Segovia 1996a), tending to become more prone to germination in conditions that would have previously maintained dormancy.

The ability of some species to remain dormant in the shade and germinate only in direct sunlight has been recognised as a characteristic uniting a group of tropical tree species of similar ecology. Swaine & Whitmore (1988) used this as the key character to define 'pioneer' species. These are the fast-growing, shade-intolerant species typically found only in gaps or other early successional sites in the forest. The more shade-tolerant species have germination that is generally not dependent on degree of shading. Swaine & Whitmore (1988) referred to these species as 'non-pioneers'. There have been few attempts to test the validity of the dichotomy proposed by Swaine & Whitmore. Raich & Gong (1990) investigated the germination of 43 Malaysian tree species in clearing, gap and forest understorey sites of 60%, 40% and 1.2% full-sun PAR. Only seven species germinated equally well in all sites. In the clearing, 16 species showed low rates of germination, or failed completely; 22 species germinated at higher rates in the understorey than the clearing and 12 species germinated better in one or other of the open sites than they did in the understorey. Some species germinated little in the understorey, but did so readily when these seeds were transferred to the clearing (Fig. 5.6). The clearing conditions killed seeds of a number of species. Raich & Gong (1990) argued that a clear dichotomy of species on germination response was not easy to discern. A more complex pattern of seed environmental responses was seen. This was also the picture to emerge from a study of 19 species from tropical West Africa (Kyereh *et al.* 1999). Only three species germinated in significantly lower proportions in complete darkness than a light treatment and only one species was found to have a germination response to variation in R : FR. All the species showed some germination in forest understorey conditions. The sample included several species that are strongly light-demanding such as *Ceiba pentandra* and *Ricinodendron heudelotii.*

A thick shell around a seed need not necessarily delay germination. In *Mezzettia parviflora* the seed has a woody covering 3–4 mm thick, derived from the middle integument (Lucas *et al.* 1991b). This protects the large seed from most attackers. Orang utans can just about open some seeds with their teeth. Yet the *Mezzettia* seedling can break out of the seed and germinate very rapidly. This is because of a special band of brittle brachysclereids (stone cells) running around the shell and a small plug through the wall at one end. This provides a built-in weakness to the shell that allows the turgor pressure of the seed to open it. Mammalian seed-eaters are not assisted by this design because the plug and band are too narrow for their teeth to exploit. Seed-eating beetles are so small that they operate at a scale where shell hardness is more important than strength. The weak band does not differ appreciably in hardness from the rest of the shell so again the seed predators cannot exploit it.

The big seeds of *Cavanillesia platanifolia* contain large quantities (27%) of mucilage (Garwood 1985). The mucilage takes up water very readily and rapidly (7 g water $g^{-1}$ fruit in 10 min). This probably assists the seed to germinate and the seedling to establish in the seasonally dry forest on Barro Colorado Island, Panama.

*Delaying germination*

Why is there this large range in time taken to germinate among species? The main advantage of waiting is to increase the likelihood of secondary dispersal and possibly to stagger germination allowing some seeds to escape unfavourable periods, such as droughts, to which they would have succumbed as seedlings. The disadvantage of waiting is seed mortality. Rapid germinators avoid seed predators by soon becoming seedlings, although some seed eaters will also attack the cotyledons of seedlings. Predator satiation may be facilitated by rapid germination because it means less time for the predators to exploit the seeds. Seeds that possess mechanisms to detect the quality of the environment and germinate or remain dormant accordingly have the benefit of both delayed and rapid germination options.

*Timing of germination*

Barro Colorado Island, Panama, has quite a marked dry season each year. The species in the rain forest showed three main patterns of timing of germination with respect to the climatic seasonality (Garwood 1983): 42% of species dispersed their seed during the dry season and remained dormant until the rains started and germination took place; 40% of species dispersed and germinated in the wet season; and 18% of species dispersed in one rainy season but did not germinate until the next.

*Germination and litter*

As already mentioned above, forest leaf litter can act as a selective filter of light, transmitting a spectral composition of reduced R : FR, which can influence the germination of seeds. Litter can also act as a physical barrier to seedling establishment. Small seedlings may not be able to push leaves out of the way and will die in the low-light conditions under the litter. For three tree species, there was a direct correlation between seed size and germination success through litter in the laboratory (Vázquez-Yanes & Orozco-Segovia 1992). Under shade-house conditions, litter layers were found negatively to influence germination and emergence of seeds of 'shade-intolerant' species more than 'shade-tolerant' ones (Molofsky & Augspurger 1992). However, there was a considerable range of susceptibility to litter suppression among the 'shade-intolerant' species. In high-light conditions, litter was actually beneficial to some species such as *Gustavia superba*. The seeds were probably cooler and moister than they would have been without litter present. Litter in gaps may facilitate colonisation by more shade-tolerant species. In a Puerto Rican forest, litter removal was found to increase seedling density significantly (Guzmán-Grajales & Walker 1991). The increase came largely from small-seeded species such as *Cecropia schreberiana* and *Chionanthus domingensis*. Large-seeded species were either unaffected by litter manipulation, or became rarer in litter-removal sites. The influence of seed size on germination and seedling establishment has been confirmed by further experiments (Everham *et al.* 1996). Litter removal and disturbance of the soil surface (scarification) have been reported as increasing the density of germinants by a factor of 2.5 in gaps in lowland dipterocarp forest in Danum Valley, Sabah, Malaysia

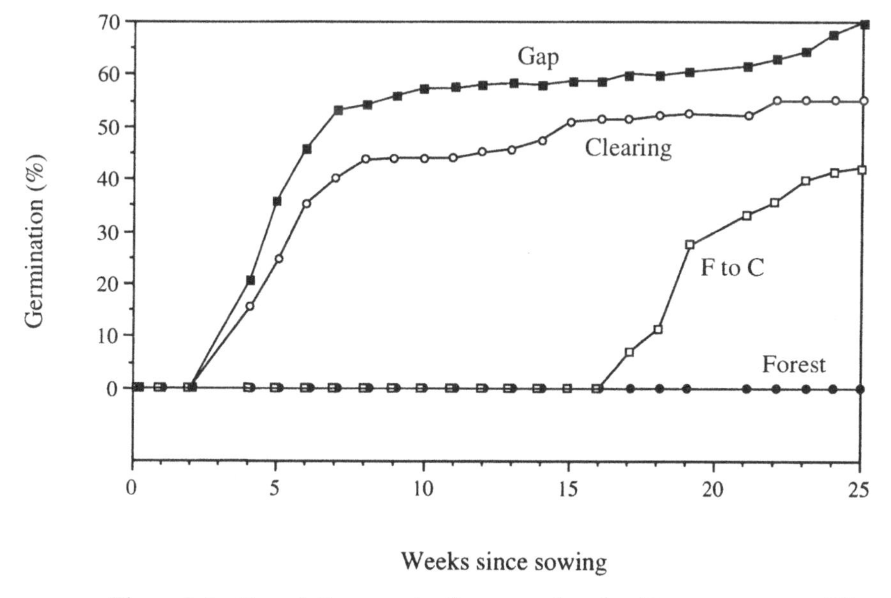


Figure 5.6 Cumulative germination over time for *Trema tomentosa* (Ulmaceae) in a forest-understorey, gap and clearing site. One basin of seeds (F to C) was moved from the forest understorey into the clearing on week 15; a second basin was left in the understorey. All basins contained 150 seeds. Note that the y-axis extends below zero to indicate clearly that no germination occurred in the forest understorey. Germination was recorded for 28 weeks, but no germination occurred after 25 weeks. Final germination was significantly different among all treatments. After Raich & Gong (1990).

(Kennedy & Swaine 1992). These germinants represented only a small fraction of the seed bank of the uppermost layer (0–5 cm) of the soil, hardly more than 5%, even with scarification. Similar results were obtained in forest in Singapore (Metcalfe & Turner 1998).

### Other influences on seedling establishment

Although it has been demonstrated that litter can influence seed germination and seedling establishment, an experimental manipulation of the understorey layer by removing all stems up to 5 cm dbh at La Selva, Costa Rica, did not appear to have any effect (Marquis *et al.* 1986). However, certain plants may be able to limit or suppress tree regeneration. For instance, dense populations of the terrestrial bromeliad *Aechmea magdalenae* significantly reduced the abundance of tree seedlings on Barro Colorado Island, Panama (Brokaw 1983).

### Germination safe sites

The establishment requirements of tree species may be narrow enough to limit the seedlings to certain definable safe sites in the forest. Examples of this include the finding that the commonest canopy tree (*Clethra occidentalis*) in the montane ridge forest of the Blue Mountains in Jamaica preferentially established (97%) on the stem bases of tree ferns (Newton & Healey 1989). Certain species, e.g. *Miconia mirabilis* and *Cecropia schreberiana*, were found to be positively associated with fallen logs in Dominica (Lack 1991). Swaine (1983) hypothesised that the stilt-roots possessed by some species (e.g. *Musanga cecropioides*) at an early stage of seedling development are an advantage in establishing on unstable sites in the forest such as fallen logs and the top of soil tip-up mounds. Tree-fall mounds are important for the regeneration of *Cecropia obtusa* in French Guiana (Riera 1985). These examples indicate the possibility of very specialised regeneration niches among tropical trees, and further observation will probably uncover more.

## Seedlings

Seedlings are perhaps the best-studied life-stage of tropical trees. This is largely for logistic reasons. Seedlings can be grown in controlled conditions with replicated manipulations that are not feasible for adult trees. Their populations are sufficiently dynamic to gather meaningful data in a matter of months or years rather than the decades required for adults. The greater accessibility of the forest floor makes physiological studies on seedlings *in situ* easier than those conducted in tree crowns. Seedlings are important from an ecological viewpoint because their relative success can influence total population size, though probably rather weakly (see Chapter 3), and spatial distribution of a species in the rain forest. However, there is a tendency to try to view seedlings as miniature trees and extrapolate from seedling characteristics to the likely behaviour of adult trees. This is relatively uncharted territory, and we have little means of judging the accuracy of such predictions at present. I have deliberately separated studies on juvenile from those on mature individuals in this book, in an attempt to make clear the relative limits of our knowledge for different stages of tree development.

A point of minor contention in forest ecology is the definition of seedling and sapling. It is accepted that seedlings are more juvenile than saplings, but the dividing line has not been fixed. Some authors prefer to restrict 'seedling' to the cotyledon-bearing stage only, but it is often employed in referring to plants up to 1 m tall or even bigger. I maintain the looser terminology.

### Seedling and sapling form

The form of newly germinated seedlings has been classified, along the lines of earlier systems, by Garwood (1996). Her system is based on cotyledon (hidden or not, foliaceous or purely storage) and germination (epigeal or hypogeal) characters and recognises five main seedling types as follows.

Phanerocotylar–epigeal–foliaceous (PEF)
Phanerocotylar–epigeal–reserve (PER)
Phanerocotylar–hypogeal–reserve (PHR)
Cryptocotylar–hypogeal–reserve (CHR)
Cryptocotylar–epigeal–reserve (CER)

Surveys of seedling types from various tropical areas show a generally similar pattern of relative abundance of the different seedling types. PEF is usually the commonest in terms of number of species; PHR and CER are relatively rare. A search for the ecological relevance of the seedling types has not been very rewarding till now (Garwood 1996). The clearest correlation is with seed size. Nearly all small seeds are phanerocotylar and epigeal; for example, 100% of seeds less than 3 mm long in a sample from Malaysia (Ng 1978). At the other end of the size spectrum the PE (F or R) seedling group becomes much rarer with only 5% of species with seeds ≥ 40 mm long in Malaysia (Ng 1978) and 19% of species with seeds ≥ 20 mm long in Gabon (Miquel 1987; Hladik & Miquel 1990). It is possible that cryptocotyly is favoured in big seeds as a means of protecting the large cotyledons or endosperm from animals.

Seedling and sapling morphology and architecture are quite varied, ranging from erect unbranched juveniles to branching plants with arching shoots. In an analysis of sapling form of common species on Barro Colorado Island, King *et al.* (1997) found that more than half of the species had

Table 5.1. *Frequency of occurrence of different general architectural forms among species with 50 or more individuals ($\geq 1$ cm dbh) in 50 ha on Barro Colorado Island*

| Architectural form | No. species | Percentage of total |
|---|---|---|
| tiers of plagiotropic branches on a vertical stem | 21 | 11 |
| plagiotropic branches (not in tiers) on a vertical stem | 33 | 18 |
| plagiotropic branches on an arching stem | 40 | 21 |
| orthotropic forms | 80 | 43 |
| intermediate forms | 12 | 6 |
| unexamined | 2 | 1 |

Data from King *et al.* (1997).

strongly orthotropic main stems (Table 5.1). Such saplings tend to have distinctly three-dimensional foliage whereas plagiotropic seedlings usually hold their leaves in flat arrays with relatively little depth of foliage on each branch (King 1998a). Leaf form is related to sapling architectural form (King & Maindonald 1999). Species with compound leaves are generally orthotropic. In species with simple leaves, orthotropic saplings tend to have larger leaves with longer petioles than plagiotropic saplings. Many species show a switch from plagiotropy as saplings to orthotropy as canopy trees. This may reflect a more efficient harvesting of the widely scattered light in the tropical rain-forest understorey by the plagiotropic sapling design.

Reviewing a large body of published data on the growth of tropical tree seedlings, Veneklaas & Poorter (1998) concluded that stem mass ratio increases with plant mass during the course of development. There is also a concomitant increase in average leaf mass per unit area (LMA), which progressively reduces the seedling leaf area ratio (LAR) and results in a trend of decreasing relative growth rate (RGR) with plant size. They argued that species follow individual allometric trajectories during the course of sapling growth. However, Kohyama & Hotta (1990) found that for nine species from the Sumatran rain forest with saplings 60–300 cm tall there were no significant interspecific differences in slope for the major allometric relationships of architectural form.

Seedling and sapling leaves can be very different from those found on the adult tree. In some species this is marked enough to be regarded as heterophylly; for example, *Scaphium macropodum* has deeply lobed leaves in juveniles, but simple leaves in mature trees (Fig. 5.7). It is clearly unlikely that small seedlings will be able to support leaves of adult size, so there is generally an increase in leaf size with age initially. However, size may peak at the sapling stage and then decline, as seen in *Cecropia obtusifolia* (Alvarez-Buylla &

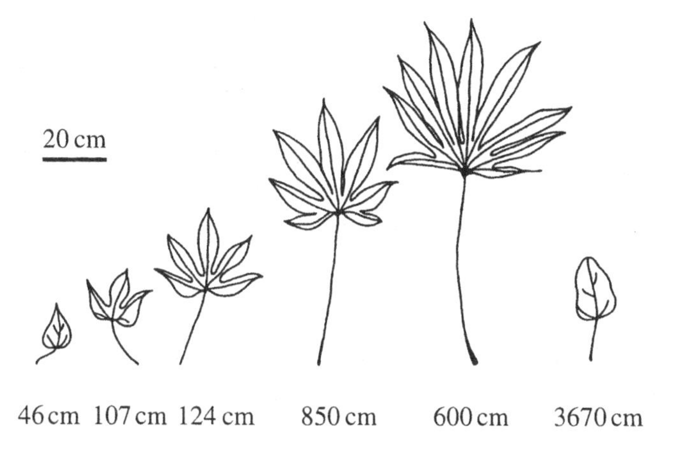


Figure 5.7 The leaf shapes of *Scaphium macropodum*. The heights of the trees from which the leaves were collected are indicated in the figure. After Yamada & Suzuki (1996).

Martínez-Ramos 1992) (Fig. 5.8). This probably reflects a maximum leaf size in saplings before they start to branch. The unbranched main stem bears very big leaves, but after branching has taken place such big leaves would tend to self-shade and could probably not be supported by the narrower axis of a secondary branch. There was a highly significant positive correlation between height at first branching and leaf length for saplings of 70 species spread across four different tropical rain-forest sites (King 1998b). Thomas & Ickes (1995) studied ontogenetic changes in leaf size in 51 species at Pasoh, Peninsular Malaysia. They found that half (26/51) exhibited larger leaves on saplings than adults. Thirteen species had larger adult leaves. These were mostly understorey treelets. Microscopic leaf-venation patterns differ between juvenile and mature trees (Roth 1996) with leaves of young trees having less dense venation with fewer vein endings and less ramification per mesh. This is probably mostly an effect of denser shade on juvenile leaves.

### Seedling dependence on seed reserves

A germinating seedling will be dependent initially on seed reserves for all its energy and nutrient requirements. The rapid development of the root will allow the seedling to become anchored and to start taking up water and nutrients. Foliaceous cotyledons can begin to photosynthesise and reduce the seedling's reliance on the seed reserves for fixed carbon. Seeds vary in the form of energy reserves they contain. Starch is common, but some seeds are rich in oils and other lipids. Lipid is a more concentrated store of energy than starch. In theory, a lipid-rich seed should produce a seedling 30% heavier than the seed (Kitajima 1996). A starchy seed of the same mass should only achieve a seedling mass 8% less than the seed mass. However, in reality plants are biochemically less efficient at utilising lipid than starch, so the advantage of lipids as seed stores is not as great as predicted, although they are still space-saving in comparison to carbohydrates. One might predict therefore that very small seeds should tend to be oilier than large ones, in order to store more energy in the tiny packet. However, wide-ranging surveys have found that big seeds tend to be oilier than small ones (Levin 1974). This is possibly to keep very energy-rich seeds within the size limits imposed by effective dispersal (Kitajima 1996).

Mineral nutrient concentrations of seeds tend to decline with seed size, both within a species (Grubb & Burslem 1998) and between species (Grubb 1996; Grubb & Coomes 1997), but not fast enough to prevent total nutrient content of seeds increasing with seed size. Grubb (1996) found that large-seeded lauraceous species dispersed by scatter-hoarding rodents in the rain forests of Queensland had higher N concentrations than the general regression of concentration against seed size would predict. He hypothesised a selection pressure in favour of higher seed N in such species as a reward for the mammalian seed dispersers.

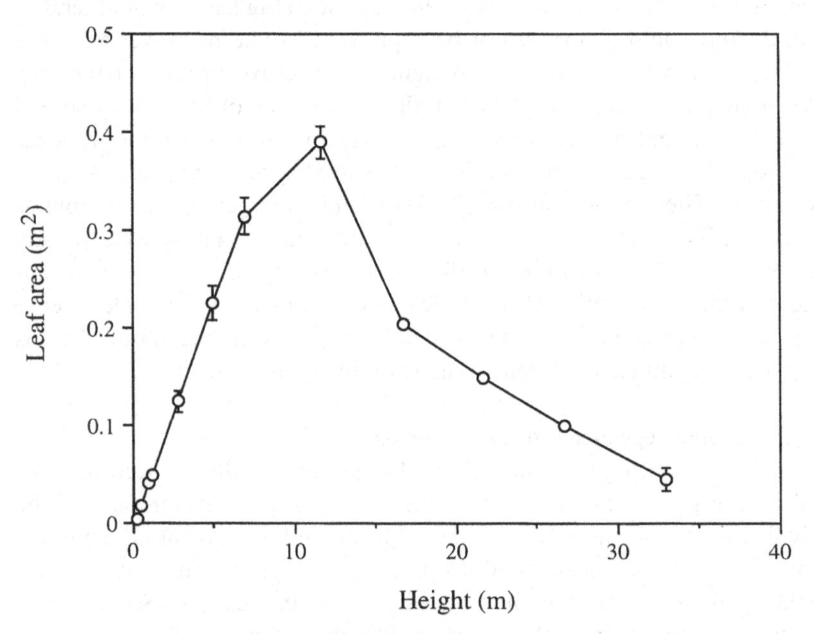


Figure 5.8 Leaf size as a function of tree height for *Cecropia obtusifolia* individuals at Los Tuxtlas, Veracruz, Mexico. After Alvarez-Buylla & Martínez-Ramos (1992).

Kitajima (1996) performed an ingenious experiment to estimate the duration of seedling dependence on seed reserves for three tree species of the Bignoniaceae. Seedlings were grown at high and low (2%) light and high and low nitrogen availability. When the growth of the low-resource treatment seedlings fell below that of the high-resource treatment it was assumed that the seed reserves were no longer the sole supplier of resources to the seedling. For all three species, the seed store of energy ran out before that of nitrogen. In the shade, the nitrogen store of the seed lasted up to 40 days. *Tabebuia rosea*, with the smallest seeds and leafy cotyledons, was reliant on seed reserves for the shortest period. It is not clear whether this was because the reserves ran out more quickly, or because photosynthesis began sooner than in the other two species. The mass-based assimilation rate of seedling cotyledons was strongly negatively related to cotyledon thickness (Kitajima 1992). Thick, largely storage, cotyledons photosynthesised little, but it may have been sufficient to balance the respiration of the storage cells, meaning that the cotyledons were not a net drain on the reserves of the seedling.

### Seedling mortality

The mortality of tree seedlings in the tropical rain forest is generally high, particularly for the new germinants. Many surveys of seedling population dynamics may well miss much of the earliest, and most intense, mortality because monitoring tiny seedlings and identifying them correctly is very difficult (Kennedy & Swaine 1992). A cohort of germinants often declines in numbers in a log-linear fashion, at least initially. That is, a straight line of negative gradient is produced by plotting seedling number on a logarithmic scale against time plotted linearly (Fig. 5.9). Li *et al.* (1996) have followed annual seedling cohorts at La Selva, Costa Rica, starting some 2–3 months after the peak of germination in the forest. The mean half-life of the multi-species cohorts was 2.49 months. Survivorship was highest for cohorts establishing during wet periods. Cohort half-lives for individual species varied from 0.37 to 40.7 months. Species such as *Cecropia obtusifolia* and *C. insignis* had very short half-lives. Large-seeded palms like *Socratea durissima* and *Welfia georgii* were among the most persistent seedlings. There was a negative relation between seedling recruitment and seedling persistence across the species (Fig. 5.10). In other words, there was no species that was very abundant as a new germinant that had a high seedling survival. Survivorship appears to improve as seedlings get larger. Among three common tree species on Barro Colorado Island, Panama, seedlings greater than 50 cm tall showed annual survival of more than 80% (De Steven 1994).

Small seedlings on the forest floor are open to may possible causes of death. These include the following.

*Breakage due to falling debris or large animals*

There is a continual rain of leaves, twigs, branches and trunks onto the forest floor. A small seedling is susceptible to damage, which may often be

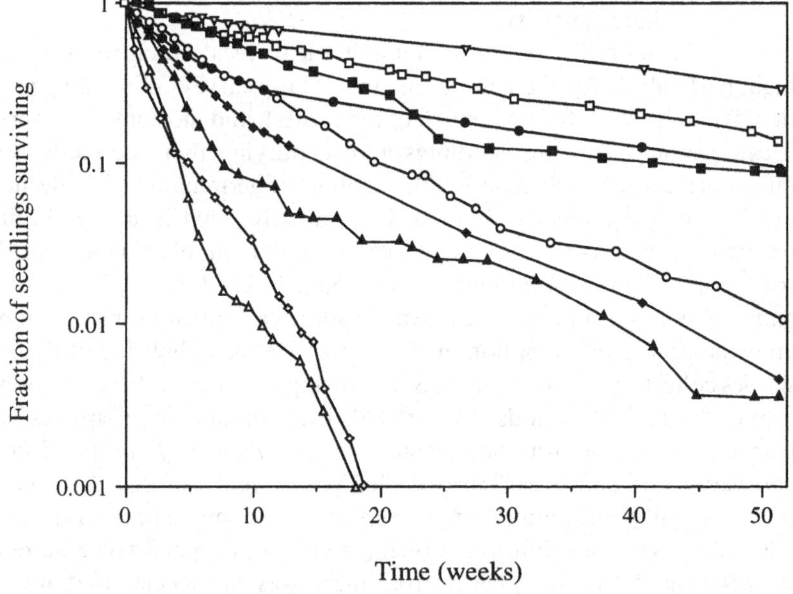


Figure 5.9 The fraction of all tagged seedlings of nine species surviving through one year under shaded conditions on Barro Colorado Island, Panama. The fraction is based on a summation of seedlings surviving at all distances from the parent. An equal number of seedlings were observed at each distance interval from the parent. After Augspurger (1984).

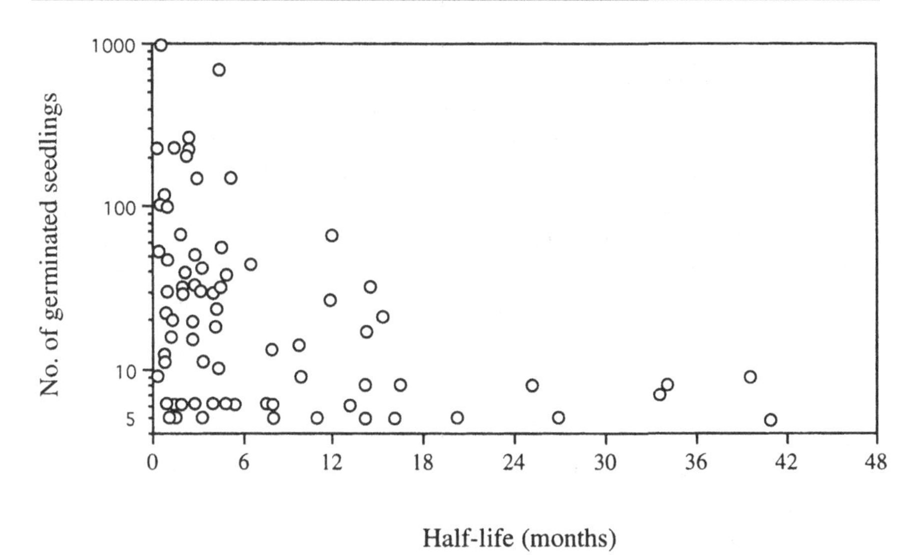


Figure 5.10 Relation between seedling abundance and survivorship among the 75 most abundant species (five or more germinated seedlings) at La Selva, Costa Rica. Longer half-lives are associated with lower seedling recruitment over the 18 month period. After Li *et al.* (1996).

fatal, from many of these falling objects (Aide 1987). In addition, large mammals (elephants, deer, pigs, human beings) will crush forest-floor vegetation as they move around. One of the major advantages to a seedling of growing larger is to become less at risk of damage by falling litter or animal activity. Clark & Clark (1989) used artificial seedlings made of plastic drinking straws and wire to assess seedling damage rates in the forest at La Selva. In one year, 82% of the seedlings were knocked over, flattened or uprooted. In the 49% of cases where the agent of destruction was identifiable, roughly half were due to falling debris and the other half to trampling and uprooting by vertebrates. Guariguata (1998) studied saplings 1–2.5 m tall of four species with an erect, unbranched architecture on Barro Colorado Island. About 3% per year of the saplings suffered damage. Over a four-year period mortality resulting from saplings being pinned to the forest floor was almost double that due to loss of the upper half of the sapling by breakage. *Alseis blackiana*, the most resilient of the species studied, was able to produce adventitious roots on the bent saplings. Interestingly, it was the most slow-growing of the four species and therefore would have the longest average residence time in the understorey.

Saplings may require to be more robustly designed than later stages to withstand falling debris; this idea fits with studies of safety factors calculated from allometric analyses of height–diameter relations (King 1996; Claussen &

Maycock 1995; Rich *et al.* 1986). However, this approach to design safety may be flawed because saplings tend to invest heavily in foliage, making crown mass a large proportion of total mass. Saplings may not therefore be as safe as they appear from trunk allometry; King (1987), employing a different approach, seems to have confirmed this. He used weights to load the crowns of treelets and saplings of canopy species at La Selva, Costa Rica. The results of the experiment led to the conclusion that the plants had stem diameters only 1.3–1.4 times the minimum diameter to prevent instability when growing in the shaded understorey. The absence of strong winds inside the forest probably allows such narrow safety limits.

*Seedling predators*

A range of different animals will attack seedlings, mostly for food, and cause their death. At the earliest stages seed-eating animals may eat the cotyledons and rob the seedling of its energy reserves. Such activity can also lead to the seedling being pulled up. Herbivorous mammals may eat the whole seedling before it becomes woody. Nest-making wild pigs are a major cause of mortality among large seedlings and saplings in Pasoh Forest, Malaysia. The pigs build a shelter of thin stems in which to raise their young and collect many saplings from the understorey for this purpose. Mammalian herbivory was observed to be a major cause of direct mortality in seedlings of *Virola nobilis* on Barro Colorado Island, and it also contributed to susceptibility to death due to drought as defoliated plants could not grow sufficiently large root systems to cope with water shortage (Howe 1990). Exclusion of mammals from seedlings in the forest reduces mortality rates (Osunkoya *et al.* 1992; Molofsky & Fisher 1993).

*Pests and disease*

Damping-off is a frequent cause of mortality in newly germinated seedlings. Damping-off is a disease caused by soil-borne fungi. It was studied, by means of field and shade-house experiments, by Augspurger & Kelly (1984). Newly germinated seeds of *Platypodium elegans* suffered density- and distance-dependent mortality from damping-off on Barro Colorado Island. Seeds placed out in the forest at four-times higher density had a greater incidence of damping-off, as did seeds nearer adult trees. Seeds of 18 species of wind-dispersed tree from Barro Colorado Island, were sown at two densities in shade houses imitating small gap (300 $\mu$mol m$^{-2}$s$^{-1}$ PAR) and understorey (17.5 $\mu$mol m$^{-2}$s$^{-1}$) conditions. Species varied considerably in their susceptibility to damping-off. Light was found to be more important than seed density in determining likelihood of damping-off, with most mortality in the deep shade. There are probably many other diseases present among populations of tropical trees, but there has been relatively little work conducted in this field. Potentially pathogenic nematodes have been reported from the roots of seedlings of *Turraeanthus africana* in Ivory Coast (Alexandre 1977) and *Dicorynia guianensis* in French Guiana (Quénéhervé *et al.* 1996).

*Drought*

The inability of young seedlings to grow very deep roots leaves them particularly susceptible to drought. High seedling mortality in dry spells has been widely reported (Turner 1990a). Mortality among seedlings of *Virola nobilis* was greatest in the understorey during the dry season on Barro Colorado Island (Fisher *et al.* 1991). Irrigation of seedlings through the drought resulted in higher survival. Seedlings in gaps survived the dry season best, despite having greater transpirational loads, because they could grow faster in the higher light of the gap and develop more extensive root systems that allowed them better access to water during the dry period.

*Shading*

The survival and growth of the seedlings of most species is reduced in the deep shade of the forest understorey. This is the topic of a later section of this chapter. For the present, suffice it to say that the low light of the forest understorey may impose such low photosynthetic rates that seedlings barely make a positive carbon gain. A few cloudy days may be sufficient to exhaust resources and kill the seedling.

*Competition*

The majority of seedlings grow in the forest understorey where they must establish despite being surrounded by many other individuals, most of which are far larger than they are. This is a highly competitive environment, and superior competitors may deprive weaker individuals of resources, reduce their growth rates and possibly even starve them altogether. Resource shortage caused by competition, and in other ways, will increase the susceptibility of seedlings to disease. Proximity to understorey palms negatively influenced survival in seedlings of two species of *Inga* planted into the forest understorey (Denslow *et al.* 1991). Coomes & Grubb (1998a) found that trenching had a positive effect on the height growth and leaf production in seedlings in the very infertile caatinga forests of Venezuela. The magnitude of response to trenching was similar in understorey and gap sites (Fig. 5.11).

Seed and seedling mortality are high for most tree species, and in combination with inadequacies in fecundity and dispersal, probably mean that many species are recruitment-limited in the forest.

**Growth and survival with respect to light**

The influence of light on the growth and survival of seedlings has been the aspect of the ecology of tropical trees most often investigated. There

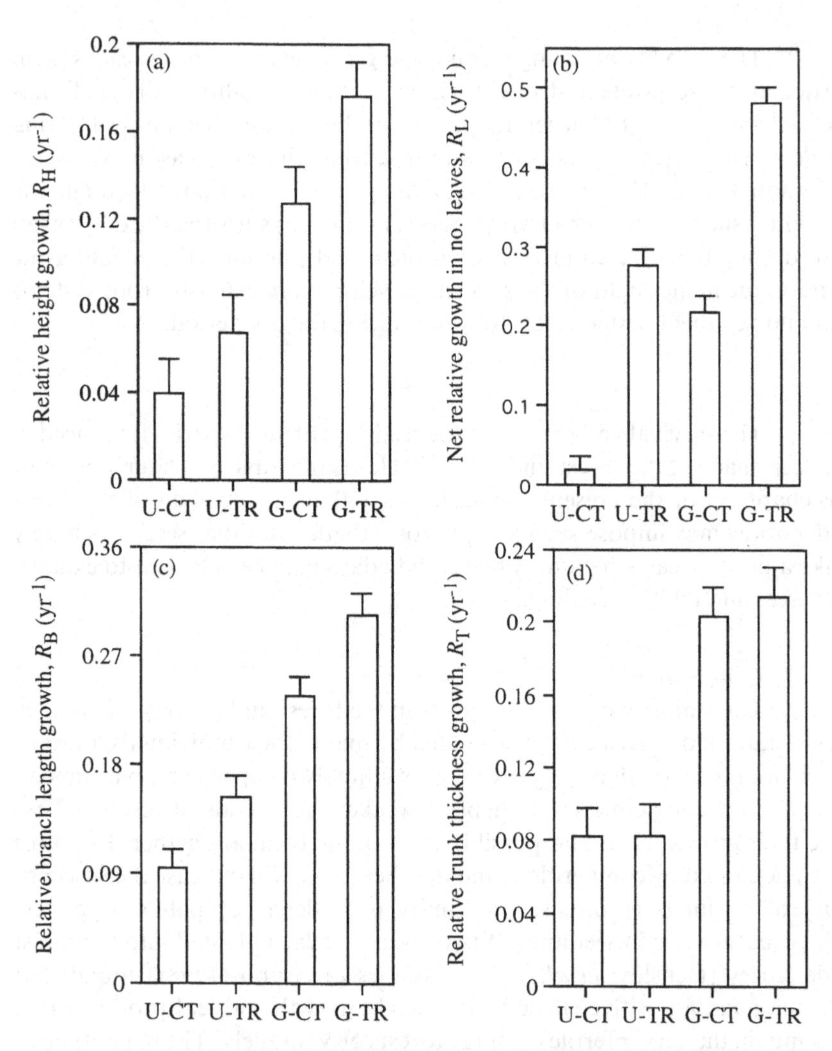


Figure 5.11 Mean relative growth rates (±1 SE) in (a) height, (b) number of leaves, (c) branch length and (d) stem thickness of trenched (TR) and untrenched (CT) saplings of 13 species, measured in understorey (U) and gaps (G), in Venezuelan caatinga. After Coomes & Grubb (1998a).

is a considerable range in the degree of 'naturalness' of such studies. Observations can be made on wild seedlings *in situ* across a variety of natural microsites in the forest. Seedlings can be planted or placed in pots into the forest, or grown in shade-houses imitating the natural microclimates of the forest, or controlled-environment chambers can be used (often well outside the tropics). Natural forest seedlings are very variable in size, age and environment. The greater degree of control and replication of artificial environments may lack the reality of the natural forest. For instance, growth chambers usually provide uniform light to plants, but in the forest the light received by a seedling on the forest floor varies almost continuously in intensity, and to some extent in spectral composition.

*Natural seedlings* in situ

Seedling growth and survival is nearly always promoted by increased light (King 1991b, 1994; Coomes & Grubb 1998b; van der Meer *et al.* 1998), except at very high levels (Zagt & Werger 1998). For example, Turner (1990b) observed that seedlings less than 1 m tall grew little under canopy shade conditions in Malaysian dipterocarp forest. There was a positive correlation between height growth and 1 $m^2$ seedling-plot light climate as estimated by using hemispherical photography. Clark *et al.* (1993) found similar results for saplings (0.5–5 m tall) of nine species at La Selva, Costa Rica, as did Dalling *et al.* (1998a) for eight out of ten pioneer species growing in gaps on Barro Colorado Island. However, shade-tolerant understorey species with seedlings that survive well in low light conditions may show little difference in survival between gap and understorey sites. For instance, Fraver *et al.* (1998) found no significant difference in survival over a four-year period for seedlings of *Protium panamense* and *Desmopsis panamensis* growing in gaps and under a continuous forest canopy in a Panamanian rain forest.

*Seedlings planted into the forest*

The range of growth possible in different light environments is exemplified by the results of an experiment by Thompson *et al.* (1988) in Queensland. Seedlings of four species were planted into tiny, medium and big gaps (0.6%, 9% and 40% full sun). After seven years, one species, *Darlingia darlingiana*, persisted in the shadiest site where its seedlings were about 15 cm tall. The largest *Toona australis* individuals in the big gap had reached 10 m in height by this time. Boot (1996) grew seedlings of Guyanan tree species in understorey and gap-edge and gap-centre sites (2.2%, 9.4% and 30.2% full-sun PAR), and performed serial harvests of seedlings in order to conduct classic growth analysis of the species. Some species could not survive in the understorey site. These tended to have the highest relative growth rate in the highest light. Most species did not show much improvement in growth rate between the gap-edge and gap-centre sites. Rooted cuttings of seven shrubby species did show faster growth in gap centres (9–23% full sun) compared with gap edges (3–11% full sun) (Denslow *et al.* 1990). Ashton *et al.* (1995) grew seedlings of four species of *Shorea* in forest gaps at three topographic positions in a forest in Sri Lanka. The valley gap was the biggest, the ridge gap the smallest and the slope gap intermediate in size. The faster growers in the big valley gap, *Shorea megistophylla* and *S. trapezifolia*, showed poorest survival in the shade, particularly on the ridge top. Another species, *S. worthingtonii*, was the best performer in the slope gap. Kobe (1999) planted small seedlings of four species of Moraceae (*sensu lato*) into various forest microsites ranging from less than 1% to 85% full sun and followed seedling survival and stem-base radial increment for one year. All the species showed a positive relationship between light availability and growth. Three (*Castilla elastica*, *Cecropia obtusifolia* and *Trophis racemosa*) also exhibited declining mortality with increasing light availability. The fourth, *Pourouma aspera*, had a peak survival at about 20% full sun. Combining growth with expectation of survival from a population of two seedlings at each light level, Kobe (1999) demonstrated that the four species neatly partitioned the light range into sections of approximately 20% full-sun increments in which each showed the best performance. The author reported that partitioning was maintained if larger seedling population sizes ($n$ = 5 or 10) were employed, although *Pourouma* steadily increased the range of light availabilities at which it was expected to grow the fastest.

*Shade-house experiments*

A common approach to studying tropical tree seedling response to light has been to grow seedlings in nurseries under different degrees of shading. Many early experiments of this type used a shading material that reduced the photon flux density incident on the plants beneath, but did not alter the spectral quality of the shade cast. Special paints or filters can be used to alter the colour as well as the quantity of light in the shade house. Thus, forest conditions can be imitated more closely, although the accuracy of the simulation is still not necessarily high.

Many experiments have been conducted where seedlings of different species were grown under a range of light conditions, usually spanning the irradiance regimes of large gaps to the forest understorey. I have summarised the results of some of these experiments by plotting growth irradiance as a percentage of full direct sun against seedling relative growth rate (Fig. 5.12). The general response to increased light availability is one of improved growth up to relatively high intensities (about 20% full sun), after which there may be

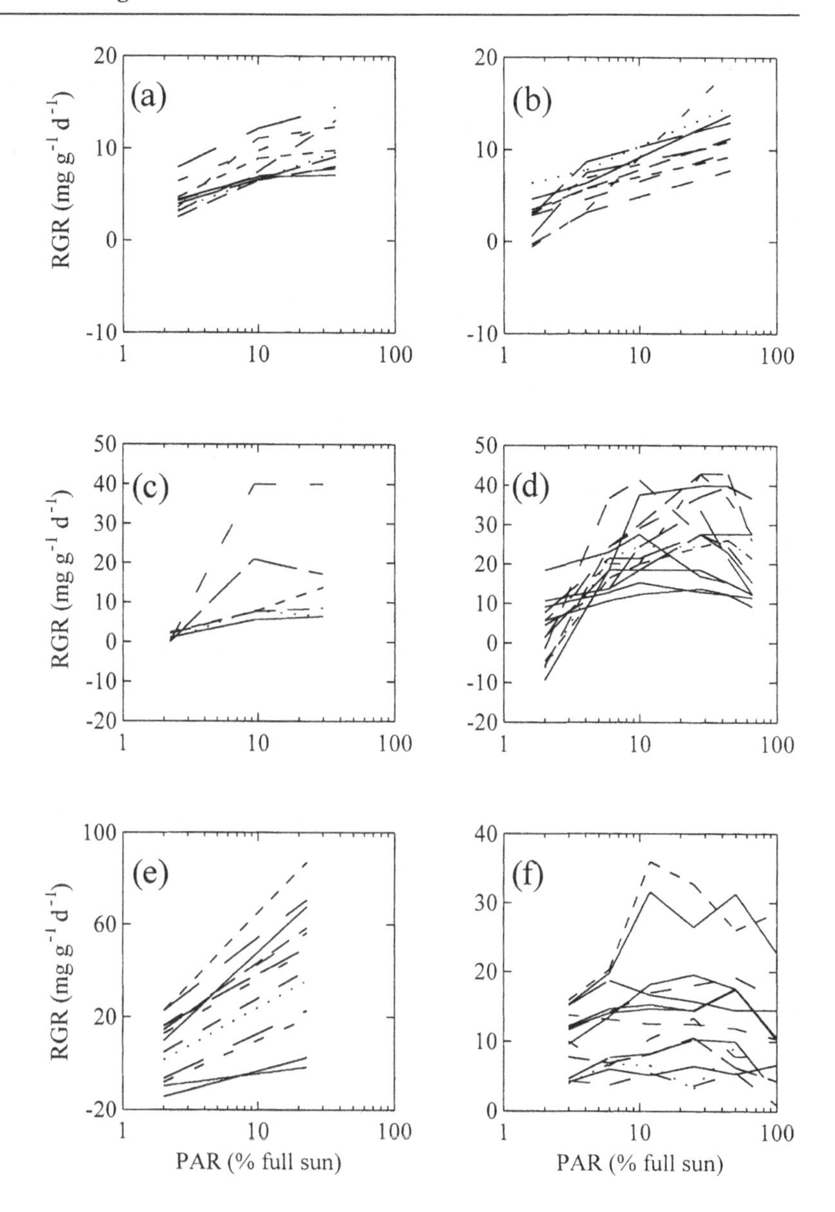


Figure 5.12 Summary of seedling growth responses to light availability (% full sun on a logarithmic scale). Each line represents a different species in the experiment. Data from (a) Osunkoya *et al.* (1994), (b) Popma & Bongers (1988), (c) Boot (1996), (d) Agyeman *et al.* (1999), (e) Kitajima (1994), (f) Poorter (1999).

levelling off, or even a decrease in performance. Most of the studies where growth declined at the highest light intensities included a treatment that involved a period of direct sunlight. Most rain-forest tree species decline in performance when exposed to such conditions. In many cases (see, for example, Kitajima 1994; Poorter 1999) there was little, if any, change in the relative performance of species between shade treatments within an experiment. Some species grow well in all light treatments, and others grow poorly in them all. Ecologists have often argued that tropical tree species may partition the forest environment by being different in their optimal conditions for growth. In terms of light, one might then expect changes in relative performance between species across different shade treatments. A very shade-tolerant species would do best in the deepest shade, a shade-intolerant in the brightest light, and intermediate species would take over at intermediate light intensities. The experimental studies reviewed here offer surprisingly little support for this concept. Of the studies summarised in Fig. 5.12, that of Agyeman *et al.* (1999) on 16 West African species showed the closest match to the predicted changes in relative performance of species across a light gradient. In the study by Popma & Bongers (1988), *Cecropia obtusifolia* switched from growing the fastest of 10 species in the high light treatment, to the slowest in the deepest shade. Denslow *et al.* (1990) found switches in relative performance between light treatments in rooted cuttings of three *Miconia* species. Another example is the study by Ashton (1995) of four species of *Shorea* from Sri Lanka. In this experiment three of the four species were the best performers in one or more shade treatment. Probable reasons for this greater variation in outcome in these compared to other studies include a relatively long time period of observations and a wide range of light treatments used including both deep shade (2% full sun or less) and some direct sun treatment.

It must be concluded that subtle differences in light availability (of say a few per cent above 3% full sun) are probably not sufficient to cause major shifts in relative performance between the seedlings of tropical tree species. The studies also show that tolerance of periods of direct full sun may also be an important factor differentiating between species.

### Seedling responses to shade

Responses in growth rate to the amount of light incident on a seedling are often reflected in a range of morphological and physiological changes. In general, seedlings show an increased allocation to leaves and reduced allocation to roots in deeper shade (Popma & Bongers 1988; Ashton 1995; Veenendal *et al.* 1996; Lee *et al.* 1997). Juvenile plants, of whatever architectural form, tend to become more spreading and less erect in shaded environments (King 1998a). Foliage display in saplings of three pioneer species growing in gaps in Mexico was found to be oriented towards intercepting the diffuse skylight rather than the direct sunshine (Ackerly & Bazzaz 1995a). The form of leaves is also responsive to shade. Seedling leaves in shade tend to be larger (Ducrey 1992; Poorter 1999), thinner (Ashton & Berlyn 1992; Strauss-Debenedetti & Berlyn 1994; Poorter 1999) and have a lower LMA (Fetcher *et al.* 1983; Veenendal *et al.* 1996; Davies 1998) and stomatal density (Fetcher *et al.* 1983; Ducrey 1992; Ashton & Berlyn 1992) than in higher irradiance. Leaf longevity may increase in the shade and leaf turnover rates reduce as leaf production drops and leaves last longer (Bongers & Popma 1990b; Ackerly & Bazzaz 1995b).

The relative growth rate (RGR) of seedlings increases with irradiance (King 1991b; Osunkoya *et al.* 1994; Veneklaas & Poorter 1998; Kitajima 1994) but may decline at the highest irradiances (Veenendal *et al.* 1996; Boot 1996). The allocation component of RGR (see Box 5.1), leaf area ratio (LAR), declines with increasing light (Fetcher *et al.* 1983; Popma & Bongers 1988; Thompson *et al.* 1992a; Osunkoya *et al.* 1994; Veneklaas & Poorter 1998) but net assimilation rate (NAR) increases more rapidly to achieve the increase in RGR with light. In their analysis of the combined data from many experiments, Veneklaas & Poorter (1998) concluded that in shade LAR was a better predictor of RGR than was NAR. In other words, what one can refer to as allocation pattern (in a broad sense to include consideration of leaf area) was more important than assimilation rates at determining growth rates in the shade. However, a number of individual experiments have shown NAR to be more influential on RGR than was LAR, even in the shade (Popma & Bongers 1988; Kitajima 1994; Boot 1996). But Kitajima (1994) did find that LAR was a better predictor of RGR than leaf-based measures of assimilation (Fig. 5.13). For 28 species with foliaceous cotyledons, LMA was the best predictor of LAR. Maintaining positive NAR in deep shade is requisite for survival under those conditions. Boot (1996) found that 'pioneers' could not maintain positive NAR in the shade and rapidly died.

Leaf-area-based maximal assimilation rates, saturating irradiances and dark respiration rates generally increase with growth irradiance in seedlings (Ramos & Grace 1990; Barker *et al.* 1997; Strauss-Debenedetti & Bazzaz 1991; Thompson *et al.* 1992b; Kitajima 1994; Davies 1998), although if very high irradiance treatments are included photosynthetic performance of some species will drop, as found by Lee *et al.* (1997). Mass-based measures of assimilation are less responsive to shade than area-based ones because LMA is generally affected more strongly than physiological processes. Compared with shade conditions, in high light young leaves are proportionately more important for carbon assimilation than old ones (Lehto & Grace 1994).

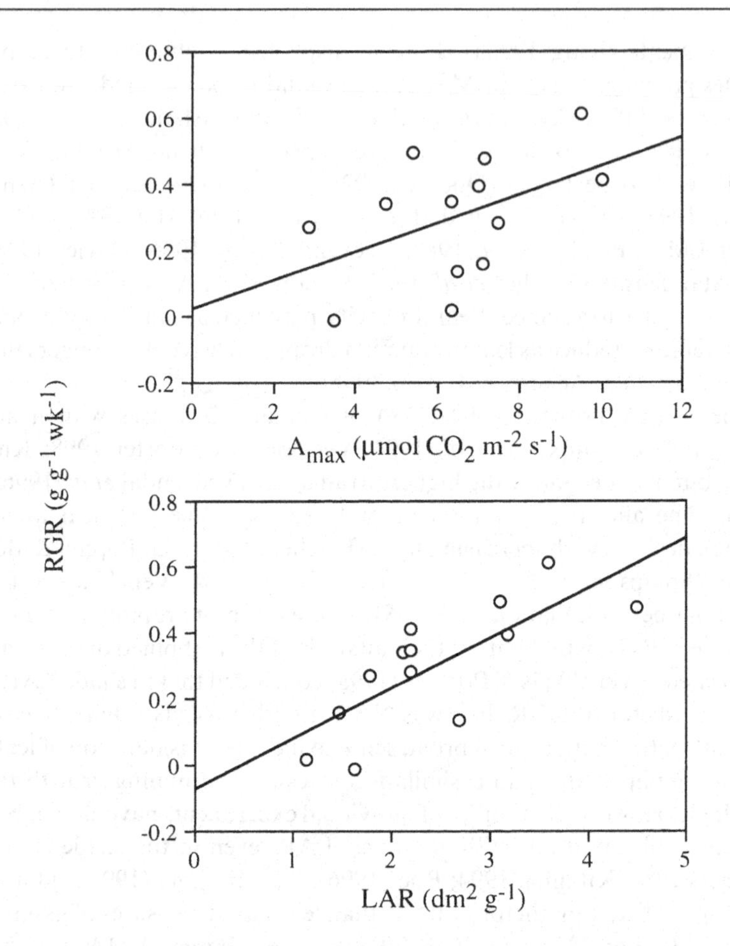


Figure 5.13 Maximal photosynthetic rate and LAR as predictors of seedling RGR for 13 species from Barro Colorado Island, Panama. After Kitajima (1994).

Leaves are the major respiratory drain on seedlings (Veneklaas & Poorter 1998), in terms of both their maintenance and their turnover, which is probably one contributory factor to the poor survival of seedlings with very high LAR in deep shade.

Careful studies of the influence of spectral quality on responses to shading have been performed on seedlings in Malaysia (Lee *et al.* 1996, 1997). Shade treatments matched for total irradiance, but with different R : FR, have been used on a range of species. These have shown that it is quantity rather than quality of PAR that has the main affect on growth and morphology of seedlings. Species were quite variable in their response to R : FR. Some showed very little response, others were quite strongly affected. Overall leaf area per unit stem length was the seedling character most influenced by spectral quality. Low R : FR often leads to internode elongation. A similar experiment on seedlings of three neotropical species (Tinoco-Ojanguren & Pearcy 1995) also showed some morphological responses to spectral quality (R : FR) but no major changes in steady-state or dynamic photosynthetic characteristics.

**BOX 5.1** Growth Analysis

Classic growth analysis is based on the equation:

$$\mathrm{RGR} = \mathrm{NAR} \times \mathrm{LAR}$$

which can be re-written as:

$$\mathrm{RGR} = \mathrm{NAR} \times \frac{\mathrm{LWR}}{\mathrm{LMA}}$$

where

RGR = relative growth rate
NAR = net assimilation rate
LAR = leaf area ratio
LWR = leaf weight ratio
LMA = leaf mass per unit area.

**Seedling shade tolerance**

The concept of shade tolerance is, like many in ecology, one that appears simple on first acquaintance, but slowly reveals a hidden complexity. Various approaches to defining shade tolerance in tropical tree seedlings have been proposed (Whitmore 1996; Kitajima 1994; Popma & Bongers 1988). These include the following.

*(a) Minimum irradiance for seedling survival*

More shade-tolerant species can survive at lower irradiances than less shade-tolerant ones. It is easy to demonstrate that certain species do die quite rapidly in deep shade but, for young seedlings, seed reserves may play an important role in surviving such conditions. Boot (1996) found that some large-seeded species could survive one year in complete darkness. Therefore, from a practical point of view, testing for minimum survival irradiance is difficult.

*(b) Length of time seedlings can survive in deep shade*

More shade-tolerant species are more persistent in deep shade. Under low irradiance a cohort of seedlings of a shade-tolerant species will reduce in numbers less quickly than that of a shade-intolerant species. The slope of the logarithmic decay for seedling populations grown in deep shade was used as an index of shade-tolerance by Augspurger (1984) and promoted by Kitajima (1994).

Points (a) and (b) are the assumptions behind using measures of average growth irradiance for juveniles in the forest as an index of shade tolerance. If the mean or median light climate for species A is brighter than for species B, then B is considered more shade-tolerant. The supposition is that the germinating seedlings started out in a similar range of light conditions, but shade-intolerant species die in the darker sites thus shifting their seedling average light conditions upwards. Alternative processes, such as light-responsive germination and rapid growth of seedlings in high light reducing the residence time of seedlings in the size-class in question, will add inaccuracy to this method, but it has been widely used (see, for example, Clark *et al.* 1993).

*(c) Growth rate in deep shade*

Popma & Bongers (1988) put forward the proposal that RGR in understorey shade conditions was the best measure of shade tolerance, arguing that ultimately growth was what mattered. Kitajima (1994) contended that survival in shade was more important than growth. This was because there would be little to be gained by growth if the seedlings eventually died, and seedlings that persist, even with negative growth, may eventually receive more light through new gap formation that will allow positive growth to begin again. The work of Kobe *et al.* (1995) on temperate forest saplings also adds weight to an emphasis on persistence. The feature of the most shade-tolerant species among the ten they studied was a lack of correlation between mortality and growth rate. Shade-tolerants could survive prolonged periods of zero growth.

*(d) Amount of PAR required in order to release a seedling from shade suppression and obtain rapid height growth*

The more shade-tolerant a species is, the less light it requires to achieve maximal growth rates. This concept has been widely applied in natural forest silviculture. It is believed that the degree of canopy opening necessary to accelerate the growth rate of juveniles is inversely proportional to the shade-tolerance of the species. This premise is based on, largely empirical, forestry research (Brown & Whitmore 1992).

I conclude that relative ability to persist in deep shade is the best measure of shade tolerance. This leads on to questions such as: what makes a species shade-tolerant? And what are the advantages of being shade-intolerant?

**The traits of shade tolerance**

The responses of seedlings to shading are similar across a wide range of species. Increased allocation to shoots, particularly leaves, the production of leaves of larger area but thinner laminas, and reduced dark respiration and maximal photosynthetic rates, make up the nearly universal shade response. Why, then, cannot all species adjust physiologically to cope with very low light availabilities? The answer seems to be that all but a few can. A small proportion of species appears to be unable to reduce respiration and leaf production and turnover rates to cope with deep shade. Such species enter a negative carbon budget, because high respiration and the inefficient turnover of leaves are not balanced by photosynthetic gains, and soon expire, although often they succumb to disease before they have respired all their resources. If the definition of shade tolerance as the relative ability of populations to persist in the deep shade of the forest understorey is accepted, then survival at low intensities of incident radiation becomes the key feature of shade tolerance. An ability to withstand damage and disease at low light availabilities is a more important trait of shade tolerance than maximising photosynthetic gains. Kitajima (1994) found that leaf gas exchange rate had few significant correlations to seedling survival in the shade for 14 species from Barro Colorado Island, Panama. Shade-tolerant species (ones more persistent in shade) had greater LMA, higher root : shoot dry mass and lower LAR in the shade than less tolerant species. This is exactly the opposite pattern to the general response to increasing shade. In other words, more shade-tolerant species show less morphological and physiological plasticity to shading than more shade-intolerant species (see, for example, Strauss-Debenedetti & Bazzaz 1991; Ducrey 1992; Strauss-Debenedetti & Berlyn 1994; Veenendal *et al.* 1996; Veneklaas & Poorter 1998). An exception to this pattern was noted by King (1991b), who found that saplings 1–2.5 m tall of species strongly associated with gaps varied little in shoot dry mass allocation to leaves with light conditions whereas shade-tolerant species increased allocation to leaves with reduced light availability.

Kitajima's explanation of the general pattern was that these are features necessary to improve survival in the deep shade. Shade-tolerant species allocate more to defence and long-term security than shade-intolerant species, which show greater plasticity in form and physiology. The shade-intolerant species end up with a highly 'risky' seedling design: very thin leaves are susceptible to herbivores, the limited root system results in sensitivity to drought, and so on. Evidence for a seedling survival versus growth rate trade-off was found among 14 pioneer species on Barro Colorado Island (Dalling *et al.* 1998a).

Shade-tolerant species, in a pattern similar to plants typical of other habitats poor in available resources, tolerate shade by harbouring their reserves, protect themselves from potential enemies and emphasise survival rather than maximal growth rates. The higher LMA of shade-tolerant species in the shade probably reflects greater leaf longevity (a longer time over which the leaf can make profits for the plant). The steeper slope for the correlation between leaf maximum assimilation rate and foliar nitrogen or phosphorus concentrations observed in shade-intolerant species (Strauss-Debenedetti & Bazzaz 1991; Raaimakers *et al.* 1995) may reflect the allocation of resources to processes other than photosynthesis, such as defence in shade-tolerant species. Pioneers typically have greater instantaneous assimilation rates per unit of nutrient in the leaf (see Chapter 2).

There may also be architectural differences between the two shade-tolerance categories. It has been argued that light-demanding species tend to be strongly orthotropic, with the leader shoot showing strong apical dominance, producing tall and thin seedling crowns. Shade-tolerants have a greater tendency to be plagiotropic, with spreading branches producing sprays of leaves and a wider crown. This dichotomy of form is exemplified by the seedlings of two dipterocarp species, *Dryobalanops lanceolata* and *Shorea leprosula* (Fig. 5.14). Kohyama (1987) named the species following these architectural strategies optimists and pessimists, respectively. This is because the orthotropic species behave in an optimistic manner, emphasising height growth into higher light conditions. The plagiotropic species are pessimistic, spreading sideways to capture more light and allow them to persist in the shade. However, Kohyama & Hotta (1990) found that tropical seedlings showed a large diversity in architecture and the pessimist–optimist dichotomy may be too simplistic to apply there.

A number of studies have reported that wood density (usually adult wood density obtained from references on timber properties) is a positive correlate of shade tolerance in juveniles (Augspurger 1984; Whitmore & Brown 1996). There are several possible reasons for this. Very fast-growing species rarely have dense wood, probably because such growth rates could not be achieved if more resources went into the wood. Dense wood may be of positive value to plants in the understorey. Wood density is correlated with resistance to decay by fungi, and probably also to wood-boring insects. Wood strength is correlated with density and the understorey is a place where wood strength may mean the difference between death and survival when a falling branch lands on a seedling.

Figure 5.14 Seedling architecture of *Shorea leprosula* and *Dryobalanops lanceolata*. *Dryobalanops* shows preferential resource allocation to the growth of a few long branches. Individuals shown have the same age and total branch length. After Zipperlen & Press (1996).

The denser, stronger wood of the understorey species would allow them to have narrower stems for a given loading. However, the generally more spreading crowns of the juveniles of understorey species might mean that they would need proportionally thicker stems for a given sapling height than large-statured species. These two factors acting in opposite directions may help explain some apparent anomalies obtained in allometric analyses of sapling form for tropical rain-forest tree species. King (1994) studied the allometric relationships of six species in Panama and found that the two species of smallest adult stature had significantly thicker stems with wider, leafier crowns at 2.5 m height than the rest. Both Thomas (1996a) and Davies *et al.* (1998) reported that saplings of species with low asymptotic height tended to have narrower stems as saplings than those of species of greater maximal size. The latter study concerned 11 species of the mostly strongly light-demanding genus *Macaranga*, where the small-statured shade-tolerants were unbranched treelets with narrow, rather than wide, crowns. Wood density differences may have allowed the small-sized species studied by Thomas not to differ significantly in trunk (*H–D*) allometry at the sapling stage.

The orthotropic unbranched, or sparsely branched sapling architecture seems to have advantages when it comes to rapid height growth. Coomes & Grubb (1998b) found that the juveniles of tall-tree species with this architecture had the fastest relative height growth rates in tree-fall gaps of species studied in the caatinga forest of Venezuela. The erect unbranched form allows greater height increment per unit increment in dry mass (King 1994).

### On being light-demanding

It has generally been assumed that the advantage of shade intolerance is faster growth at high light intensities. When considered across the complete shade tolerance spectrum there is evidence for a tradeoff between growth rate and survival (Kitajima 1994; Dalling *et al.* 1998b; Veneklaas & Poorter 1998). However, the view that shade tolerance and slow maximal growth rates, or the reverse, fast maximal growth and shade intolerance, are necessarily strongly coupled has been questioned (Grubb 1996). Over a narrower range of shade tolerance there may be a greater scatter of maximal growth rates. For example, among six large-tree species at La Selva there was no correlation between survival and growth rate (Clark & Clark 1992). However, the general trend appears to be one of faster growth for shade-intolerant species, but with a spread of points about any regression line.

### The relevance of shade tolerance

That species differ in shade tolerance cannot be doubted, and that the species lying at the two ends of the axis differ considerably is readily demonstrated, but what about small differences? Is a difference of a few positions in the ranking of shade tolerance of species across the community sufficient to provide separate ecological niches for the species concerned? We are still a very long way from being able to answer such a question definitively, but available evidence points to little likelihood of this being the case. Many of the seedling growth studies find that there is little or no change in the rank of seedling performance between species across a range of light climates (Veneklaas & Poorter 1998). Relatively few experiments have clearly demonstrated a switch in best performance between one species and another under different shade treatments. The study of Whitmore & Brown (1996) of dipterocarp seedlings in different sized gaps in the Danum Valley, Sabah, Malaysia, is a case in point. An experiment was initiated through making canopy gaps of various sizes over natural populations of dipterocarp seedlings. This was done in order to see which species would perform best in which gap size, and whether the received wisdom of foresters was correct in predicting that the light-demanders would do better in the large gaps and the more shade-tolerant in the small gaps. There were three common species among the seedlings, the shade-tolerant (as judged by observation of foresters and wood density) *Hopea nervosa* and the light-demanding *Parashorea malaanonan* and *Shorea johorensis*. Over the first 40 months of the observations initial seedling

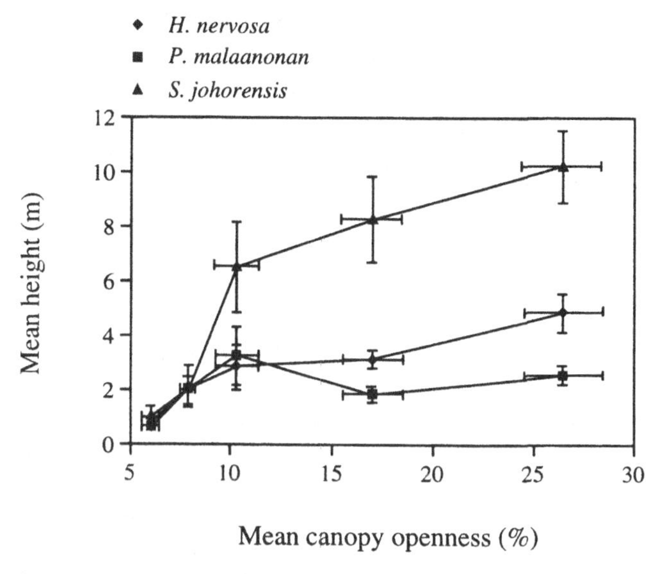


Figure 5.15 Mean height of all individuals of the three common species in gaps of increasing size at 77 months at Danum Valley, Sabah, Malaysia. Only seedlings that survived the full 77 months of the experiment are used to calculate means. Error bars represent 95% confidence intervals around the mean values. After Whitmore & Brown (1996).

size at gap formation was the most reliable predictor of seedling performance across all the gaps. This meant *Hopea nervosa* was doing best as it had larger seedlings initially. Being shade-tolerant, its seedlings could persist and grow slowly in the forest understorey, allowing them a head start compared with the light-demanders when the gaps were formed. Clearly, this head start is maintained for a long period. However, by 77 months the light-demanders, particularly *Shorea johorensis*, had caught up and overtaken the *Hopea nervosa* saplings (Fig. 5.15). *Parashorea malaanonan* was handicapped in its performance by a susceptibility to herbivores. Thus it took more than six years for the predicted differences in shade tolerance to come into effect. It is important to note that being overtopped by the light-demanders may not mean failure on the part of the shade-tolerants. It may represent another phase of persistence in the shade until the light climate improves again. The comparatively long span of time it takes for differences in shade tolerance to become apparent in the relative performance allows time for other factors to come into play. In the example from Sabah, the influence of herbivores prevented the faster-growing *Parashorea malaanonan* exploiting its inherent advantage in the larger gaps. Dalling *et al.* (1998b) found no evidence of gap-partitioning by size among pioneer species on Barro Colorado Island. Brown & Jennings (1998) concluded that the expected niche differentiation of tropical tree species by gap size has not been substantiated. They argued that this might reflect technical problems with categorising gap light climates. The lack of experimental evidence for major changes in relative performance between species for small changes in illumination would seem to me to make it unlikely that more precise measurements of light *in situ* would alone be more revelatory in terms of comparative ecology.

**Sudden changes in light availability**

Any point in the forest is subject to considerable variation in the quantity and quality of photosynthetically active radiation it receives. The forest understorey probably shows the greatest short-term magnitude in variation, because of the occasional bright sunflecks set against the generally low irradiance of the forest interior. Despite typically being of short duration, their brightness means that sunflecks contribute a large proportion of the total daily irradiance. Physiological investigations have been conducted to investigate the ability of plants growing in the forest to exploit sunflecks. What has generally been found is that a leaf that has not received a sunfleck for a matter of hours is quite slow to respond to a sunfleck and does not achieve very high maximal photosynthetic rates. However, if another sunfleck occurs soon afterwards, the performance in the second sunfleck is considerably better than in the first. In effect, the initial sunfleck induces a state of better readiness for subsequent ones, but the induced state gradually declines over time. Among the limited studies of photosynthesis under rapidly changing light in tropical tree seedlings there is no clear evidence that shade-tolerant species are better at exploiting sunflecks. Zipperlen & Press (1997) compared the photosynthetic performance of *Dryobalanops lanceolata* and *Shorea leprosula* seedlings under sunfleck conditions at both low and high background irradiances. The more light-demanding *Shorea leprosula* actually showed more rapid induction to sunflecks, higher maximal photosynthetic rates during them, and longer maintenance of the induced state. *Dryobalanops lanceoalata* maintained a higher degree of induction over the short term, so with its lower dark respiration rate, it probably made more efficient use of rapidly fluctuating light regimes. Similar results were found in a comparison of small saplings of the shade-tolerant *Dipteryx panamensis* and the shade-intolerant *Cecropia obtusifolia* in Costa Rica (Poorter & Oberbauer 1993). Stomatal conductance increases sigmoidally when a sunfleck hits a leaf and slowly declines thereafter (Chazdon & Pearcy 1986; Tinoco-Ojanguren & Pearcy 1992; Zipperlen & Press 1997). The stomata respond such that a relatively constant internal partial pressure of carbon dioxide is maintained.

However, in some cases oscillations in stomatal conductance have been observed after a sunfleck (Zipperlen & Press 1997). This is thought to be due to over-compensation by the stomata, which leads to a damped oscillation in the rate of carbon fixation.

Another major change in light climate, but over a different temporal scale, is gap formation. If the tree shading juveniles dies, or is blown over, or loses a limb there can be a sudden large increase in the daily totals and the maximum instantaneous solar radiation received. High direct sunlight poses two problems to these seedlings. Firstly, the massive increase in the radiation load on the leaves is likely to lead to much higher leaf temperatures, particularly if transpiration cannot rise sufficiently to dissipate the extra heat absorbed. Secondly, the high fluxes of light are liable to cause photoinhibition of photosynthesis in the seedlings. The seedlings of many tropical rain-forest tree species do not grow well in open conditions. Saplings of the shade-tolerant dipterocarp *Neobalanocarpus heimii* showed lower maximal photosynthetic rates in an open site than in a gap site, whereas those of *Shorea leprosula* and *Macaranga gigantea* increased under full sun (Ishida *et al.* 1999a).

Leaf scorching and bleaching, and often leaf drop, are typical symptoms of gap formation over previously shaded seedlings. Shade-acclimated *Shorea curtisii* seedlings exhibited leaf scorch within a few days of being moved into large-gap conditions (Turner & Newton 1990). During artificial gap formation in dipterocarp forest in Borneo it was found that small seedlings suffered more extensive exposure damage than large ones (Brown & Whitmore 1992). This was probably because the self-shading achieved by the large seedlings protected some of the leaves from direct insolation. The small seedlings were not self-shading. The key to acclimation by juveniles to a more brightly lit microclimate is the growth of new leaves of a more 'sun-type' physiology (Newell *et al.* 1993). Seedlings of early-successional species acclimate more quickly than late-successional ones (Fetcher *et al.* 1987) because of faster leaf turnover.

The alternative change of an increase in degree of shading, such as when canopy trees grow outward and fill up a gap, may also be physiologically disturbing to a plant. In experiments where seedlings were moved factorially between light treatments it has been found that movement from high to low light required the longer acclimation time than a stepwise increase in illumination (Popma & Bongers 1991; Osunkoya & Ash 1991). A sudden reduction in light availability means that leaves with high respiration rates have to operate at below the compensation point for much of the time. Sun leaves are often shed in this circumstance as otherwise they would remain a carbon drain on the plant.

### Water

The tropical rain forest is, as its name suggests, a generally wet place, and not one where studies of plant water relations readily spring to mind. However, many tropical rain forests do have dry seasons, and even the wettest of climates are liable to dry spells at intervals. During a dry spell small seedlings, with their limited root systems, are probably the woody plants most susceptible to drought. Reviewing the literature on factors influencing the species composition of lowland tropical rain forest, Sollins (1998) drew the conclusion that soil physical properties were most commonly correlates of community variation. Physical properties are often related to ease of drainage of a soil. Tree species in a forest often appear to specialise in microhabitats related to soil drainage, e.g. freely draining sites, or those possessing a tendency to waterlogging. The exclusion of species from sites of the wrong drainage type may operate through seed and seedling mortality. Therefore investigations of relative tolerance of drought or waterlogging in juvenile stages may be highly relevant to understanding the ecological interactions of species in the rain forest.

The limited number of investigations of tropical tree seedling drought tolerance shows a very varied pattern of water use and response to water shortage between species. There was a twofold range in leaf water-use efficiency among seedlings of ten species of dipterocarp (Press *et al.* 1996). Droughting seedlings by using controlled watering and competing grass plants produced a range of physiological responses among three species in Singapore (Burslem *et al.* 1996). However, all three (*Antidesma cuspidatum*, *Hopea griffithii* and *Vatica maingayi*) showed a reduction in total leaf area, an increase in LMA and relative allocation to roots (particularly fine lateral roots) under simulated drought compared with control plants. Droughted *Hopea* seedlings managed to maintain shoot water potentials quite close to the control plants, whereas the sparingly watered *Vatica* seedlings showed markedly more negative water potentials than the controls. Reekie & Wayne (1992) subjected the seedlings of three 'pioneer' tree species to 5 d droughting cycles. *Piper auritum* and *Cecropia obtusifolia* showed little physiological response to droughting until the pot soil water content dropped to about 20%. At this point stomatal conductance, photosynthetic rates and canopy display area fell dramatically. *Trichospermum mexicanum* showed a more gradual response to declining soil water content. The first two species are heavily dependent on the turgor of foliage to maintain canopy display. This may explain the physiological behaviour observed in these species.

Huc *et al.* (1994) used a carbon-stable-isotope technique to estimate the average leaf internal partial pressure of carbon dioxide ($c_i$) and long-term water-use efficiency of saplings of five species growing in plantation stands in

French Guiana. They calculated that the three 'pioneer' species had significantly higher average $c_i$ but lower water-use efficiency because of generally higher stomatal conductance than the two 'late-stage' species. The pioneers had higher specific hydraulic conductance that allowed them to maintain low water potentials in the leaves and keep up high rates of transpiration and hence carbon fixation. The susceptibility to cavitation brought about by the high hydraulic conductance was mitigated by sensitive stomata that closed when the soil water content started to drop.

Tyree *et al.* (1998b) studied the hydraulic conductance of seedlings of five neotropical tree species. The conductance of all the species increased in proportion to the total leaf area of the growing seedlings such that the leaf specific conductance remained relatively constant. Two 'pioneer' species had finer root systems than three 'shade-tolerant' ones, with a greater root area and root length per unit root-system dry mass. However, this difference among species seemed to diminish as the seedlings got bigger. The pioneers had higher root and shoot hydraulic conductances per unit leaf area (also found by Becker *et al.* (1999b) for saplings in Brunei). They were therefore able to maintain less negative water potentials for a given transpiration rate and a higher conductance per unit dry mass of investment. The pioneers appear to be designed for rapid uptake and transport of water to maintain high transpiration. The cost may be a root system susceptible to below-ground herbivores (few woody roots) and a xylem system liable to cavitation.

Excess water is common in many tropical forest types, and most dryland forests will contain hollows where water collects and which will contain waterlogged soils during wet periods of the year. At the extreme certain Amazonian forests are completely submerged, foliage and all, during the flood peaks. Many tropical tree species must therefore be tolerant of waterlogging. Little research has been done on this, but adaptations must include an ability to respire anaerobically in the root system and a tolerance of species of ion such as $Mn^{2+}$ that are normally toxic to unadapted plants. In Guyana, *Mora excelsa* is characteristically associated with forest sites with a high water table and gleyed or mottled soils, whereas its congener, *M. gongrijpii*, is rare in such sites. The seeds and seedlings of *M. excelsa* were shown to be more tolerant of prolonged flooding than those of *M. gonggrijpii* (ter Steege 1994b). The latter grew faster under favourable conditions, indicating that tolerance of waterlogging may have a physiological cost.

### Mineral nutrients

Seedlings are generally the subjects of studies on the mineral nutrition of tropical trees. There appears to be a relationship between light-demandingness and growth response to increases in soil nutrient availability.

Typically fast-growing species have been shown to respond readily to increased soil fertility by faster growth, but more shade-tolerant species often do not respond (Turner *et al.* 1993a; Burslem *et al.* 1995; Raaimakers & Lambers 1996; Veenendaal *et al.* 1996), although positive results have sometimes been reported (see, for example, Gunatilleke *et al.* 1997). Seedlings not showing a growth response to increased nutrient availability may still take up more nutrients as reflected in higher concentrations in plant parts (Fig. 5.16) (Raaimakers & Lambers 1996; Veenendaal *et al.* 1996). It may be that shade-tolerant species that are adequately mycorrhizal have a sufficient

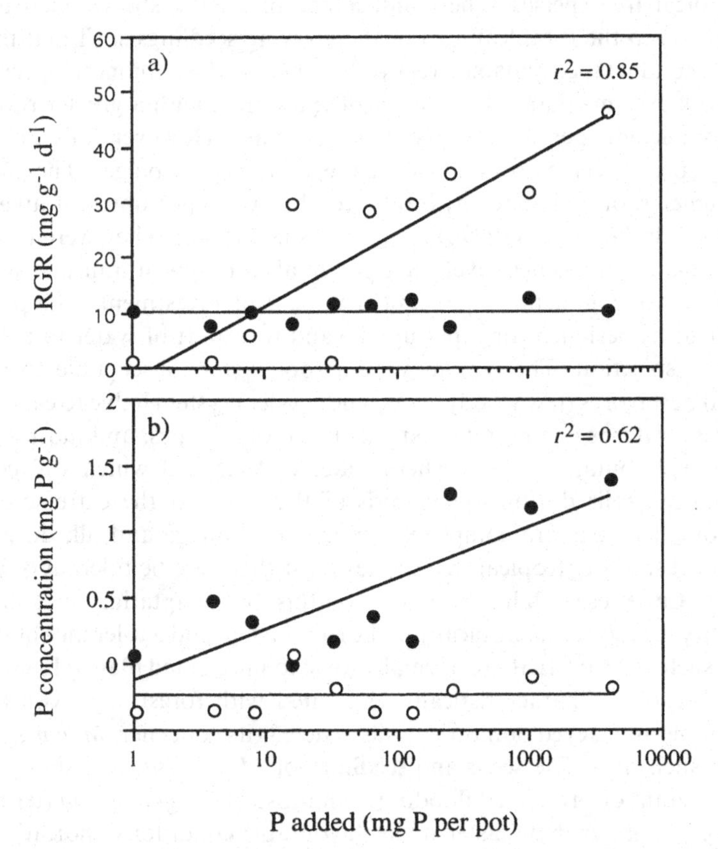


Figure 5.16 Relation between (a) relative growth rate (RGR, mg $g^{-1} d^{-1}$) and (b) P concentration (mg P $g^{-1}$ (plant dry mass) and P added to the plants (mg), for *Tapirira obtusa* (open circles) 192 days after germination, and *Lecythis corrugata* (closed circles) 220 days after germination. The RGR of *L. corrugata* and P concentration of *T. obtusa* showed no significant relation with the amount of P added per pot; the lines represent their means. After Raaimakers & Lambers (1996).

supply of nutrients and do not need any more to maintain their fairly limited growth rates.

There have been some dramatic demonstrations of the importance of mycorrhizas for growth of seedlings. Alexander *et al.* (1992) showed that adding mycorrhizal inoculum to potted seedlings of *Paraserianthes falcataria* and *Parkia speciosa* had a very marked effect on seedling growth, much more so than adding a large dose of phosphate to the growing medium. Janos (1980) found that potted seedlings of 23 out of 28 species (more than half trees) from La Selva, Costa Rica, showed significant positive growth responses to inoculation with a source of VAM fungi. Seedlings of *Beilschmiedia pendula* inoculated with VAM showed enhanced RGR in comparison with uninoculated plants (Lovelock *et al.* 1996). However, seedlings in the forest are nearly always infected with mycorrhizas. Light availability, and other factors, may influence the degree of mycorrhizal infection and the fungal symbiont involved, as has been demonstrated for ectomycorrhizal seedlings of *Shorea parvifolia* in Borneo (Ingleby *et al.* 1998).

A major implication of research on mycorrhizas is that experimental manipulations of soil fertility conducted on seedlings in pots are only realistic if the seedlings are infected with the mycorrhizas typical of the field situation.

Morphological responses to increased nutrient availability generally include a reduced dry mass allocation to roots, an increase in LAR and a reduction in LMA (Veenendal *et al.* 1996). Leaf turnover tends to increase as leaf production rate goes up and leaf longevity declines (Ackerly & Bazzaz 1995b).

Using a liquid culture technique, Osaki *et al.* (1997) were able to demonstrate a positive response to aluminium addition to the nutrient medium in the seedling growth of the aluminium accumulators *Melastoma malabathricum* and *Melaleuca cajuputi*. Growth was poor without Al in the medium. *Melastoma malabathricum* may have a physiological requirement for the element, although Osaki *et al.* (1997) hypothesised that the apparent need for Al may stem from the requirement for aluminium–phosphorus precipitates at the root surface to allow phosphorus uptake.

There is some circumstantial evidence that magnesium availability may sometimes limit seedling growth in forest soil (Burslem *et al.* 1996, for the Al-accumulating *Antidesma cuspidatum*; Gunatilleke *et al.* 1997). It is possible that competition with aluminium ions is what limits uptake of magnesium from the soil by the plant. Osaki *et al.* (1997) found that *Melastoma malabathricum* in liquid culture took up less Mg and Ca when grown at high Al. However, further studies showed no interaction between growth responses to Al and Mg in liquid culture (Watanabe *et al.* 1997). It therefore seems less likely that it is abundance of free Al ions that is producing the Mg response.

**Defence in juveniles**

In what has become a classic study, Coley (1983) compared rates of herbivory on saplings 1–2 m tall of 46 species growing in gaps on Barro Colorado Island, Panama. An index of leaf toughness (ability of the lamina to resist penetration by a blunt rod) was the best negative correlate of rate of leaf area loss across the sample. The concentration of phenolic compounds was not correlated with herbivory rate. Mature leaves of the 22 'pioneer' species in the sample were grazed, on average, at six times the rate of the 24 'persistent' (shade-tolerant) species. The pioneers had significantly shorter leaf life spans on average (6.8 months, range 3.9–21 months) than the persistents (21.7, 7.3–35 months) (Coley 1988), with less tough leaves containing less fibre and phenolic compounds but more nitrogen and water (Coley 1983). Coley (1988) found a positive correlation between leaf life length and abundance of quantitative defences, notably condensed tannins and fibre concentrations in leaves of the saplings on Barro Colorado Island (Fig. 5.17).

In an experiment on seedlings of the fast-growing *Cecropia peltata*, Coley (1986) found that the foliar concentration of condensed tannins was negatively correlated with rate of herbivory when compared between individuals. However, there was a cost to tannin production: a reduced growth rate. Rates of leaf production were negatively correlated with tannin concentration. In a similar study, using rooted cuttings of the shade-tolerant shrub *Psychotria horizontalis* (Sagers & Coley 1995), it was found that plants with tough leaves and high tannin concentrations suffered less herbivory. Growth was negatively correlated with tannin concentration but did not show a significant relationship with leaf 'toughness'.

Small seedlings may be more heavily defended chemically than large ones. Coley (1987) reported a 20% reduction in mean condensed-tannin concentrations in saplings of *Cecropia* after one year of growth, and significantly greater palatability to invertebrate herbivores of leaves from large saplings of eight *Cecropia* species in Peru, than those from small saplings. Presumably there is a switch from chemically to physically based defences with ontogeny.

Fast-growing species may be able to compensate for losses to herbivores by regrowth rather than investing heavily in defence. Lim & Turner (1996) demonstrated that small seedlings of *Trema tomentosa* and *Macaranga heynei* subjected to a 50% loss of leaf area had no significant difference in dry mass from control individuals after 9 weeks. Fast growth and high rates of leaf turnover are probably important factors in this rapid recovery. A third species, *Dillenia suffruticosa*, could not compensate. Larger seedlings of *Heliocarpus appendiculatus* growing in a large gap at Los Tuxtlas did fall behind control plants in growth when artificially defoliated at levels as low as 25% (Núñez-Farfán & Dirzo 1991). The defoliation treatments also reduced

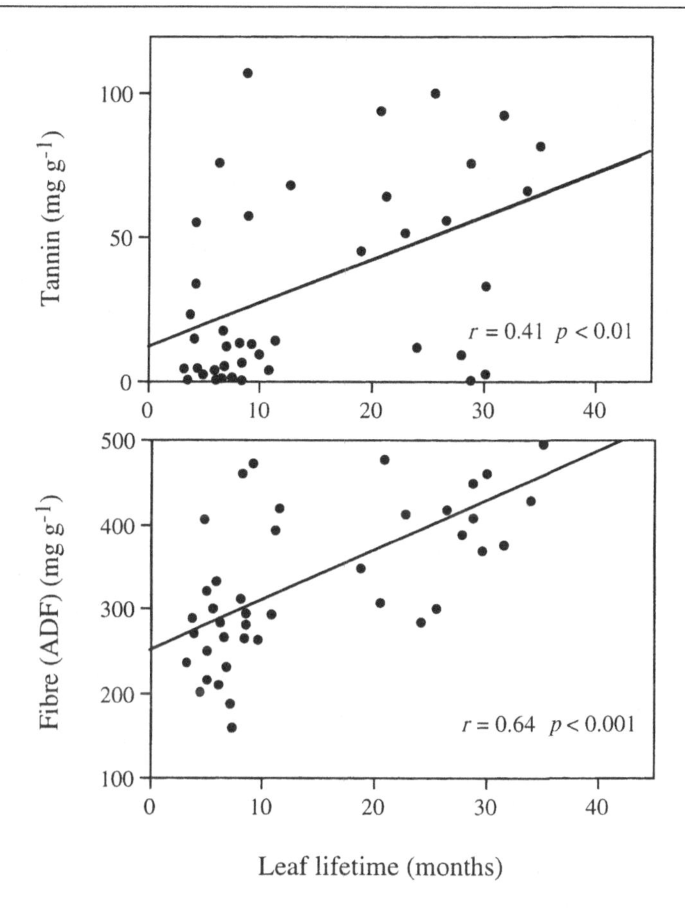


Figure 5.17 Fibre and condensed-tannin concentrations as a function of leaf lifetimes for tree species on Barro Colorado Island, Panama ($n = 41$). Fibre content was measured by using the acid–detergent method and tannin content by using the BuOH–HCl assay for proanthocyanin-based tannins. After Coley (1988).

seedling survival. Seedlings of two *Shorea* species 21 months old were more resistant to defoliation in terms of survival (Becker 1983); 25% leaf-area loss did not increase mortality, and complete defoliation was required significantly to reduce survival. The more shade-tolerant *Shorea maxwelliana* seedling cohorts survived for longer after 100% defoliation than those of *S. leprosula*, possibly because of lower respiration rates.

It is interesting to compare the results of two studies of the influence of light and soil fertility on allocation to chemical defence in seedlings of tropical trees. In one study, Nichols-Orians (1991) investigated tannin concentrations in leaves of the Central American legume *Inga oerstediana*. In the other, Höft *et al.* (1996) measured foliar alkaloid content of *Tabernaemontana pachysiphon*, a small tree from Africa. Increased soil nutrient availability led to reduced tannin concentrations in the *Inga* leaves, but increased alkaloid concentrations in those of *Tabernaemontana.* Conversely, increased light availability increased tannin content in *Inga* foliage, but reduced alkaloids in that of *Tabernaemontana.* It is the nitrogen content of alkaloid molecules that probably provides the explanation to this paradoxical pattern of defence investment. Under high-light conditions plant growth is likely to be limited by nutrient availability and there will be excess photosynthate to devote to carbon-based defences such as tannins, but nitrogen will be required for growth, so allocation to alkaloids should be reduced. At high soil fertility carbon assimilation will tend to limit growth, so there will be nitrogen available for alkaloid production, but less carbon will be available for tannins. This control of allocation by physiological processes related to growth limitation has been referred to as the carbon–nutrient balance hypothesis (Bryant *et al.* 1983). The results of the two studies referred to above support the hypothesis, as do similar experimental manipulations of degree of shade where foliar condensed tannin concentrations were measured in seedlings of *Cecropia peltata* (Coley 1987) and rooted cuttings of two *Miconia* species (Denslow *et al.* 1990). However, there is one further observation of interest to consider from the study of Höft *et al.* (1996). The *Tabernaemontana* seedlings had higher alkaloid concentrations at high nutrient availability, but allocated a higher proportion of leaf nitrogen to alkaloids at low soil fertility. This would seem to indicate that materials used for defence are not entirely surplus to other requirements.

Folgarait & Davidson (1994, 1995) studied the influence of light and soil nutrient availability on seedlings of six Peruvian myrmecophytic *Cecropia* species. In all species, more light increased the dry-mass production of glycogen-rich Müllerian bodies (but not lipid-rich pearl bodies) and foliar phenolic concentrations. High nutrient availability led to increased production of ant-food bodies but reduced leaf tannin concentrations. The three species characteristic of riverbanks and landslides grew faster than the three species typical of tree-fall gaps in the forest. The former tended to invest less in defence (fewer Müllerian bodies per unit leaf area, and less tough leaves) and initiated myrmecophytism later in development.

In conclusion, studies on seedlings and saplings of tropical trees were influential in the development of current theories of plant defence. In general, juveniles of pioneer species are defended less well than those of climax species, particularly in terms of leaf toughness and fibre concentration. They are also more attractive to herbivores and pathogens because of their high nitrogen concentrations. The strong emphasis on growth in the physiology of pioneers favours low investment in defence.

# 6

# Classificatory systems for tropical trees

## Height at maturity

Forest ecologists have generally understated the rather obvious distinction between species in their height at reproductive maturity. Species are often divided into height classes in ecological analyses, but it seems more with the purpose of comparing like with like within the stature groups, than of making comparisons among the groups. However, some general trends do emerge from the literature, and these are summarised in Table 6.1 as a series of characteristics of small-statured species in comparison to those of larger size at maturity.

The factor that appears to determine mature height is the size at onset of reproduction. Small-statured tree species start reproduction at smaller size (Thomas 1996b; Davies & Ashton 1999). Thomas (1996b) found this both in absolute terms and in size relative to asymptotic height. Allocation to reproduction probably requires a compromise in height growth rate, and hence these trees are left behind in the height growth race by the taller-growing species. Thomas (1996b) found that the change in the slope of the *H–D* regression from linear to asymptotic for a species generally coincided with the onset of reproduction.

Reproduction at small size need not necessarily imply reproduction at an earlier age. A very shade-tolerant understorey tree may grow very slowly and so be as old as, if not older than, a canopy tree that grew up in a gap. However, it is likely that on average small-statured trees do reproduce earlier in life than large-statured ones. Certainly, within a species, the time taken to reach reproductive size is likely to be proportional to that size. So an individual with genes for delayed reproduction might reach a larger size before it started to reproduce. The advantage of doing so would probably be to increase reproductive output per reproductive event because of larger size and increased light availability, but in comparison to other individuals in the population it would have foregone several episodes of reproduction to reach larger size before it began. If the risk of mortality is relatively high in the understorey, as it probably is because of falling debris and scarce resources,

Table 6.1. *Characteristics of small-statured species of tropical rain-forest trees in comparison with large-statured ones*

| Characteristic | Method of comparison | Total number of species included in study | Reference |
|---|---|---|---|
| denser wood | negative correlation between stem peripheral wood density and asymptotic height | 38 | Thomas (1996a) |
| earlier reproduction | positive correlation between both absolute and relative height at onset of reproductive maturity and asymptotic height | 37 | Thomas (1996b) |
| | positive correlation between absolute (but not relative) height at onset of reproduction and asymptotic height | 11 | Davies & Ashton (1999) |
| smaller seeds | negative correlation between seed mass and asymptotic height | 11 | Davies & Ashton (1999) |
| slower growth | positive correlation between average adult growth rate and asymptotic height | 38 | Thomas (1996a) |
| lower maximal assimilation rate | positive correlation between $A_{max}$ for saplings growing in the shade and asymptotic height | 38 | Thomas & Bazzaz (1999) |
| more shade tolerant | positive correlation between average crown illumination index and asymptotic height | 11 | Davies & Ashton (1999) |
| higher mortality rate | species from smaller adult size classes had significantly higher mortality for stems 1–9.9 cm dbh | 194 | Condit *et al.* (1995) |
| | species from large adult size classes had significantly lower mortality for stems ≥ 10 cm dbh | 128 | Condit *et al.* (1995) |
| | mortality of stems ≥ 10 cm dbh for understorey species higher than for larger size classes | 22 | Korning & Balslev (1994) |

then delaying reproduction may risk death before reproduction can begin. However, if the delay in reproduction brings proportionally greater rewards, that is improved total lifetime reproductive output for the individual, then continued growth will be favoured. An emphasis on growth may also allow better survival in the understorey because of a quicker passage time through the shadier, lower layers of the forest. There is evidence that understorey species do suffer higher average rates of mortality than those of larger stature (Table 6.1), even when stems of the same size class are compared. This probably indicates that reproduction is in itself a survival risk. For instance, the production of a superficial root system in small-statured species may, as Becker & Castillo (1990) suggest, provide effective nutrient uptake for reproduction but compromise the ability of those species to obtain water deeper in the soil during dry periods and hence reduce the probability of survival.

If species of small stature do have higher mortality rates than those of typically larger size, then for all species in the forest to maintain roughly uniform population densities through time the rate of recruitment to populations of small-statured species must be higher than to those of greater mature height. In other words, species of the understorey are predicted to have more dynamic populations than those of the canopy. Kohyama (1993, 1994) used computer models based on one-sided competition – bigger individuals suppress smaller ones through competition for light – to demonstrate stable co-existence in multi-species stands founded on a fundamental tradeoff between maximal height and reproductive output such that per capita recruitment of smaller species was greater than large ones. His forest architecture hypothesis predicts that:

1. Taller forests can support more species.
2. Faster demographic rates allow equilibrium to be reached more quickly.
3. High rates of tree growth allow more species to live together.

These predictions fit patterns of forest diversity on a large scale; for example, tropical rain forests are more diverse than typically shorter, temperate ones. But there is still much scope for testing the applicability of the model to real forests rather than to those generated by computers. Kohyama (1996) has demonstrated the predicted maximal height–recruitment tradeoff for species in the temperate rain forests of Yakushima, Japan (Fig. 6.1), but there has been no explicit test of this hypothesis in tropical forests. It is possible to look for circumstantial evidence of the tradeoff in other published results. A definite negative trend between recruitment rate and projected life span based on growth simulation (Fig. 6.2) was found for species from La Selva, Costa Rica (Lieberman *et al.* 1985a). However, visual inspection of recruitment rate data from Australia (Connell *et al.* 1984) (Fig. 6.3) and Malaysia (Okuda *et al.* 1997) gives no hint of greater rates of recruitment for understorey species. Favrichon (1994) found virtually no correlation between per capita recruitment and maximum diameter or height for 122 species from French Guiana. A confounding factor in the success of recruitment may be species' relative abundance if compensatory mortality is occurring. This will need to be taken into account in any test of the forest architecture hypothesis.

It is tempting to link stature with shade tolerance, and imply that understorey species are more shade-tolerant than are those of the canopy. This is true to the extent that almost inevitably large individual trees will be less shade-tolerant than small individuals because large trees have a greater proportion of non-photosynthetic tissue and hence a greater respiratory load. In the forest, the increased light availability with height means that this physiological handicap is not a drawback. Therefore, when comparing species in shade tolerance we need to compare individuals of similar size. The understorey species will be accompanied by juveniles of the canopy trees that in theory could be of equal, if not greater, shade tolerance. However, if the

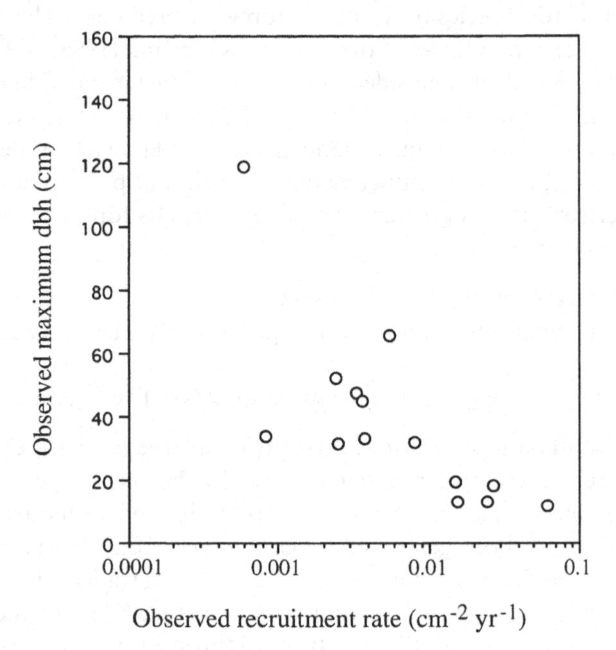


Figure 6.1 Relation between the per capita recruitment rate (the observed recruitment rate divided by the basal area of the species) to dbh of 2 cm and above and the observed maximum dbh, for the 14 most abundant species in a warm-temperate rain forest on the island of Yakushima, southern Japan. After Kohyama (1996).

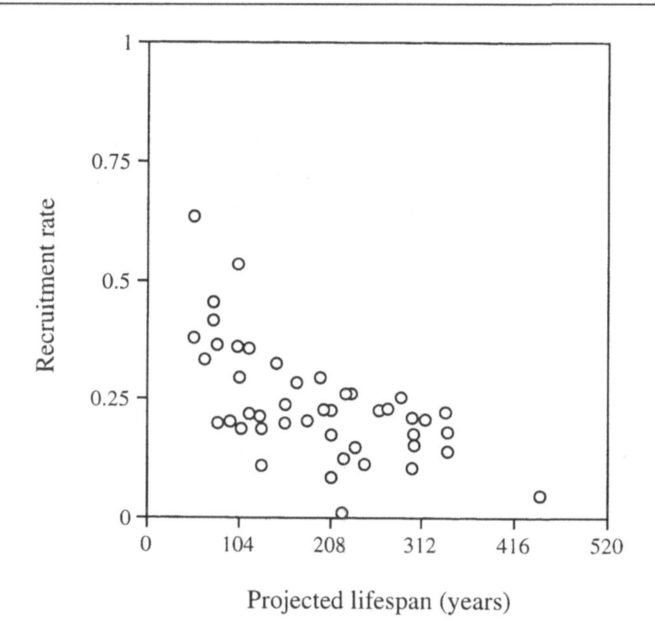


Figure 6.2 Relation between rate of recruitment and species longevity for trees at La Selva, Costa Rica. Recruitment defined as proportion of live stems present in 1982 that were not enumerated in 1969. After Lieberman *et al.* (1985).

degree of acclimation to light climate is finite because of physiological limitations it would be predicted that small-statured species should generally be more shade-tolerant than larger-sized species (Thomas & Bazzaz 1999). Thomas & Bazzaz (1999) found that maximal photosynthetic rates (whether expressed on area, mass or leaf nitrogen bases) of saplings growing in the forest understorey were positively correlated with asymptotic height. Saplings of understorey species tended to exhibit higher assimilation rates at low light intensities than those of canopy species especially when expressed on a leaf mass basis. Among 11 sympatric species of *Macaranga* there was a positive correlation between crown illumination index and asymptotic height (Davies & Ashton 1999). These results confirm the supposition that across species increasing adult stature is associated with reductions in shade tolerance.

There has been speculation about changes in the shade tolerance of species with developmental stage, not just in line with the likely decrease in shade tolerance with size because of increased respiratory load. These putative ontogenetic variations in shade tolerance between species have been grouped into different 'temperaments' by Oldeman & van Dijk (1991). They recognised both species that became less shade-tolerant with age and those that became more so. These groupings are based on field observations and foresters' experience. A species becoming more shade-tolerant with age is surprising. This temperament is usually exemplified by species that require gaps for establishment, but can be overtopped and survive as a sapling beneath a canopy. Examples of this ecological group are anecdotal, with no accurate measurements of light climate to confirm casual observation. It

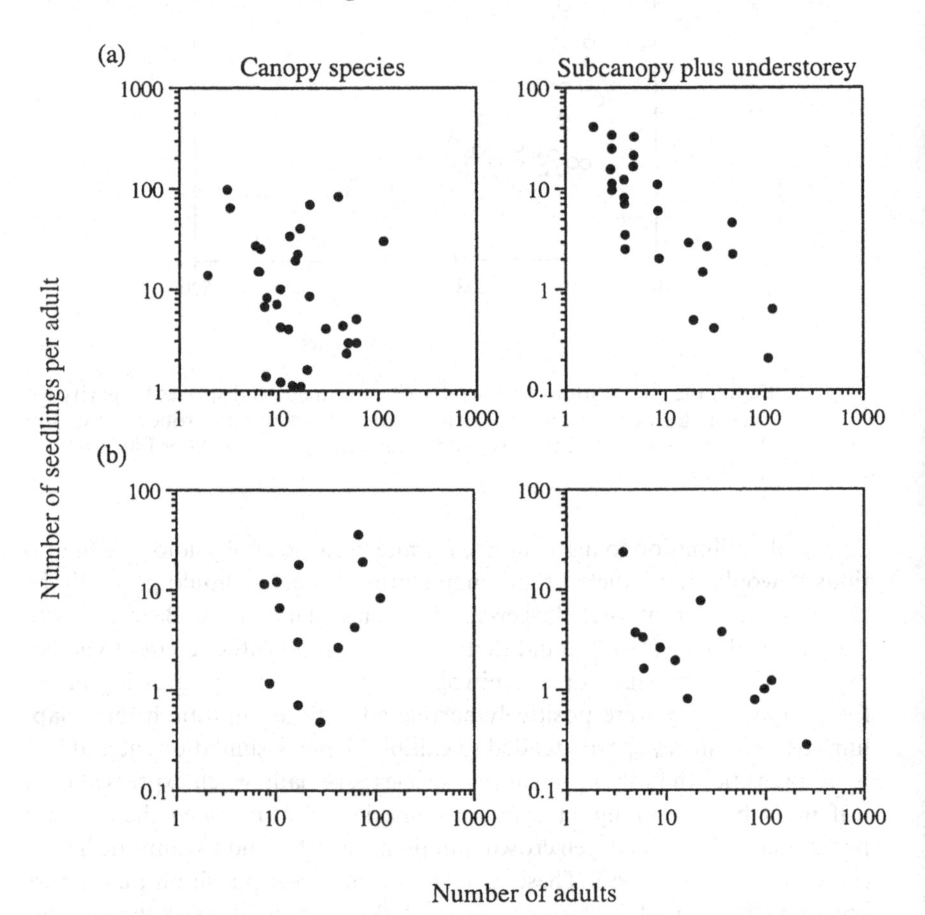


Figure 6.3 Seedling recruitment in relation to adult abundance for two sites in tropical Queensland. Seedling recruitment of species of various adult abundances on Area 1 (a) and Area 2 (b). Seedlings per adult in relation to the number of adults of the same species. Recruitment is the total number of new seedlings of a species appearing at all censuses between 1965 and 1981. All species were included that had at least two adults, a total basal area of at least 0.01 $m^2$ and at least 10 seedlings. After Connell *et al.* (1984).

seems difficult to understand why a seedling or sapling should not possess the physiological latitude available to an older individual. Smaller size should actually make them more flexible, particularly in allocation pattern, but possibly also less resistant to pests, pathogens and other vicissitudes of the understorey.

**BOX 6.1.** The nomenclature of the pioneer–climax dichotomy

Those fast-growing light-demanding species characteristic of large gaps in the forest, as well as forest edges and clearings, are the ecological group most frequently recognised by those attempting ecological classification of tropical trees. Various names have been coined for the group which reflect different aspects of their ecology. 'Pioneer' is perhaps the most widely used, and with 'early-succession' and 'secondary', refers to the abundance of such species in the early stages of secondary succession. 'Nomads' (van Stenos 1956) is a name suggesting the tendency of successive generations of individuals to move around the forest as they can only establish in new gaps. 'Gamblers' (Oldeman & van Dijk 1991) alludes to the high-risk way of life of these species with high fecundity and high mortality.

The other species have been associated with later stages of succession, often referred to as 'climax' or 'late-successional'. This implies incorrectly that 'pioneers' are not found in climax forest. As all forests are being subjected to disturbance at frequent intervals, gap-demanding species are part of the tropical forest climax landscape. In addition, their seeds are often common in the soil seed bank throughout the forest. The 'climax' species group does not include 'pioneers' that are found in the climax forest. Denomination by subtraction was therefore resorted to by Swaine & Whitmore (1988) who invented the term 'non-pioneers' for the others. The Classics provided the antinym for van Steenis's 'nomads'. His rather poetic term 'dryads' has not found favour in the ecological literature. In recognition of the difficult existence of the juvenile stages of these species Oldeman & van Dijk (1991) employed the term 'strugglers'.

## The pioneer–climax dichotomy

The pioneer–climax axis or dichotomy (Box 6.1) among tropical tree species has been associated with many features of their biology as summarised in Table 6.2. There has been a strong tendency to rely on anecdotal accounts, impressions from observation in the field and the application of circular arguments (*Cecropia* and *Macaranga* are pioneers, therefore features of these genera are characteristics of pioneers) to support these points. Other studies generally rely upon an act of faith by the reader that the designation of species to ecological categories (e.g. pioneer versus climax) is correct (Clark & Clark 1992). Ideally, we would have an independent measure of position along the pioneer–climax axis against which to judge other character traits. As there is no agreed index of pioneering ability, the best approach is to search for consistently correlated (but not autocorrelated) traits that covary in a manner predicted by our understanding of the pioneer–climax concept. Perhaps the best example of this is the use of correlation between first-axis score of a principal components analysis of growth and demographic data for species and sapling leaf longevities (Condit *et al.* 1996a). This demonstrated a strong link between short-lived leaves and fast-growing species with high population turnover rates, in other words reliable evidence that pioneers tended to have leaves that lived for a shorter period than did those of climax species. Unfortunately such robust analyses are rare and we must still resort to less critical methods to analyse the nature of the pioneer–climax dichotomy.

It is informative to analyse the propositions put forward in Table 6.2 to consider the amount and strength of evidence in their favour.

**Diagnostic characters**

Swaine & Whitmore (1988) defined pioneer and 'non-pioneer' groups on seed germination characteristics. They argued that pioneers are those species that are only found in gaps because their seeds do not germinate elsewhere on the forest floor. Climax species will germinate under canopy shade, as well as in gaps. A degree of vagueness is inherent in the Swaine & Whitmore dichotomy because they provide no minimum gap size or period of direct sunlight that is required for pioneer germination. This germination-dependent definition has been questioned by several workers (Alvarez-Buylla & Martínez-Ramos 1992), and even the original authors indicate that modifications are required:

> 'The definition of pioneers offered by Swaine & Whitmore (1988) was too crude and needs more precise formulation.' Kennedy & Swaine (1992)

> 'It can now be seen that the Swaine and Whitmore dichotomy is a necessary but not a sufficient description of the variety of tree autecology found in nature. It gives overemphasis to just two features, germination and establishment.' Whitmore (1996)

> 'Light-mediated germination is relatively rare among these forest trees, even among pioneers, so that the working definition of a pioneer should be seen to depend more on a species' ability to survive in forest shade.' Kyereh *et al.* (1999)

Table 6.2. *Characters proposed as part of the pioneer syndrome*

Those given in **bold** are ones for which I consider there is good evidence of a strong link to the pioneer syndrome. Those given in *italics* have little or no evidence of an association. Others have some evidence but require better studies to confirm their importance as features of the pioneer syndrome. In general climax species exhibit the antithetical traits to those listed for pioneers, with the exception of (1) where climax species have gap-independent germination. The opposite of (12) is a positive or reverse-J stand table, and of (32) is recalcitrant seeds.

*Diagnostic*
1. Seeds only germinate in canopy gaps open to the sky and which receive some full sunlight
2. Not found beneath a closed forest canopy, i.e. shade-intolerant

*Reproduction*
3. **Seeds small**
4. Often a large number of seeds per fruit
5. Reproductive effort high: seeds produced in large numbers
6. Reproduction begins from early in life
7. Once begun, reproduction is more or less continuous
8. Seeds dispersed over long distances by animals or the wind

*Demography*
9. **Seeds abundant in the soil seed bank**
10. **Juvenile mortality high**
11. Trees short-lived
12. Negative stand table

*Growth and Form*
13. **Growth rapid**
14. Growth continuous and indeterminate
15. Branching relatively sparse with few orders
16. **Wood pale and of low density**
17. Leaves generally large
18. **Leaves short-lived (and hence with a high turnover)**
19. **Leaves thin with low mass per unit area**
20. Defences generally poor
21. Mutualisms with defensive ants relatively common
22. Root system highly branched and superficial (Whitmore)/*deep* (Bazzaz)
23. Root:shoot ratio low
24. Morphological plasticity high
25. Germination mostly epigeal
26. Cotyledons often foliaceous

*Physiology*
27. **Photosynthetic rate high**
28. **Respiration rate high**
29. **Light compensation point high**
30. *Nitrate is the main form of N used*
31. Physiological plasticity high
32. Seeds often orthodox

*General*
33. **Strongly associated with disturbance**

After Swaine & Whitmore (1988), Bazzaz (1991), Whitmore (1998).

Surprisingly, there have been few wide-ranging tests of the Swaine & Whitmore germination hypothesis. Raich & Gong (1990) found a reasonable agreement between expectations based on field observation of the ecology of species and the germination characteristics. However, some pioneers did show low frequencies of germination under canopy shade. This was also found in *Cecropia obtusifolia* by Alvarez-Buylla & Martínez-Ramos (1992), which led them to question a dichotomy based solely on seed germination. The recent research of Kyereh *et al.* (1999) on West African species has cast even more doubt on the applicability of germination as a diagnostic character. They found that a majority of species, including accepted pioneers, will germinate in the dark and all show some germination in forest understorey conditions. The general shade intolerance of pioneer species (see below) must act as a selection pressure in favour of mechanisms that maintain seed dormancy under shaded conditions, but it is clear that not all pioneer species have acquired dormancy mechanisms.

Because of both physiological limitations and the presence of predators, seeds will not survive indefinitely in the dormant state in tropical forest soils, so a loss of dormancy with time might be expected. The occurrence of some pioneer seedlings in shady environments (for example, Kennedy & Swaine (1992) report the occurrence of some *Macaranga hypoleuca* and *Endospermum peltatum* seedlings in very tiny gaps at the Danum Valley, Sabah) may be the result of such 'last-chance' germination.

Pioneer individuals can occasionally be found in heavily shaded sites within the forest, but the chances of mortality for such individuals are high and in general pioneers are associated with sites of above-average brightness of illumination (Clark & Clark 1992; Lieberman *et al.* 1995; Clark *et al.* 1996; Davies *et al.* 1998). A principal components analysis summarising the results of an experiment where seedlings of 16 West African tree species were grown at a range of light intensities found the first principal component differentiating the species to be strongly linked to RGR at 2% full sun (Agyeman *et al.* 1999). The pioneers tended to show negative growth rates in deep shade, and the authors suggested that seedling RGR under forest understorey conditions may be the most practicable way to estimate the position of a species on the pioneer–climax axis.

### Reproduction

The general supposition in the ecological literature is that pioneer species are characterised by the production of very large numbers of small seeds from early in life. As has already been outlined, the comparison of seed size between pioneer and climax species is a contentious issue, although it must be remembered that the debating point is the adaptive value of either small seed size in pioneers, or large seed size in climax species, not the fact that pioneers do tend to have smaller seeds on average than climax species. This has been demonstrated many times (Foster & Janson 1985; Primack & Lee 1991; Ibarra-Manríquez & Oyama 1992; Favrichon 1994; Hammond & Brown 1995; Metcalfe & Grubb 1995). Pioneers also tend to have large numbers of seeds per fruit (Opler *et al.* 1980; Prevost 1983; Ibarra-Manríquez & Oyama 1992). The fecundity of pioneer species can indeed be very high. Individuals of *Cecropia obtusifolia* produce $1.4 \times 10^4$ to $1.4 \times 10^7$ seeds annually (Alvarez-Buylla & Martínez-Ramos 1990) and *Solanum rugosum*, a short-lived, early-successional shrub in French Guiana, has been estimated to have an average lifetime fecundity of 2.3 million seeds (Prevost 1983). Pioneers begin reproduction early in life; for instance, *Cecropia obtusifolia* has been recorded as flowering at three years of age (Alvarez-Buylla & Martínez-Ramos 1992), but this may not represent the onset of maturity at smaller relative size compared with climax species. Thomas (1996b) reported *Macaranga gigantea* to have a relative size at the onset of maturity very similar to the average for 37, mostly shade-tolerant, tree species at Pasoh, Malaysia. In Mexico, Ibarra-Manríquez & Oyama (1992) found that 30 pioneer species had significantly longer flowering and fruiting periods on average than 109 'persistent' species.

Zoochory and anemochory are certainly not confined to pioneer species. As might be expected (see Chapter 4), wind-dispersal was shown to be commoner among tall pioneers than short ones in French Guiana (Favrichon 1994). Small seeds are often inferred to have an advantage of increased dispersal distance, but evidence that small-seeded pioneers gain greater dispersal distances than large-seeded climax species is wanting. It is possible that the large numbers of seeds produced by pioneers increase the likelihood of some seeds being dispersed long distances. Dalling *et al.* (1998a) did find that small-seeded pioneer species on Barro Colorado Island showed a seedling distribution pattern among gaps less dependent on adult distribution than larger-seeded species of similar ecology.

### Demography

The abundance of pioneer seeds in the soil seed bank of tropical rain forests is well documented (Garwood 1989). Such species generally dominate in terms of numbers of individuals, although some sites remote from major disturbance have been found to contain few seeds of extreme pioneer trees (see, for example, Kennedy & Swaine 1992). The tiny seedlings of pioneers are susceptible to damage and death from many causes, resulting in high mortality rates. Seedlings of three gap-demanders (*Ochroma pyramidale*, *Cochlospermum vitifolium* and *Cavanillesia platanifolia*) showed very high susceptibility to death from damping-off under shade conditions (Augspurger & Kelly 1984). The three named species with the lowest seedling-cohort half-life in a study at La Selva, Costa Rica, were *Cecropia obtusifolia*, *C. insignis* and *Goethalsia meiantha*. They halved in population size in well under one month (Li *et al.* 1996). *C. obtusifolia* was also found to have a very high juvenile mortality at Los Tuxtlas, Mexico, where more than 99% of seeds and seedlings died in the first year (Alvarez-Buylla & Martínez-Ramos 1992), and a high mortality (48% $yr^{-1}$) for stems 1–10 cm dbh at Barro Colorado Island, Panama (Condit *et al.* 1995). All the species with high mortality rates on Barro Colorado Island had recruitment strongly associated with gaps in the canopy (Condit *et al.* 1995).

Estimates of the longevities of tropical tree species generally show pioneer species to have short life spans in comparison with climax species (Lieberman *et al.* 1985a; Condit *et al.* 1995; Martínez-Ramos & Alvarez-Buylla 1998). For instance, *Cecropia sciadophylla* was estimated to take 54 years to grow from 10 cm dbh to maximum size in Ecuador (Korning & Balslev 1994), faster than 21 other common species.

Pioneer species tend not to have the reverse-J stand table (size–frequency distribution) typical of most tree species in the forest, but have relatively fewer individuals in the smaller size classes. In an analysis from the 50 ha plot on Barro Colorado Island, extreme pioneers such as *Cecropia insignis* and *Zanthoxylum belizense* were found to exhibit such negative stand tables (Hubbell & Foster 1987; see Fig. 6.4), whereas shade-tolerant species had the reverse-J distribution. Negative stand tables have generally been interpreted as a failure of adequate regeneration and high juvenile mortality by the species concerned and an indication that the species' population is in decline within the stand. However, Condit *et al.* (1998) have shown both theoretically and with real data from the 50 ha plot on Barro Colorado Island that rates of population size change and mortality have only a weak effect on size–frequency distributions in populations. Much more influential is growth rate of small size classes. Pioneers have negative stand tables because their juveniles grow rapidly and soon leave the small size classes, whereas climax species realise very slow growth rates among small size classes and hence accumulate large numbers of individuals in the lower intervals.

### Growth and form

The net assimilation rate of some tropical woody pioneer species can come close to that of herbaceous *Helianthus annuus* (Whitmore & Gong 1983). Pioneer species can often achieve height growth rates in excess of 1 m per annum, and values of more than 5 m per year have been recorded (Alexandre 1989; Bazzaz 1991). *Cecropia sciadophylla* had a mean diameter growth rate nearly twice as fast as the second fastest species among 22

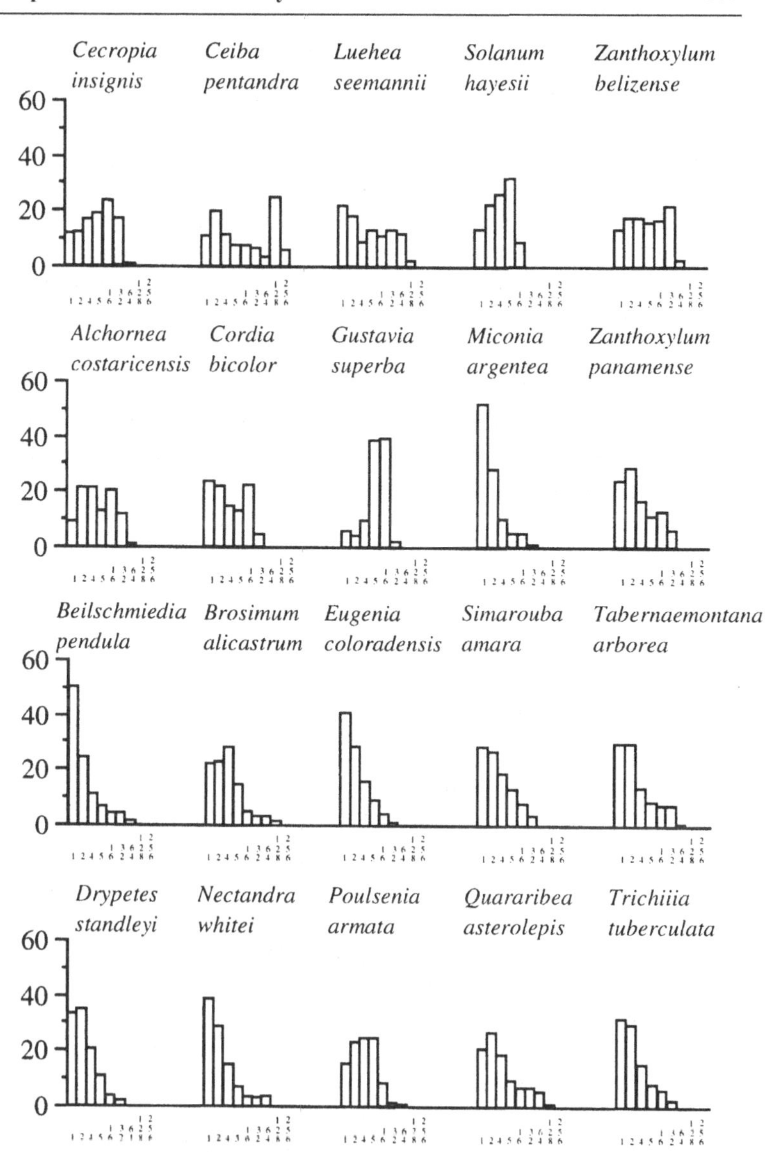


Figure 6.4 Diameter distributions ($\log_2$-scale) for 20 species from 50 ha on Barro Colorado Island. The upper row represents 'sun' species, the second row 'partial sun' species, the third row 'indifferent' species and the bottom row 'shade' species. After Hubbell & Foster (1987).

monitored in Amazonian Ecuador (Korning & Balslev 1994). Species of *Macaranga* had the highest mean growth rates in long-term inventory plots at Sungei Menyala and Bukit Lagong in Peninsular Malaysia (Manokaran & Kochummen 1994).

Pioneers are frequently characterised by ecologists as exhibiting continuous and indeterminate growth. However, published accounts providing reliable evidence of this are wanting. Roux's model and Rauh's model tend to be common among early successional species (de Foresta 1983; Vester & Cleef 1998), but the functional significance of this remains unclear. Pioneers are more likely to be orthotropic in general architecture and leaf display than climax species (King & Maindonald 1999), at least as saplings.

Many authors have emphasised the soft and light wood of pioneer species (Swaine & Whitmore 1988; Alexandre 1989; Bazzaz 1991). Certainly well-known genera of pioneers such as *Cecropia*, *Macaranga*, *Musanga* and *Trema* typically have very low timber densities. In Borneo, species from secondary forest stands 2–6 years old were found to have stem specific gravity in the range 0.28–0.34 (mean 0.31), considerably lower than nearby primary forest where the range was 0.54–0.62 (mean 0.58) (Suzuki 1999). Adult wood density has been shown to be a positive correlate of juvenile shade tolerance (Augspurger 1984). Low-density wood is relatively weak, and there is growing evidence that pioneer species are poor at re-sprouting after breakage of the trunk or major branches (Putz & Brokaw 1989; Negrelle 1985). Possible contributory factors to this limited ability to recover from major damage are a low allocation by the plant to stored carbohydrates and the paucity of dormant buds (Zimmerman *et al.* 1994).

Popma *et al.* (1992) found that obligate gap species had higher mean lamina area and lower LMA than gap-dependent and gap-independent species at Los Tuxtlas. Leaves of pioneers have shorter life spans than those of more shade-tolerant species (Coley 1983; Coley & Aide 1991). Condit *et al.* (1996a) used the first-axis scores of a principal components analysis of growth and demographic data for species on Barro Colorado Island, in conjunction with Coley's (1983) leaf longevity data, to show that all pioneers had leaf life lengths of less than 1 year whereas those of shade-tolerant species were mostly above 1 year. Short leaf life span is linked to high foliar nutrient concentrations and low LMA (Reich *et al.* 1995).

In a review of published data, Coley & Aide (1991) found that tropical gap species had leaves of lower condensed tannin and crude fibre concentration on average than shade-tolerant species. Pioneers also generally have leaves of low toughness (Coley & Aide 1991; Choong *et al.* 1992). Low fibre concentration and relatively high nitrogen content make the foliage of pioneers highly palatable to herbivores. Schupp & Feener (1991) showed that gap-dependent species were significantly more likely to be ant-defended than shade-tolerant ones on Barro Colorado Island. Important pioneer genera such as *Cecropia*, *Macaranga*, *Piper* and *Ochroma* all contain myrmecophytic species.

There have been few detailed studies of root-system form in tropical trees, and it is not possible to conclude that there is a consistent difference between pioneer and climax species from published data. Tyree *et al.* (1998b) found that seedlings of two pioneer species had finer root systems than those of three climax species, with significantly higher root length and area per unit dry mass of root system.

The relatively large morphological and physiological plasticity of pioneer species has been demonstrated by a number of studies, mostly concerning seedlings (see, for example, Strauss-Debenedetti & Bazzaz 1991; Ducrey 1992; Strauss-Debenedetti & Berlyn 1994; Veenendal *et al.* 1996; Veneklaas & Poorter 1998).

The common occurrence of epigeal germination and the possession of foliaceous phanerocotylar seedings are probably a pioneer character linked to the generally small seed size of this ecological group. The link between seed size and seedling form is quite well established (Garwood 1996; Osunkoya 1996).

**Physiology**

The physiological characterisation of pioneer species has consistently emphasised high rates of photosynthesis, particularly on a per unit dry mass basis, high dark respiration rates and high light compensation points (Medina 1986; Eschenbach *et al.* 1998). Probably the best demonstration of the link between leaf form and physiology and successional status is the work of Davies (1998) on nine species of *Macaranga* growing in Lambir Hills National Park, Sarawak. He found that there were significant correlations between seedling LMA and mass-based maximal assimilation rate and the first-axis scores of a principal components analysis of life-history characteristics of the species. The principal component represented an index of pioneering with a significant positive correlation to $A_{max}$ and a negative one to LMA. Seedlings of *Cecropia obtusifolia* and *Ficus insipida* were shown to have higher maximal photosynthetic rates than those of three more shade-tolerant species of the Moraceae (*sensu lato*), and also to have greater physiological amplitude in response to different growth conditions (Strauss-Debenedetti & Bazzaz 1991).

The only evidence available that points to an importance of nitrate as an N source for pioneers is the higher nitrate reductase activity in the leaves of pioneers compared with those of climax species (Fredeen *et al.* 1991; Turner 1995b). However, in general these activities are still low. Most tropical soils are acidic and it is likely that ammonium rather than nitrate is the main form of nitrogen available to plant roots. In liquid culture, *Melastoma malabathricum* seedlings grew better with ammonium rather than nitrate as the N source (Watanabe *et al.* 1998), although this may have been brought about by the solubilisation of greater amounts of phosphorus in the more acidic medium containing $NH_4^+$. Further research is necessary, but it seems likely that tropical rain-forest pioneers are well able to use ammonium as their primary source of nitrogen.

**General**

The terms 'pioneer' and 'early successional' indicate the clear perceptual link ecologists have made between such species and succession-initiating disturbance. That pioneers do, in fact, dominate secondary forest in the tropics can hardly be doubted, although species characteristic of secondary succession on very infertile soils appear not to possess all the 'typical' pioneer traits (Sim *et al.* 1992).

The small, frequently photoblastic or thermoblastic, seeds of pioneers are unlikely to germinate successfully from under litter, or a layer of soil, even when in gaps. Therefore, pioneers are often strongly associated with soil disturbance. Davies *et al.* (1998) found that among eleven species of *Macaranga* the more shade-tolerant species had a smaller proportion of stems found in sites with evidence of earlier soil disturbance.

In the 50 ha plot on Barro Colorado Island, Panama, the density of saplings (1–3.9 cm dbh) of both pioneer and climax species was significantly higher in gaps (areas with canopy less than 5 m tall estimated on a 5 m × 5 m grid) (Hubbell *et al.* 1999). The relative abundance of pioneers increased with gap size but species richness of pioneers did not when a correction for area sampled was included. This confirms that gaps are important sites of recruitment for a wide range of species and that big gaps favour pioneers more than climax species. But it seems, on Barro Colorado Island at least, that there is not a strong partitioning of the upper end of the light-availability gradient by pioneer species. If there were, one would expect increasing species richness of pioneers per unit gap area with gap size because big gaps would contain the most light-demanding species in their centres and the less light-demanding species at the edges. There may be some big-gap specialists present but their number is too small to influence the diversity–gap size relationship.

If disturbance favours pioneer species, then more frequently, or more intensely, disturbed forests should have a greater abundance of pioneer individuals, and possibly also a higher diversity of pioneer species. A study often cited as a demonstration of this phenomenon was that by Whitmore (1989) on Kolombangara, an island in the Solomons. Using a pseudo-quantitative pioneer index based on observations of the requirements of the commonest 12 species for seedling establishment and for onward growth of saplings, Whitmore found that the more cyclone-prone north coast of the island had a higher stand pioneer index value than the less disturbed west coast. However, more detailed analysis of the long-term data for the period including the direct hit of cyclone Annie in 1967 showed that forest stands showed relatively little change in the rank order of species abundance after the major disturbance event because there was a positive correlation between mortality during the hurricane and recruitment after it among the species (Burslem & Whitmore 1996). It now seems likely that a history of more intense anthropogenic disturbance on the north coast is what has led to the greater abundance of pioneer species there.

## Conclusion

There is a considerable body of evidence to support the existence of a pioneer–climax axis among tropical tree species. However, there is little support for any discrete clustering of species along the axis. Any classification of species into groups has to be done on more or less arbitrary break points. In a primary rain-forest community, it seems that the number of extreme pioneer species is relatively small. Only 6 out of 108 common species on Barro Colorado Island appeared to be strongly pioneering in demographic characteristics (Welden *et al.* 1991). Species of this guild appeared to be absent altogether from the lowland dipterocarp forest at Pasoh, Peninsular Malaysia (Condit *et al.* 1999).

In addition to the distinction between pioneer and climax groups being arbitrary, it must also be remembered that the association of the character traits with species is not universal. A particular pioneer may have many of the features listed in Table 6.2, but is not likely to possess them all. For instance, many pioneer species possess relatively small leaves. The converse is also true: some climax species show a few pioneer characteristics.

The pioneer–climax axis appears to be a combination of two other axes. These are the *r*- versus *K*-selection axis of reproductive effort and allocation (MacArthur & Wilson 1967; Pianka 1970), and an axis of shade tolerance. Pioneers are typically *r*-selected, beginning reproduction early in life and producing large numbers of seeds, often almost continuously. Climax species, in comparison, delay reproduction and produce smaller seed crops discontinuously. Offspring (seed) size was included in the original distinction between the *r*- and *K*-selected groups, and pioneers do tend to have relatively small seeds. However, phylogenetically controlled analyses have not found a consistent clear-cut distinction within clades between light-demanders and shade-tolerators in seed size (see page 185).

In order to reproduce rapidly trees will need high levels of resource availability; therefore it is not surprising that *r*-selected species are light-demanding. The *r–K* axis and the shade-tolerance axis appear to run parallel, at least when stretched across their complete ranges. However, in the middle of the ranges there is probably considerable scatter. Species of intermediate shade tolerance probably show a wide variation in reproductive characteristics and vice versa. Another important point is that there are no clear discontinuities of the position of species along the axes. The recognition of groups is a fairly arbitrary choice, and is more important as a means of emphasising the direction of change along the axis than accurately circumscribing a particular ecological or evolutionary pathway. Some authors have chosen to break the axis into even more groups; for example, five groups were recognised by Oldeman & van Dijk (1991). I believe it would be difficult to classify species consistently with such a system, unless one quantifiable measure, such as wood density, is taken as the agreed yardstick and arbitrary groups are defined in those terms.

### Combining the two axes

The height at maturity and the pioneer–climax axes can be combined to produce a more detailed classification. They appear to be reasonably independent axes, with both pioneers and climax species covering a wide range of statures. If we recognise the pioneer and climax groups then each can be divided into height classes. This, in effect, is what Shugart (1987) did in his two by two classification of forest trees on their gap requirement for regeneration and propensity to form gaps when they die. The four groups of species recognised on growth rate and longevity by Lieberman *et al.* (1990) (see Chapter 3) form a similar 2 × 2 table where longevity substitutes for size at maturity and growth rate replaces regeneration requirement or shade-tolerance. Favrichon (1994) also recognised two main axes differentiating species in the forests of French Guiana, based on ordination of a large number of characteristics for those species. The axes he referred to as potential size and heliophily, the latter being basically potential maximum growth rate. These different systems can be resolved into one, as shown in Fig. 6.5. I use the terms understorey, canopy, small pioneer and large pioneer for the four species groups such a system recognises. Favrichon (1994, 1998) included three size divisions in his classification system, but could not find a consistent distinction in degree of heliophily between the biggest trees. Thus he differentiated five ecological groups, the above four plus emergents. It is possible that in French Guiana there are few pioneers of very large stature, but it also seems likely that the range in shade tolerance among species diminishes with adult

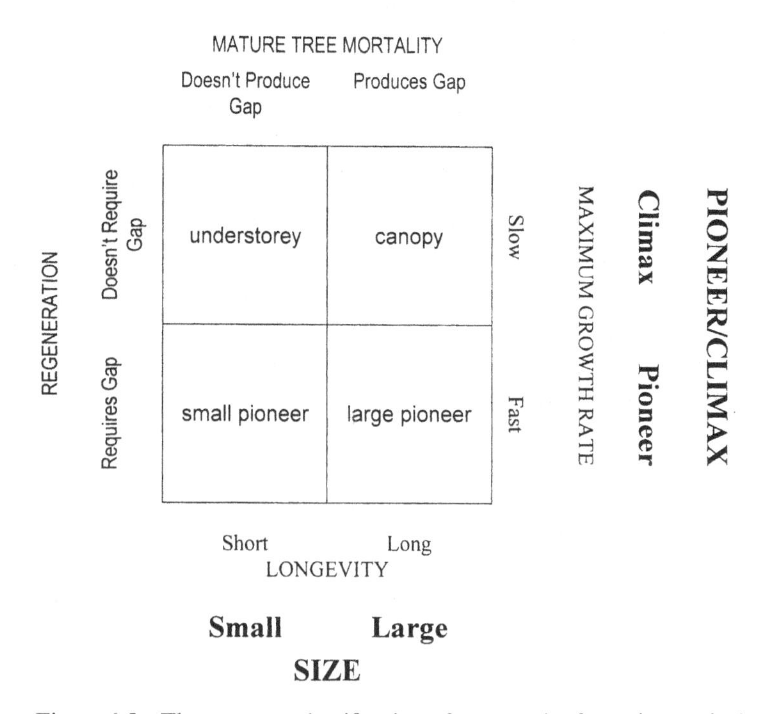


Figure 6.5 The two-way classification of tree species from the tropical rain forest based on size at maturity and the pioneer–climax axis. The top and left headings to the central box are based on Shugart's system (Shugart 1987); the bottom and right inner headings are based on the system of Lieberman *et al.* (1990). Favrichon (1994) recognised the axes as potential size and maximum growth rate (heliophily). The names for the four ecological groups do not follow these authors.

stature. Further studies are needed to confirm whether emergents are consistently different in general ecology from canopy-top species thoughout the tropics.

**Long-lived pioneers or late-secondary species**

The pioneers of large stature are a group that has received some attention. They are often presented as an intermediate group on the pioneer–climax axis (Alexandre 1989) referred to variously as large or long-lived pioneers, or late-secondary species. They are characterised as species that can grow quickly to large size, require gaps to regenerate and are relatively intolerant of shade. They reproduce freely, but not as prolifically as the smaller pioneers do. Clearly if they are to grow to large size, and some of the large pioneers such as *Ceiba pentandra* can attain enormous heights and girths, these species cannot allocate all their resources to reproduction from small size. Therefore, they will not be as strongly *r*-selected as their smaller-statured pioneer brethren. Foresters take great interest in the large pioneers because they have characteristics that frequently make them excellent timber trees. Their fast growth and light but strong wood make them superior to the slow-growing and heavy-timbered shade-tolerant climax species for the purpose of growing timber. Large pioneers are often anemochorous (Alexandre 1989). Ashton & Hall (1992) noted large pioneer species to be more diverse and abundant on more nutrient-rich soils than relatively infertile sites in northwestern Borneo.

A number of authors have made reference to a group of species of very large stature that also require gaps for regeneration (Newbery & Gartlan 1996; Condit *et al.* 1998). These so-called 'giant invaders' are frequently characterised by very strongly negative stand tables. Explanations for this lack of regeneration have generally referred to either changes in disturbance regime where the emergent trees are the relics of intense disturbance in the past, or regeneration tied to very sporadic disturbance events after which the new recruits grow very rapidly and soon reach large size. A known example of the latter is the mahogany *Swietenia macrophylla* that in Amazonian Bolivia has been shown to regenerate successfully only on large erosion gullies (Gullison *et al.* 1996). The relative abundance of giant invaders in the rain forests of West Africa (Alexandre 1989; Newbery & Gartlan 1996) has tantalised many ecologists into seeking an explanation.

### Parallels between pioneers and shade-tolerant understorey species

There appear to be some parallels in the evolutionary pressures exerted on pioneers and shade-tolerant species of small stature. In both, the relatively high risk of mortality favours early reproduction. Although this will represent a stronger pressure in pioneers, it is probable that both ecological classes require relatively high recruitment rates to maintain populations. Therefore, whereas they may have very different physiology and growth form, an emphasis on reproduction may be a linking thread between the two groups. It is possible that the evolution from light-demanding to shade-tolerant physiology, or vice versa, may not be that difficult. Grubb & Metcalfe (1996) considered it possible that some of the small-seeded shade-tolerant species they studied in tropical Australia had evolved from light-demanders. Similarly, Mulkey *et al.* (1993) identified the pioneer-like features of Central American *Psychotria furcata*, an understorey shrub that flowers in the shade, particularly in comparison with two other *Psychotria* species of more typical shade-tolerant demeanour.

## Where does this leave us?

The sorts of ecological classification outlined above are syntheses of our knowledge concerning the characteristics of tree species in the rain forest. They are valuable tools for communicating information if used critically. Exceptions to the generalisations abound, but only the naïve expect ecology to produce absolute solutions. Further subdivisions of the major classes by height, light requirements or other characters are possible, and lead to the recognition of guilds of species of similar life history, regeneration requirements and physiology. An important question to ask is how consistently can species be assigned to guilds across different forests, or by different workers using the same system of classification. The answer is probably that it depends on how much information is available for the species concerned. If data on growth and demography, or even stand tables, are to hand then classification may be possible, but an ecologist entering a tropical rain forest outside the handful of well-studied sites will not have this information and must rely on educated guesses for classifying the tree flora.

One of my objectives when I set out to write this book was to see if there was a new synthesis of the comparative ecology of tropical trees awaiting discovery amidst the voluminous literature on tropical rain forests. I have failed to find a new synthesis. This is very likely to be a reflection of my own shortcomings, but I also believe that it is testament to the complexity of the ecology and evolution of the tropical rain-forest tree community. Scientists like simple, elegant solutions to problems, but rain forests just do not work that way.

# REFERENCES


Ackerly, D.D. & Bazzaz, F.A. 1995a. Seedling crown orientation and interception of diffuse radiation in tropical forest gaps. *Ecology* **76**: 1134–46.

Ackerly, D.D. & Bazzaz, F.A. 1995b. Leaf dynamics, self-shading and carbon gain in seedlings of a tropical pioneer tree. *Oecologia* **101**: 289–98.

Adler, G.H. & Kestell, D.W. 1998. Fates of neotropical tree seeds influenced by spiny rats (*Proechimys semispinosus*). *Biotropica* **30**: 677–81.

Agrawal, A.A. 1997. Do leaf domatia mediate plant-mite mutualism? An experimental test of the effects on predators and herbivores. *Ecological Entomology* **22**: 371–6.

Agrawal, A.A. & Rutter, M.T. 1998. Dynamic anti-herbivore defense in ant-plants: the role of induced responses. *Oikos* **83**: 227–36.

Agyeman, V.K., Swaine, M.D. & Thompson, J. 1999. Responses of tropical forest tree seedlings to irradiance and the derivation of a light response index. *Journal of Ecology* **87**: 815–27.

Aide, T.M. 1987. Limbfalls: a major cause of sapling mortality for tropical forest plants. *Biotropica* **19**: 284–5.

Aide, T.M. 1993. Patterns of leaf development and herbivory in a tropical understory community. *Ecology* **74**: 455–66.

Alexander, I. 1989. Mycorrhizas in tropical forests. Pp. 169–188 in *Mineral nutrients in tropical forest and savanna ecosystems* (ed. J. Proctor). Blackwell Scientific Publications, Oxford.

Alexander, I., Norani Ahmad & Lee, S.S. 1992. The role of mycorrhizas in the regeneration of some Malaysian forest trees. *Philosophical Transactions of the Royal Society of London* B **335**: 379–88.

Alexandre, D.Y. 1977. Régeneration naturelle d'un arbre caractéristique de la forêt équatoriale de Côte-d'Ivoire: *Turraeanthus africana* Pellegr. *Oecologia Plantarum* **12**: 241–62.

Alexandre, D.Y. 1980. Caractere saisonnier de la fructification dans une forêt hygrophile de Cote-d'Ivoire. *Revue d'Ecologie (La Terre et la Vie)* **34**: 335–59.

Alexandre, D.Y. 1989. *Dynamique de la régénération naturelle en forêt dense de Côte d'Ivoire.* ORSTOM, Paris.

Allen, E.B., Allen, M.F., Helm, D.J., Trappe, J.M., Molina, R. & Rincon, E. 1995. Patterns and regulation of mycorrhizal plant and fungal diversity. *Plant and Soil* **170**: 47–62.

Altshuler, D.L. 1999. Novel interactions of non-pollinating ants with pollinators and fruit consumers in a tropical forest. *Oecologia* **119**: 600–6.

Alvarez-Buylla, E.R. & Martínez-Ramos, M. 1990. Seed bank versus seed rain in the regeneration of a tropical pioneer tree. *Oecologia* **84**: 314–25.

Alvarez-Buylla, E.R. & Martínez-Ramos, M. 1992. Demography and allometry of *Cecropia obtusifolia*, a neotropical pioneer tree – an evalution of the climax-pioneer paradigm for tropical rain forests. *Journal of Ecology* **80**: 275–90.

Alvarez-Buylla, E.R., García-Barrios, R., Lara-Moreno, C. & Martínez-Ramos, M. 1996. Demographic and genetic models in conservation biology: applications and perspectives for tropical rain forest tree species. *Annual Review of Ecology and Systematics* **27**: 387–421.

Andrade, J.L., Meinzer, F.C., Goldstein, G., Holbrook, N.M., Cavelier, J., Jackson, P. & Silvera, K. 1998. Regulation of water flux through trunks, branches and leaves in trees of a lowland forest. *Oecologia* **115**: 463–71.

Andresen, E. 1999. Seed dispersal by monkeys and the fate of dispersed seeds in a Peruvian rain forest. *Biotropica* **31**: 145–58.

Angiosperm Phylogeny Group 1998. An ordinal classification for the families of flowering plants. *Annals of the Missouri Botanical Garden* **85**: 531–53.

Appanah, S. 1982. Pollination of androdioecious *Xerospermum intermedium* Radlk. (Sapindaceae) in a rain forest. *Biological Journal of the Linnean Society* **18**: 11–34.

Appanah, S. 1985. General flowering in the climax rain forests of South-East Asia. *Journal of Tropical Ecology* **1**: 225–40.

Armbruster, W.S. 1984. The role of resin in angiosperm pollination: ecological and chemical considerations. *American Journal of Botany* **71**: 1149–60.

Armbruster, W.S. & Webster, G.L. 1979. Pollination of two species of *Dalechampia* (Euphorbiaceae) in Mexico by Euglossine bees. *Biotropica* **11**: 278–83.

Armstrong, J.E. 1997. Pollination by deceit in nutmeg (*Myristica insipida*, Myristicaceae): floral display and beetle activity at male and female trees. *American Journal of Botany* **84**: 1266–74.

Aronow, L. & Kerdel Vegas, F. 1965. Seleno-cystathionine, a pharmacologically active factor in seeds of *Lecythis ollaria*. *Nature* **205**: 1185–6.

Ashton, M.S. 1995. Seedling growth of co-occurring *Shorea* species in the simulated light environments of a rain forest. *Forest Ecology and Management* **72**: 1–12.

Ashton, P.M.S. & Berlyn, G.P. 1992. Leaf adaptations of some *Shorea* species to sun and shade. *New Phytologist* **121**: 587–96.

Ashton, P.M.S., Gunatilleke, C.V.S. & Gunatilleke, I.A.U.N. 1995. Seedling survival and growth of four *Shorea* species in Sri Lankan rainforest. *Journal of Tropical Ecology* **11**: 263–79.

Ashton, P.S. & Hall, P. 1992. Comparisons of structure among mixed dipterocarp forests of north-western Borneo. *Journal of Ecology* **80**: 459–81.

Ashton, P.S., Givnish, T.J. & Appanah, S. 1988. Staggered flowering in the Dipterocarpaceae: new insights into floral induction and the evolution of mast fruiting in the aseasonal tropics. *American Naturalist* **132**: 44–66.

Asner, G.P. & Goldstein, G. 1997. Correlating stem biomechanical properties of Hawaiian canopy trees with hurricane wind damage. *Biotropica* **29**: 145–50.

Atger, C. & Edelin, C. 1994. Premières données sur l'architecture comparée des systèmer racinaires et caulinaires. *Canadian Journal of Botany* **72**: 963–75.

Auffenberg, W. 1988. *Gray's monitor lizard*. University of Florida Press.

Augspurger, C.K. 1981. Reproductive synchrony of a tropical shrub: experimental studies on effects of pollinators and seed predators on *Hybanthus prunifolius* (Violaceae). *Ecology* **62**: 775–88.

Augspurger, C.K. 1983a. Seed dispersal of the tropical tree, *Platypodium elegans*, and the escape of its seedlings from fungal pathogens. *Journal of Ecology* **71**: 759–71.

Augspurger, C.K. 1983b. Offspring recruitment around tropical trees: changes in cohort distance with time. *Oikos* **40**: 189–96.

Augspurger, C.K. 1984. Light requirements of neotropical tree seedlings: a comparative study of growth and survival. *Journal of Ecology* **72**: 777–95.

Augspurger, C.K. 1986. Morphology and dispersal potential of wind-dispersed diaspores of neotropical trees. *American Journal of Botany* **73**: 353–63.

Augspurger, C.K. & Franson, S.E. 1988. Input of wind-dispersed seeds into light-gaps and forest sites in a Neotropical forest. *Journal of Tropical Ecology* **4**: 239–52.

Augspurger, C.K. & Hogan, K.P. 1983. Wind dispersal of fruits with variable seed number in a tropical tree (*Lonchocarpus pentaphyllus*: Leguminosae). *American Journal of Botany* **70**: 1031–7.

Augspurger, C.K. & Kelly, C.K. 1984. Pathogen mortality of tropical tree seedlings: experimental studies of the effects of dispersal distance, seedling density, and light conditions. *Oecologia* **61**: 211–17.

Augspurger, C.K. & Kitajima, K. 1992. Experimental studies of seedling recruitment from contrasting seed distributions. *Ecology* **73**: 1270–84.

Bailey, I.W. & Sinnott, E.W. 1916. The climatic distribution of certain types of angiosperm leaves. *American Journal of Botany* **3**: 24–39.

Baker, D.D. & Schwintzer, C.R. 1990. Introduction. Pp. 1–13 in *The biology of* Frankia *and actinorhizal plants* (ed. C.R. Schwintzer & J.D. Tjepkema). Academic Press, London.

Baker, H.G. & Baker, I. 1983. Floral nectar sugar constituents in relation to pollinator type. Pp. 117–41 in *Handbook of experimen-*

*tal pollination biology* (ed. C.E. Jones & R.J. Little). Scientific and Academic Editions, New York.

Baker, H.G., Baker, I. & Hodges, S.A. 1998. Sugar composition of nectars and fruits consumed by birds and bats in the tropics and subtropics. *Biotropica* **30**: 559–86.

Barajas-Morales, J. 1985. Wood structural differences between trees of two tropical forests in Mexico. *IAWA Bulletin* **6**: 355–64.

Barker, M. & Becker, P. 1995. Sap flow rate and sap nutrient content of a tropical rain forest canopy species, *Dryobalanops aromatica*, in Brunei. *Selbyana* **16**: 201–11.

Barker, M.G. & Booth, W.E. 1996. Vertical profiles in a Bruneian rain forest: II. Leaf characteristics of *Dryobalanops aromatica*. *Journal of Tropical Forest Science* **9**: 52–66.

Barker, M.G., Press, M.C. & Brown, N.D. 1997. Photosynthetic characteristics of dipterocarp seedlings in three tropical rain forest light environments: a basis for niche partitioning? *Oecologia* **112**: 453–63.

Bawa, K.S. 1990. Plant-pollinator interactions in tropical rain forests. *Annual Review of Ecology and Systematics* **21**: 399–422.

Bawa, K.S. & Crisp, J.E. 1980. Wind-pollination in the understorey of a rain forest in Costa Rica. *Journal of Ecology* **68**: 871–6.

Bawa, K.S., Bullock, S.H., Perry, D.R., Coville, R.E. & Grayum, M.H. 1985a. Reproductive biology of tropical lowland rain forest trees. Part 2. Pollination systems. *American Journal of Botany* **72**: 346–56.

Bawa, K.S., Perry, D.R. & Beach, J.H. 1985b. Reproductive biology of tropical lowland rain forest trees. Part 1. Sexual systems and incompatability mechanisms. *American Journal of Botany* **72**: 331–45.

Bazzaz, F.A. 1991. Regeneration of tropical forests: physiological responses of pioneer and secondary species. Pp. 91–118 in *Rain forest regeneration and management* (ed. A. Gómez-Pompa, T.C. Whitmore & M. Hadley). UNESCO, Paris.

Beach, J.H. 1984. The reproductive biology of the peach or 'pejibayé' palm (*Bactris gasipaes*) and a wild congener (*B. porschiana*) in the Atlantic lowlands of Costa Rica. *Principes* **28**: 107–19.

Beard, J.S. 1946. The natural vegetation of Trinidad. *Oxford Forestry Memoirs* **20**: 1–152.

Beattie, A. 1989. Myrmecotrophy: plants fed by ants. *Trends in Ecology and Evolution* **4**: 172–6.

Becker, P. 1983. Effects of insect herbivory and artificial defoliation on survival of *Shorea* seedlings. Pp. 241–52 in *Tropical rain forest: ecology and management* (ed. S.L. Sutton, T.C. Whitmore & A.C. Chadwick). Blackwell Scientific Publications, Oxford.

Becker, P. 1996. Sap flow in Bornean heath and dipterocarp forest trees during wet and dry periods. *Tree Physiology* **16**: 295–9.

Becker, P. & Castillo, A. 1990. Root architecture of shrubs and saplings in the understorey of a tropical moist forest in lowland Panama. *Biotropica* **22**: 242–9.

Becker, P., Norhartini Sharbini & Razali Yahya 1999a. Root architecture and root : shoot allocation of shrubs and saplings in two lowland tropical forests: implications for life-form composition. *Biotropica* **31**: 93–101.

Becker, P., Tyree, M.T. & Tsuda, M. 1999b. Hydraulic conductances of angiosperms versus conifers: similar transport sufficiency at the whole-plant level. *Tree Physiology* **19**: 445–52.

Bennett, R.N. & Wallsgrove, R.M. 1994. Secondary metabolites in plant defence mechanisms. *New Phytologist* **127**: 617–33.

Bentley, B.L. & Carpenter, E.J. 1984. Direct transfer of newly-fixed nitrogen from free-living epiphyllous microorganisms to their host plant. *Oecologia* **63**: 52–6.

Béreau, M., Gazel, M. & Garbaye, J. 1997. Les symbioses mycorhiziennes des arbres de la forêt tropicale humide de Guyane française. *Canadian Journal of Botany* **75**: 711–16.

Bernal, R. & Ervik, F. 1996. Floral biology and pollination of the dioecious palm *Phytelephas seemannii* in Colombia: An adaptation to Staphylinid beetles. *Biotropica* **28**: 682–96.

Bigelow, S.W. 1993. Leaf nutrients in relation to stature and life form in tropical rain forest. *Journal of Vegetation Science* **4**: 401–8.

Bittrich, V. & Amaral, M.C.E. 1996a. Pollination biology of *Symphonia globulifera* (Clusiaceae). *Plant Systematics and Evolution* **200**: 101–10.

Bittrich, V. & Amaral, M.C.E. 1996b. Flower morphology and pollination biology of some *Clusia* species from Gran Sabana (Venezuela). *Kew Bulletin* **51**: 681–94.

Blate, G.M., Peart, D.R. & Leighton, M. 1998. Post-dispersal predation on isolated seeds: a comparative study of 40 tree species in a Southeast Asian rainforest. *Oikos* **82**: 522–38.

Bloom, A.J., Chapin, F.S. & Mooney, H.A.

1985. Resource limitation in plants – an economic analogy. *Annual Review of Ecology and Systematics* **16**: 363–92.

Bolan, N.S. 1991. A critical review on the role of mycorrhizal fungi in the uptake of phosphorus by plants. *Plant and Soil* **134**: 189–207.

Bongers, F. & Popma, J. 1988. Is exposure-related variation in leaf characteristics of tropical rain forest species adaptive? Pp. 191–200 in *Plant form and vegetation structure* (ed. M.J.A. Werger, P.J.M. van der Aart, H.J. During & J.T.A. Verhoeven). SPB Academic Publishing, The Hague.

Bongers, F. & Popma, J. 1990a. Leaf characteristics of the tropical rain forest flora of Los Tuxtlas, Mexico. *Botanical Gazette* **151**: 354–65.

Bongers, F. & Popma, J. 1990b. Leaf dynamics of seedlings of rain forest species in relation to canopy gaps. *Oecologia* **82**: 122–7.

Bongers, F., Popma, J., Meave del Castillo, J. & Carabias, J. 1988. Structure and floristic composition of the lowland rain forest of Los Tuxtlas, Mexico. *Vegetatio* **74**: 55–80.

Bongers, F. & Sterck, F.J. 1998. Architecture and development of rainforest trees: responses to light. Pp. 125–62 in *Dynamics of tropical communities* (ed. D.M. Newbery, H.H.T. Prins & N.D. Brown). Blackwell Science, Oxford.

Boot, R.G.A. 1996. The significance of seedling size and growth rate of tropical rain forest tree seedlings for regeneration in canopy openings. Pp. 267–83 in *The ecology of tropical forest tree seedlings* (ed. M.D. Swaine). UNESCO, Paris.

Boshier, D.H., Chase, M.R. & Bawa, K.S. 1995. Population genetics of *Cordia alliodora* (Boraginaceae), a neotropical tree. 3. Gene flow, neighborhood, and population structure. *American Journal of Botany* **82**: 484–90.

Breitsprecher, A. & Bethel, J.S. 1990. Stem-growth periodicity of trees in a tropical wet forest of Costa Rica. *Ecology* **71**: 1156–64.

Brewer, S.W. & Rejmánek, M. 1999. Small rodents as significant dispersers of tree seeds in a Neotropical forest. *Journal of Vegetation Science* **10**: 165–74.

Brokaw, N.V.L. 1983. Groundlayer dominance and apparent inhibition of tree regeneration by *Aechmea magdalenae* (Bromeliaceae) in a tropical forest. *Tropical Ecology* **24**: 194–200.

Brown, E.D. & Hopkins, M.J.G. 1995. A test of pollinator specificity and morphological convergence between nectarivorous birds and rainforest tree flowers in New Guinea. *Oecologia* **103**: 89–100.

Brown, N.D. & Jennings, S. 1998. Gap-size niche differentiation by tropical rainforest trees: a testable hypothesis or a broken-down bandwagon? Pp. 79–93 in *Dynamics of tropical communities* (ed. D.M. Newbery, H.H.T. Prins & N.D. Brown). Blackwell Science, Oxford.

Brown, N.D. & Whitmore, T.C. 1992. Do dipterocarp seedlings really partition tropical rain forest gaps? *Philosophical Transactions of the Royal Society, London* B **335**: 369–78.

Brown, W.H. 1919. Vegetation of the Philippine Mountains. *Bureau of Science Publications* **13**: 1–434.

Bruijnzeel, L.A. 1989. Nutrient cycling in moist tropical forests: the hydrological framework. Pp. 383–415 in *Mineral nutrients in tropical forest and savanna ecosystems* (ed. J. Proctor). Blackwell Scientific Publications, Oxford.

Bruijnzeel, L.A. & Proctor, J. 1995. Hydrology and biogeochemistry of tropical montane cloud forests: what do we really know? Pp. 38–78 in *Tropical montane cloud forests* (ed. L.S. Hamilton, J.O. Juvik & F.N. Scatena). Springer-Verlag, New York.

Bruijnzeel, L.A., Waterloo, M.J., Proctor, J., Kuiters, A.T. & Kotterink, B. 1993. Hydrological observations in montane rain forests on Gunung Silam, Sabah, Malaysia, with special reference to the 'Massenerhebung' effect. *Journal of Ecology* **81**: 145–67.

Brummitt, R.K. 1992. *Vascular plant families and genera*. Royal Botanic Gardens, Kew.

Bryant, J.P., Chapin, F.S. & Klein, D.R. 1983. Carbon/nutrient balance of boreal plants in relation to vertebrate herbivory. *Oikos* **40**: 357–68.

Buchmann, S.L. 1983. Buzz pollination in angiosperms. Pp. 73–113 in *Handbook of experimental pollination biology* (ed. C.E. Jones & R.J. Little). Scientific and Academic Editions, New York.

Buchmann, S.L. 1987. The ecology of oil flowers and their trees. *Annual Review of Ecology and Systematics* **18**: 343–69.

Buchmann, S.L. & Buchmann, M.D. 1981. Anthecology of *Mouriri myrtilloides* (Melastomataceae: Memecyleae), an oil flower in Panama. *Biotropica* **13** (suppl.): 7–24.

Burger, W.C. 1980. Why are there so many kinds of flowering plants in Costa Rica? *Brenesia* **17**: 371–88.

Burkey, T.V. 1994. Tropical tree species diversity: a test of the Janzen–Connell model. *Oecologia* **97**: 533–40.

Burslem, D.F.R.P. & Whitmore, T.W. 1996. A long-term record of forest dynamics from the Solomon Islands. Pp. 121–31 in *Biodiversity and the dynamics of ecosystems* (ed. I.M. Turner, C.H. Diong, S.S.L. Lim & P.K.L. Ng). DIWPA, Kyoto.

Burslem, D.F.R.P., Grubb, P.J. & Turner, I.M. 1995. Responses to nutrient addition among shade-tolerant tree seedlings of lowland tropical rain forest in Singapore. *Journal of Ecology* **83**: 113–22.

Burslem, D.F.R.P., Grubb, P.J. & Turner, I.M. 1996. Responses to simulated drought and elevated nutrient supply among shade-tolerant tree seedlings of lowland tropical forest in Singapore. *Biotropica* **28**: 636–48.

Cain, S.A., de Oliveira Castro, G.M., Pires, J.M. & da Silva, N.T. 1956. Applications of some phytosociological techniques to Brazilian rain forest. *American Journal of Botany* **43**: 911–41.

Canny, M.J. 1995. A new theory for the ascent of sap – cohesion supported by tissue pressure. *Annals of Botany* **75**: 343–57.

Canny, M.J. 1997a. Vessel contents of leaves after excision – a test of Scholander's assumption. *American Journal of Botany* **84**: 1217–22.

Canny, M.J. 1997b. Vessel contents during transpiration – embolisms and refilling. *American Journal of Botany* **84**: 1223–30.

Carey, E.V., Brown, S., Gillespie, A.J.R. & Lugo, A.E. 1994. Tree mortality in mature lowland tropical moist and tropical lower montane moist forests of Venezuela. *Biotropica* **26**: 255–65.

Carlquist, S. 1988. *Comparative wood anatomy*. Springer-Verlag, Berlin. 436 pp.

Carthew, S.M. & Goldingay, R.L. 1997. Non-flying mammals as pollinators. *Trends in Ecology and Evolution* **12**: 104–8.

Cavelier, J. & Goldstein, G. 1989. Leaf anatomy and water relations in tropical elfin cloud forest tree species. Pp. 243–53 in *Structural and functional responses to environmental stresses* (ed. K.H. Kreeb, H. Richter & T.M. Hinckley). SPB Academic, The Hague.

Chambers, J.Q., Higuchi, N. & Schimel, J.P. 1998. Ancient trees in Amazonia. *Nature* **391**: 135–6.

Chan, H.T. & Appanah, S. 1980. Reproductive biology of some Malaysian dipterocarps. I. Flowering biology. *Malaysian Forester* **43**: 132–43.

Chapman, C.A. & Chapman, L.J. 1996. Frugivory and the fate of dispersed and non-dispersed seeds of six African tree species. *Journal of Tropical Ecology* **12**: 491–504.

Chapman, C.A., Kaufman, L. & Chapman, L.J. 1998. Buttress formation and directional stress experienced during critical phases of tree development. *Journal of Tropical Ecology* **14**: 431–9.

Chapman, L.J., Chapman, C.A. & Wrangham, R.W. 1992. *Balanites wilsoniana*: elephant dependent dispersal? *Journal of Tropical Ecology* **8**: 275–83.

Charles-Dominique, P., Atramentowicz, M., Gérard, C.-D. H., Hladik, A., Hladik, C.M. & Prévost, M.F. 1981. Les mammiferes frugivores arboricoles nocturnes d'une forêt guyanaise: inter-relations plantes-animaux. *Revue d'Ecologie (La Terre et la Vie)* **35**: 341–5.

Chase, M.R., Moller, C., Kesseli, R. & Bawa, K.S. 1996. Distant gene flow in tropical trees. *Nature* **383**: 398–9.

Chawla, A.S. & Kapoor, V.K. 1997. *Erythrina* alkaloids. Pp. 37–49 in *Plant and fungal toxicants* (ed. J.P.F. D'Mello). CRC Press, Boca Raton.

Chazdon, R.L. 1988. Sunflecks and their importance to forest understorey plants. *Advances in Ecological Research* **18**: 1–63.

Chazdon, R.L. 1991. Effects of leaf and ramet removal on growth and reproduction of *Geonoma congesta*, a clonal understorey palm. *Journal of Ecology* **79**: 1137–46.

Chazdon, R.L. & Field, C.B. 1987. Determinants of photosynthetic capacity in six rainforest *Piper* species. *Oecologia* **73**: 222–30.

Chazdon, R.L. & Pearcy, R.W. 1986. Photosynthetic responses to light variation in rainforest species I. Induction under constant and fluctuating light conditions. *Oecologia* **69**: 517–23.

Chazdon, R.L., Pearcy, R.W., Lee, D.W. & Fetcher, N. 1996. Photosynthetic responses of tropical forest plants to contrasting light environments. Pp. 5–55 in *Tropical forest plant ecophysiology* (ed. S.S. Mulkey, R.L. Chazdon & A.P. Smith). Chapman & Hall, New York.

Chenery, E.M. & Sporne, K.R. 1976. A note on the evolutionary status of aluminium-accumulators among dicotyledons. *New Phytologist* **76**: 551–4.

Choong, M.F. 1996. What makes a leaf tough and how this affects the pattern of *Castanop-*

*sis fissa* leaf consumption by caterpillars. *Functional Ecology* **10**: 668–74.
Choong, M.F., Lucas, P.W., Ong, J.S.Y., Pereira, B., Tan, H.T.W. & Turner, I.M. 1992. Leaf fracture toughness and sclerophylly: their correlations and ecological implications. *New Phytologist* **121**: 597–610.
Cintra, R. & Horna, V. 1997. Seed and seedling survival of the palm *Astrocaryum murumuru* and the legume tree *Dipteryx micrantha* in gaps in Amazonian forest. *Journal of Tropical Ecology* **13**: 257–77.
Cipollini, M.L. & Levey, D.J. 1998. Secondary metabolites as traits of ripe fleshy fruits: a response to Eriksson and Ehrlén. *American Naturalist* **152**: 908–11.
Clark, D.A. & Clark, D.B. 1984. Spacing dynamics of a tropical rain forest tree: evaluation of the Janzen–Connell model. *American Naturalist* **124**: 769–88.
Clark, D.A. & Clark, D.B. 1992. Life history diversity of canopy and emergent trees in a neotropical rain forest. *Ecological Monographs* **62**: 315–44.
Clark, D.A. & Clark, D.B. 1994. Climate-induced annual variation in canopy tree growth in a Coast Rican tropical rain forest. *Journal of Ecology* **82**: 865–72.
Clark, D.A., Clark, D.B. & Grayum, M.H. 1992. Leaf demography of a neotropical rain forest cycad, *Zamia skinneri* (Zamiaceae). *American Journal of Botany* **79**: 28–33.
Clark, D.B. & Clark, D.A. 1989. The role of physical damage in the seedling mortality regime of a neotropical rain forest. *Oikos* **55**: 225–30.
Clark, D.B. & Clark, D.A. 1991. The impact of physical damage on canopy tree regeneration in tropical rain forest. *Journal of Ecology* **79**: 447–57.
Clark, D.B. & Clark, D.A. 1996. Abundance, growth and mortality of very large trees in neotropical lowland rain forest. *Forest Ecology and Management* **80**: 235–44.
Clark, D.B., Clark, D.A. & Rich, P.M. 1993. Comparative analysis of microhabitat utilization by saplings of nine tree species in neotropical rain forest. *Biotropica* **25**: 397–407.
Clark, D.B., Clark, D.A., Rich, P.M., Weiss, S. & Oberbauer, S.F. 1996. Landscape-scale evaluation of understory light and canopy sytructure: methods and application in a neotropical lowland rain forest. *Canadian Journal of Forest Research* **26**: 747–57.
Claussen, J.W. & Maycock, C.R. 1995. Stem allometry in a North Queensland tropical rainforest. *Biotropica* **27**: 421–6.
Coley, P.D. 1983. Herbivory and defensive characteristics of tree species in a lowland tropical forest. *Ecological Monographs* **53**: 209–33.
Coley, P.D. 1986. Costs and benefits of defense by tannins in a neotropical tree. *Oecologia* **70**: 238–41.
Coley, P.D. 1987. Patrones en las defensas de las plantas: ¿porqué los herbívoros ciertas especies? *Revista de Biología Tropical* **35** (suppl. 1): 151–64.
Coley, P.D. 1988. Effects of plant growth rate and leaf lifetime on the amount and type of anti-herbivore defense. *Oecologia* **74**: 531–6.
Coley, P.D. & Aide, T.M. 1989. Red coloration of tropical leaves: a possible antifungal defence. *Journal of Tropical Ecology* **5**: 293–300.
Coley, P.D. & Aide, T.M. 1991. Comparison of herbivory and plant defenses in temperate and tropical broad-leaved forests. Pp. 25–49 in *Plant–animal interactions: evolutionary ecology in tropical and temperate regions* (ed. P.W. Price, T.M. Lewinsohn, G.W. Fernandes & W.W. Benson). John Wiley & Sons, New York.
Coley, P.D. & Barone, J.A. 1996. Herbivory and plant defenses in tropical forests. *Annual Review of Ecology and Systematics* **27**: 305–35.
Coley, P.D., Bryant, J.P. & Chapin, F.S. 1985. Resourve availability and plant anti-herbivore defense. *Science* **230**: 895–9.
Coley, P.D. & Kursar, T.A. 1996. Anti-herbivore defenses of young tropical leaves: physiological constraints and ecological trade-offs. Pp. 305–36 in *Tropical forest plant ecophysiology* (ed. S.S. Mulkey, R.L. Chazdon & A.P. Smith). Chapman & Hall, New York.
Comstock, J.P. 1999. Why Canny's theory doesn't hold water. *American Journal of Botany* **86**: 1077–81.
Condit, R. 1995. Research in large, long-term tropical forest plots. *Trends in Ecology and Evolution* **10**: 18–22.
Condit, R., Ashton, P.S., Manokaran, N., LaFrankie, J.V., Hubbell, S.P. & Foster, R.B. 1999. Dynamics of the forest communities at Pasoh and Barro Colorado: comparing two 50–ha plots. *Philosopical Transactions of the Royal Society of London* B **354**: 1739–48.
Condit, R., Hubbell, S.P. & Foster, R.B. 1992a. Short-term dynamics of a neotropical forest. *BioScience* **42**: 822–8.
Condit, R., Hubbell, S.P. & Foster, R.B. 1992b.

Recruitment near conspecific adults and the maintenance of tree and shrub diversity in a neotropical forest. *American Naturalist* **140**: 261–86.

Condit, R., Hubbell, S.P. & Foster, R.B. 1993. Identifying fast-growing native trees from the neotropics using data from a large, permanent census plot. *Forest Ecology and Management* **62**: 123–43.

Condit, R., Hubbell, S.P. & Foster, R.B. 1995. Mortality rates of 205 neotropical tree and shrub species and the impact of a severe drought. *Ecological Monographs* **65**: 419–39.

Condit, R., Hubbell, S.P. & Foster, R.B. 1996a. Assessing the response of plant functional types to climatic change in tropical forests. *Journal of Vegetation Science* **7**: 405–16.

Condit, R., Hubbell, S.P., LaFrankie, J.V., Sukumar, R., Manokaran, N, Foster, R.B. & Ashton, P.S. 1996b. Species–area and species–individual relationships for tropical trees: a comparison of three 50–ha plots. *Journal of Ecology* **84**: 549–62.

Condit, R., Sukumar, R., Hubbell, S.P. & Foster, R.B. 1998. Predicting population trends from size distributions: a direct test in a tropical tree community. *American Naturalist* **152**: 495–509.

Connell, J.H. 1971. On the role of natural enemies in preventing competitive exclusion in some marine animals and in rain forest trees. Pp. 298–312 in *Dynamics of populations* (ed. P.J. den Boer & G.R. Gradwell). PUDOC, Wageningen.

Connell, J.H. & Lowman, M.D. 1989. Low-diversity tropical rain forests: some possible mechanisms for their existence. *American Naturalist* **134**: 88–119.

Connell, J.H., Tracey, J.G. & Webb, L.J. 1984. Compensatory recruitment, growth, and mortality as factors maintaining rain forest tree diversity. *Ecological Monographs* **54**: 141–61.

Coomes, D.A. & Grubb, P.J. 1996. Amazonian caatinga and related communities at La Esmeralda, Venezuela: forest structure, physiognomy and floristics, and control by soil factors. *Vegetatio* **122**: 167–91.

Coomes, D.A. & Grubb, P.J. 1998a. Responses of juvenile trees to above- and belowground competition in nutrient-starved Amazonian rain forest. *Ecology* **79**: 768–82.

Coomes, D.A. & Grubb, P.J. 1998b. A comparison of 12 tree species of Amazonian caatinga using growth rates in gaps and understorey, and allometric relationships. *Functional Ecology* **12**: 426–35.

Corlett, R.T. 1996. Characteristics of vertebrate-dispersed fruits in Hong Kong. *Journal of Tropical Ecology* **12**: 819–33.

Corlett, R.T. 1998. Frugivory and seed dispersal by vertebrates in the Oriental (Indomalayan) region. *Biological Reviews of the Cambridge Philosophical Society* **73**: 413–48.

Croat, T.B. 1979. The sexuality of the Barro Colorado Island flora (Panama). *Phytologia* **42**: 319–48.

Crome, F.H.J. & Irvine, A.K. 1986. 'Two bob each way': the pollination and breeding system of the Australian rain forest tree *Syzygium cormiflorum* (Myrtaceae). *Biotropica* **18**: 115–25.

Crook, M.J. & Ennos, A.R. 1998. The increase in anchorage with tree size of the tropical tap rooted tree *Mallotus wrayi* King (Euphorbiaceae). *Annals of Botany* **82**: 291–6.

Crook, M.J., Ennos, A.R. & Banks, J.R. 1997. The function of buttress roots: a comparative study of the anchorage systems of buttressed (*Aglaia* and *Nephelium ramboutan* species) and non-buttressed (*Mallotus wrayi*) tropical trees. *Journal of Experimental Botany* **48**: 1703–16.

Cunningham, S.A. 1995. Ecological constraints in fruit initiation by *Calyptrogyne ghiesbreghtiana* (Arecaceae): floral herbivory, pollen availability, and visitation by pollinating bats. *American Journal of Botany* **82**: 1527–36.

Cunningham, S.A. 1997. The effect of light environment, leaf area, and stored carbohydrates on inflorescence production by a rain forest understorey palm. *Oecologia* **111**: 36–44.

Dalling, J.W., Hubbell, S.B. & Silvera, K. 1998a. Seed dispersal, seedling establishment and gap partitioning among tropical pioneer trees. *Journal of Ecology* **86**: 674–89.

Dalling, J.W., Swaine, M.D. & Garwood, N.C. 1998b. Dispersal patterns and seed bank dynamics of pioneer trees in moist tropical forest. *Ecology* **79**: 564–78.

Dalling, J.W. & Wirth, R. 1998. Dispersal of *Miconia argentea* seeds by the leaf-cutting ant *Atta colombica*. *Journal of Tropical Ecology* **14**: 705–10.

D'Arcy, W.G. 1973. *Correliana* (Myrsinaceae), a new palmoid genus of the tropical rain forest. *Annals of the Missouri Botanical Garden* **60**: 442–8.

Davies, A.G., Bennett, E.L. & Waterman, P.G. 1988. Food selection by two South-East Asian colobine monkeys (*Presbytis rubicunda* and *Presbytis melalophos*) in relation to plant chemistry. *Biological Journal of the Linnean Society* **34**: 33–56.

Davies, S.J. 1998. Photosynthesis of nine pioneer *Macaranga* species from Borneo in relation to life history. *Ecology* **79**: 2292–308.

Davies, S.J. & Ashton, P.S. 1999. Phenology and fecundity in 11 sympatric pioneer species of *Macaranga* (Euphorbiaceae) in Borneo. *American Journal of Botany* **86**: 1786–95.

Davies, S.J., Palmiotto, P.A., Ashton, P.S., Lee, H.S. & LaFrankie, J.V. 1998. Comparative ecology of 11 sympatric species of *Macaranga* in Borneo: tree distribution in relation to horizontal and vertical resource heterogeneity. *Journal of Ecology* **86**: 662–73.

Dayanandan, S., Attygalla, D.N.C., Abeygunasekera, A.W.W.L., Gunatilleke, I.A.U.N. & Gunatilleke, C.V.S. 1990. Phenology and floral morphology in relation to pollination of some Sri Lankan dipterocarps. Pp. 103–33 in *Reproductive ecology of tropical forest plants* (ed. K.S. Bawa & M. Hadley). UNESCO, Paris.

de Alwis, D.P. & Abeynayake, K. 1980. A survey of mycorrhizae in some forest trees of Sri Lanka. Pp. 146–53 in *Tropical mycorrhiza research* (ed. P. Mikola). Clarendon Press, Oxford.

de Castro, F., Williamson, G.B. & Moraes de Jesus, R. 1993. Radial variation in the wood specific gravity of *Joannesia princeps*: the roles of age and diameter. *Biotropica* **25**: 176–82.

de Foresta, H. 1983. Le spectre architectural: application à l'étude des relations entre architecture des arbres et écologie forestière. *Bulletin du Muséum d'Histoire Naturelle, Adansonia* **5**: 295–302.

de Foresta, H. & Kahn, F. 1984. Un systeme racinaire adventif dans un tronc creux d'*Eperua falcata* Aubl. *Revue d'Ecologie (la Terre et la Vie)* **39**: 347–50.

de la Fuente, M.A.S. & Marquis, R.J. 1999. The role of ant-tended extrafloral nectaries in the protection and benefit of a Neotropical rainforest tree. *Oecologia* **118**: 192–202.

De Steven, D. 1988. Light gaps and long-term seedling performance of a neotropical canopy tree (*Dipteryx panamensis*, Leguminosae). *Journal of Tropical Ecology* **4**: 407–11.

De Steven, D. 1994. Tropical tree seedling dynamics: recruitment patterns and ther population consequences for three canopy species in Panama. *Journal of Tropical Ecology* **10**: 369–83.

De Steven, D. & Putz, F.E. 1984. Impact of mammals on early recruitment of a tropical canopy tree, *Dipteryx panamensis*, in Panama. *Oikos* **43**: 207–16.

den Outer, R.W. 1993. Evolutionary trends in secondary phloem anatomy of trees, shrubs and climbers from Africa (mainly Ivory Coast). *Acta Botanica Neerlandica* **42**: 269–87.

den Outer, R.W. & van Veenendaal, W.L.H. 1976. Variation in wood anatomy of species with a distribution covering both rain forest and savanna areas of the Ivory Coast, West-Africa. *Leiden Botanical Series* **3**: 182–95.

Denslow, J.S. & Gomez Dias, A.E. 1990. Seed rain to tree-fall gaps in a neotropical rain forest. *Canadian Journal of Forest Research* **20**: 642–8.

Denslow, J.S., Newell, E.A. & Ellison, A.M. 1991. The effects of understorey palms and cyclanths on the growth and survival of *Inga* seedlings. *Biotropica* **23**: 225–34.

Denslow, J.S., Schultz, J.C., Vitousek, P.M. & Strain, B.R. 1990. Growth responses of tropical shrubs to treefall gap environments. *Ecology* **71**: 165–79.

Detienne, P. & Chanson, B. 1996. L'éventail de la densité du bois des feuillus. *Bois et Forêts des Tropiques* **250**: 19–30.

Dew, J.L. & Wright, P. 1998. Frugivory and seed dispersal by four species of primates in Madagascar's eastern rain forest. *Biotropica* **30**: 425–37.

Dickinson, T.A. & Tanner, E.V.J. 1978. Exploitation of hollow trunks by tropical trees. *Biotropica* **10**: 231–3.

D'Mello, J.P.F. 1995. Toxicity of non-protein amino acids from plants. Pp. 145–53 in *Amino acids and their derivatives in higher plants* (ed. R.M. Wallsgrove). SEB Seminar Series 56. Cambridge University Press, Cambridge.

Dobat, K. & Peikert-Holle, T. 1985. *Blüten und Fledermäuse. Bestäubung durch Fledermäuse und Flughunde (Chiropterophilie).* Kramer, Frankfurt.

Doley, D., Yates, D.J. & Unwin, G.L. 1987. Photosynthesis in an Australian rainforest tree, *Argyrodendron peralatum*, during the rapid development and relief of water deficits in the dry season. *Oecologia* **74**: 441–9.

Doligez, A. & Joly, H. 1997. Mating system of *Carapa procera* (Meliaceae) in the French Guiana tropical forest. *American Journal of Botany* **84**: 461–70.

Dolph, G.E. & Dilcher, D.L. 1980. Variation in leaf size with respect to climate in Costa

Rica. *Biotropica* **12**: 91–9.

Donoghue, M.J. & Ackerly, D.D. 1996. Phylogenetic uncertainties and sensitivity analysis in comparative biology. *Philosopical Transaction of the Royal Society of London* B **351**: 1241–9.

Dowsett-Lemaire, F. 1988. Fruit choice and seed dissemination by birds and mammals in the evergreen forests of upland Malawi. *Revue d'Ecologie (La Terre et la Vie)* **43**: 251–85.

Dubost, G. 1984. Comparison of the diets of frugivorous forest ruminants of Gabon. *Journal of Mammalogy* **65**: 298–316.

Ducrey, M. 1992. Variation in leaf morphology and branching pattern of some tropical rain forest species from Guadeloupe (French West Indies) under semi-controlled light conditions. *Annales des Sciences Forestieres* **49**: 553–70.

Eguiarte, L.E., Pérez-Nasser, N. & Piñero, D. 1992. Genetic structure, outcrossing rate and heterosis in *Astrocaryum mexicanum* (tropical palm): implications for evolution and conservation. *Heredity* **69**: 217–28.

Ellison, A.M., Denslow, J.S., Loiselle, B.A., Brenés M., D. 1993. Seed and seedling ecology of neotropical Melastomataceae. *Ecology* **74**: 1733–49.

Emmons, L.H. 1991. Frugivory in treeshrews (*Tupaia*). *American Naturalist* **138**: 642–49.

Endress, P.K. 1994. *Diversity and evolutionary biology of tropical flowers*. Cambridge University Press, Cambridge.

Ennos, A.R. 1993a. The function and formation of buttresses. *Trends in Ecology and Evolution* **8**: 350–1.

Ennos, A.R. 1993b. The scaling of root anchorage. *Journal of Theoretical Biology* **161**: 61–75.

Ennos, A.R. 1995. Development of buttresses in rainforest trees: the influence of mechanical stress. Pp. 293–301 in *Wind and trees* (ed. M.P. Coutts & J. Grace). Cambridge University Press, Cambridge.

Ennos, A.R. 1997. Wind as an ecological factor. *Trends in Ecology and Evolution* **12**: 108–11.

Eriksson, O. & Ehrlén, J. 1998. Secondary metabolites in fleshy fruits: are adaptive explanations needed? *American Naturalist* **152**: 905–7.

Ervik, F. & Bernal, R. 1996. Floral biology and insect visitation of the monoecious palm *Prestoea decurrens* on the Pacific Coast of Colombia. *Principes* **40**: 86–92.

Ervik, F. & Feil, J.P. 1997. Reproductive biology of the monoecious understory palm *Prestoea schultzeana* in Amazonian Ecuador. *Biotropica* **29**: 309–17.

Eschenbach, C., Glauner, R., Kleine, M., Kappen, L. 1998. Photosynthesis rates of selected tree species in lowland dipterocarp rainforest of Sabah, Malaysia. *Trees* **12**: 356–65.

Estrada, A. & Coates-Estrada, R. 1991. Howler monkeys (*Alouatta palliata*), dung beetles (Scarabaeidae) and seed dispersal: ecological interactions in the tropical rain forest of Los Tuxtlas, Mexico. *Journal of Tropical Ecology* **7**: 459–74.

Everham, E.M., Myster, R.W. & Van-DeGenachte, E. 1996. Effects of light, moisture, temperature, and litter on the regeneration of five tree species in the tropical montane forest of Puerto Rico. *American Journal of Botany* **83**: 1063–8.

Ewers, F.W., Carlton, M.R., Fisher, J.B., Kolb, K.J. & Tyree, M.T. 1997. Vessel diameters in roots versus stems of tropical lianas and other growth forms. *IAWA Journal* **18**: 261–79.

Fægri, K. & van der Pijl, L. 1979. *The principles of pollination ecology*. Third edition. Pergamon Press, Oxford. 244 pp.

Farrell, B.D., Dussourd, D.E. & Mitter, C. 1991. Escalation of plant defense: do latex and resin canals spur plant diversification? *American Naturalist* **138**: 881–900.

Favrichon, V. 1994. Classification des espèces arborées en groupes fonctionnels en vue de la réalisation d'un modèle de dynamique de peuplement en forêt guyanaise. *Revue d'Ecologie (La Terre et la Vie)* **49**: 379–403.

Favrichon, V. 1998. Apports d'un modèle démographique plurispécifique pour l'étude des relations diversité/dynamique en forêt tropicale guyanaise. *Annales des Sciences Forestieres* **55**: 655–69.

Fetcher, N., Oberbauer, S.F., Rojas, G. & Strain, B.R. 1987. Efectos del régimen de luz sobre la fotosíntesis y el crecimiento en plántulas de árboles de un bosque lluvioso tropical de Costa Rica. *Revista de Biología Tropical* **35** (suppl. 1): 97–110.

Fetcher, N., Strain, B.R., Oberbauer, S.F. 1983. Effects of light regime on the growth, leaf morphology, and water relations of seedlings of two species of tropical trees. *Oecologia* **58**: 314–18.

Fiala, B. & Linsenmair, K.E. 1995. Distribution and abundance of plants with extrafloral nectaries in the woody flora of a lowland primary forest in Malaysia. *Biodiversity and*

*Conservation* **4**: 165–82.
Fiala, B., Maschwitz, U. & Tho, Y.P. 1991. The association between *Macaranga* trees and ants in South-east Asia. Pp. 263–309 in *Ant-plant interactions* (ed. C.R. Huxley & D.F. Cutler). Oxford University Press, Oxford.
Fineran, B.A. 1991. Root hemi-parasitism in the Santalales. *Botanische Jahrbucher für Systematik, Pflanzengeschichte und Pflanzengeographie* **113**: 271–5.
Fisher, B.L., Howe, H.F. & Wright, S.J. 1991. Survival and growth of *Virola surinamensis* yearlings: water augmentation in gap and understorey. *Oecologia* **86**: 292–7.
Fisher, J.B. 1982. A survey of buttresses and aerial roots of tropical trees for presence of reaction wood. *Biotropica* **14**: 56–61.
Fitter, A.H. & Moyersoen, B. 1996. Evolutionary trends in root-microbe symbioses. *Philosophical Transaction of the Royal Society of London* B **351**: 1367–75.
Fleming, T.H. 1988. *The short-tailed fruit bat. A study in plant–animal interactions*. University of Chicago Press, Chicago.
Fleming, T.H. & Heithaus, E.R. 1981. Frugivorous bats, seed shadows, and the structure of tropical forests. *Biotropica* (suppl.) **13**: 45–53.
Folgarait, P.J. & Davidson, D.W. 1994. Antiherbivore defenses of myrmecophytic *Cecropia* under different light regimes. *Oikos* **71**: 305–20.
Folgarait, P.J. & Davidson, D.W. 1995. Myrmecophytic *Cecropia*: antiherbivore defenses under different nutrient treatments. *Oecologia* **104**: 189–206.
Forget, P.-M. 1989. La régénération naturelle d'une espèce autochore de la forêt Guyanaise: *Eperua falcata* Aublet (Caesalpiniaceae). *Biotropica* **21**: 115–25.
Forget, P.-M. 1990. Seed-dispersal of *Vouacapoua americana* (Caesalpiniaceae) by caviomorph rodents in French Guiana. *Journal of Tropical Ecology* **6**: 459–68.
Forget, P.-M. 1992a. Regeneration ecology of *Eperua grandiflora* (Caesalpiniaceae), a rodent-dispersed tree species in French Guiana. *Biotropica* **24**: 146–56.
Forget, P.-M. 1992b. Seed removal and seed fate in *Gustavia superba* (Lecythidaceae). *Biotropica* **24**: 408–14.
Forget, P.-M. 1993. Post-dispersal predation and scatterhoarding of *Dipteryx panamensis* (Papilionaceae) seeds by rodents in Panama. *Oecologia* **94**: 255–61.
Forget, P.M. & Milleron, T. 1991. Evidence for secondary dispersal by rodents in Panama. *Oecologia* **87**: 596–9.
Forget, P.-M., Milleron, T. & Feer, F. 1998. Patterns in post-dispersal seed removal by neotropical rodents and seed fate in relation to seed size. Pp. 25–49 in *Dynamics of tropical communities* (ed. D.M. Newbery, H.H.T. Prins & N.D. Brown). Blackwell Science, Oxford.
Forget, P.-M., Munoz, E. & Leigh, E.G. 1994. Predation by rodents and bruchid beetles on seeds of *Scheelea* palms on Barro Colorado Island, Panama. *Biotropica* **26**: 420–6.
Foster, M.S. & Delay, L.S. 1998. Dispersal of mimetic seeds of three species of *Ormosia* (Leguminosae). *Journal of Tropical Ecology* **14**: 389–411.
Foster, R.B. 1977. *Tachigalia versicolor* is a suicidal neotropical tree. *Nature* **268**: 624–6.
Foster, S.A. 1986. On the adaptive value of large seeds for tropical moist forests: a review and synthesis. *Botanical Review* **52**: 260–99.
Foster, S.A. & Janson, C.H. 1985. The relationship between seed size and establishment conditions in tropical woody plants. *Ecology* **66**: 773–80.
Fragoso, J.M.V. 1997. Tapir-generated seed shadows: scale-dependent patchiness in the Amazon rain forest. *Journal of Ecology* **85**: 519–29.
Frankel, S. & Berenbaum, M. 1999. Effects of light regime on antioxidant content of foliage in a tropical forest community. *Biotropica* **31**: 422–9.
Fraver, S., Brokaw, N.V.L. & Smith, A.P. 1998. Delimiting the gap phase in the growth cycle of a Panamanian forest. *Journal of Tropical Ecology* **14**: 673–81.
Fredeen, A.L. & Field, C.B. 1991. Leaf respiration in *Piper* species native to a Mexican rainforest. *Physiologia Plantarum* **82**: 85–92.
Fredeen, A.L., Griffin, K. & Field, C.B. 1991. Effects of light quantity and quality and soil nitrogen status on nitrate reductase activity in rainforest species of the genus *Piper*. *Oecologia* **86**: 441–6.
Gale, N. & Barfod, A.S. 1999. Canopy tree mode of death in a western Ecuadorian rain forest. *Journal of Tropical Ecology* **15**: 415–36.
Ganzhorn, J.U. 1992. Leaf chemistry and the biomass of folivorous primates in tropical forests. *Oecologia* **91**: 540–7.
Garber, P.A. 1986. The ecology of seed dispersal in two species of callitrichid primates (*Saguinus mystax* and *Seguinus fuscicollis*).

*American Journal of Primatology* **10**: 155–70.
Gartner, B.L. 1989. Breakage and regrowth of *Piper* species in rain forest understory. *Biotropica* **21**: 303–7.
Garwood, N.C. 1983. Seed germination in a seasonal tropical forest in Panama: a community study. *Ecological Monographs* **53**: 159–81.
Garwood, N.C. 1985. The role of mucilage in the germination of Cuipo, *Cavanillesia platanfolia* (H. & B.) H.B.K. (Bombacaceae), a tropical tree. *American Journal of Botany* **72**: 1095–105.
Garwood, N.C. 1989. Tropical soil seed banks: a review. Pp. 149–209 in *Ecology of soil seed banks* (ed. M.A. Leck, V.T. Parker & R.L. Simpson). Academic Press, San Diego.
Garwood, N.C. 1996. Functional morphology of tropical tree seedlings. Pp. 59–129 in *The ecology of tropical forest tree seedlings* (ed. M.D. Swaine). UNESCO, Paris.
Garwood, N.C. & Lighton, J.R.B. 1990. Physiological ecology of seed respiration in some tropical species. *New Phytologist* **115**: 549–58.
Gaume, L., McKey, D. & Anstett, M.-C. 1997. Benefits conferred by 'timid' ants: active anti-herbivore protection of the rainforest tree *Leonardoxa africana* by the minute ant *Petalomyrmex phylax*. *Oecologia* **112**: 209–16.
Gautier-Hion, A., Duplantier, J.-M., Quris, R., Feer, F., Sourd, C., Decoux, J.-P., Dubost, G., Emmons, L., Erard, C., Hecketsweiler, P., Moungazi, A., Roussilhon, C. & Thiollay, J.-M. 1985. Fruit characters as a basis of fruit choice and seed dispersal in a tropical forest vertebrate community. *Oecologia* **65**: 324–37.
Gautier-Hion, A. & Maisels, F. 1994. Mutualism between a leguminous tree and large African monkeys as pollinators. *Behavioural Ecology and Sociobiology* **34**: 203–10.
Gautier-Hion, A. & Michaloud, G. 1989. Are figs always keystone resources for tropical frugivorous vertebrates? A test in Gabon. *Ecology* **70**: 1826–33.
Gavin, D.G. & Peart, D.R. 1997. Spatial structure and regeneration of *Tetramerista glabra* in peat swamp rain forest in Indonesian Borneo. *Plant Ecology* **131**: 223–31.
Gavin, D.G. & Peart, D.R. 1999. Vegetative life history of a dominant rain forest canopy tree. *Biotropica* **31**: 288–94.
Gentry, A.H. 1969. A comparison of some leaf characteristics of tropical dry forest and tropical wet forest in Costa Rica. *Turrialba* **19**: 419–28.
Gentry, A.H. 1982. Patterns of neotropical plant species diversity. *Evolutionary Biology* **15**: 1–84.
Gershenzon, J. 1994. Metabolic costs of terpenoid accumulation in higher plants. *Journal of Chemical Ecology* **20**: 1281–328.
Gilbert, G.S., Hubbell, S.P. & Foster, R.B. 1994. Density and distance-to-adult effects of a canker disease of trees in a moist tropical forest. *Oecologia* **98**: 100–8.
Gitay, H., Noble, I.R. & Connell, J.H. 1999. Deriving functional types for rain-forest trees. *Journal of Vegetation Science* **10**: 641–50.
Gittleman, J.L. & Luh, H.-K. 1992. On comparing comparative methods. *Annual Review of Ecology and Systematics* **23**: 383–404.
Givnish, T.J. 1984. Leaf and canopy adaptations in tropical forests. Pp. 51–84 in *Physiological ecology of plants in the wet tropics* (ed. E. Medina, H.A. Mooney & C. Vázquez-Yánes). Junk, The Hague.
Givnish, T.J. 1987. Comparative studies of leaf form: assessing the relative roles of selective pressures and phylogenetic constraints. *New Phytologist* **106** (suppl. 1): 131–60.
Givnish, T.J. 1999. On the causes of gradients in tropical tree diversity. *Journal of Ecology* **87**: 193–210.
Givnish, T.J., Sytsma, K.J., Smith, J.F. & Hahn, W.J. 1994. Thorn-like prickles and heterophylly in *Cyanea*: adaptation to extinct avian browsers in Hawaii? *Proceedings of the National Academy of Sciences of the United States of America* **91**: 2810–14.
Goldstein, G., Andrade, J.L., Meinzer, F.C., Holbrook, N.M., Cavelier, J., Jackson, P. & Celis, A. 1998. Stem water storage and diurnal patterns of water use in tropical forest canopy trees. *Plant, Cell and Environment* **21**: 397–406.
Gorchov, D.L., Cornejo, F., Ascorra, C.F. & Jaramillo, M. 1995. Dietary overlap between frugivorous birds and bats in the Peruvian Amazon. *Oikos* **74**: 235–50.
Gould, E. 1978. Foraging behavior of Malaysian nectar-feeding bats. *Biotropica* **10**: 184–93.
Gould, S.J. & Lewontin, R.C. 1979. The spandrels of San Marco and the Panglossian paradigm: a critique of the adaptationist programme. *Proceedings of the Royal Society of London* B **205**: 581–98.
Goulding, M. 1980. *The fishes and the forest.* University of California Press, Berkeley.
Gribel, R., Gibbs, P.E. & Quieróz, A.L. 1999.

Flowering phenology and pollination biology of *Ceiba pentandra* (Bombacaceae) in Central Amazonia. *Journal of Tropical Ecology* **15**: 247–63.

Grieg, N. 1993. Predispersal seed predation on five *Piper* species in tropical rainforest. *Oecologia* **93**: 412–20.

Grubb, P.J. 1974. Factors controlling the distribution of forest-types on tropical mountains: new facts and a new perspective. Pp. 13–46 in *Altitudinal zonation in Malesia* (ed. J. R. Flenley). *University of Hull Geography Department Miscellaneous Series* No. 16.

Grubb, P.J. 1977. Control of forest growth and distribution on wet tropical mountains: with special reference to mineral nutrition. *Annual Review of Ecology and Systematics* **8**: 83–107.

Grubb, P.J. 1996. Rainforest dynamics: the need for new paradigms. Pp. 215–33 in *Tropical rainforest research – current issues* (ed. D.S. Edwards, W.E. Booth & S.C. Choy). Kluwer Academic Publishers, Dordrecht.

Grubb, P.J. 1998a. A reassessment of the strategies of plants which cope with shortages of resources. *Perspectives in Plant Ecology, Evolution and Systematics* **1**: 1–29.

Grubb, P.J. 1998b. Seeds and fruits of tropical rainforest plants: interpretation of the range in seed size, degree of defence and flesh/seed quotients. Pp. 1–24 in *Dynamics of tropical communities* (ed. D.M. Newbery, H.H.T. Prins & N.D. Brown). Blackwell Science, Oxford.

Grubb, P.J. & Burslem, D.F.R.P. 1998. Mineral nutrient concentrations as a function of seed size within seed crops: implications for competition among seedlings and defence against herbivory. *Journal of Tropical Ecology* **14**: 177–85.

Grubb, P.J. & Coomes, D.A. 1997. Seed mass and nutrient content in nutrient-starved tropical rainforests in Venezuela. *Seed Science Research* **7**: 269–80.

Grubb, P.J., Lloyd, J.R., Pennington, T.D. & Whitmore, T.C. 1963. A comparison of montane and lowland rainforest in Ecuador. I. The forest structure, physiognomy and florisitics. *Journal of Ecology* **51**: 567–601.

Grubb, P.J. & Metcalfe, P.J. 1996. Adaptation and inertia in the Australian tropical lowland rain-forest flora: contradictory trends in intergeneric comparisons of seed size in relation to light demand. *Functional Ecology* **10**: 512–20.

Grubb, P.J., Metcalfe, D.J., Grubb, E.A.A. & Jones, G.D. 1998. Nitrogen-richness and protection of seeds in Australian tropical rainforest: a test of plant defence theory. *Oikos* **82**: 467–82.

Grubb, P.J. & Stevens, P.F. 1985. *The forests of the Fatima Basin and Mt Kerigomna, Papua New Guinea with a review of montane and subalpine rainforests in Papuasia*. Australian National University, Canberra.

Grünmeier, R. 1990. Pollination by bats and non-flying mammals of the African tree *Parkia bicolor* (Mimosaceae). *Memoirs of the New York Botanical Garden* **55**: 83–104.

Guariguata, M.R. 1998. Response of forest tree saplings to experimental mechanical damage in lowland Panama. *Forest Ecology and Management* **102**: 103–11.

Guariguata, M.R. & Gilbert, G.S. 1996. Interspecific variation in rates of trunk wound closure in a Panamanian lowland forest. *Biotropica* **28**: 23–9.

Gullison, R.E., Panfil, S.N., Strouse, J.J. & Hubbell, S.P. 1996. Ecology and management of mahogany (*Swietenia macrophylla* King) in the Chimanes Forest, Beni, Bolivia. *Botanical Journal of the Linnean Society* **122**: 9–34.

Gunatilleke, C.V.S., Gunatilleke, I.A.U.N., Perera, G.A.D., Burslem, D.F.R.P., Ashton, P.M.S. & Ashton, P.S. 1997. Reponses to nutrient addition among seedlings of eight closely related species of *Shorea* in Sri Lanka. *Journal of Ecology* **85**: 301–11.

Gunatilleke, C.V.S., Perera, G.A.D., Ashton, P.M.S. & Gunatilleke, I.A.U.N. 1996a. Seedling growth of *Shorea* section *Doona* (Dipterocarpaceae) in soils from topographically different sites of Sinharaja rain forest in Sri Lanka. Pp. 245–65 in *The ecology of tropical forest tree seedlings* (ed. M.D. Swaine). UNESCO, Paris.

Gunatilleke, I.A.U.N., Ashton, P.M.S., Gunatilleke, C.V.S. & Ashton, P.S. 1996b. An overview of seed and seedling ecology of *Shorea* (section *Doona*) Dipterocarpaceae. Pp. 81–102 in *Biodiversity and the dynamics of ecosystems* (ed. I.M. Turner, C.H. Diong, S.S.L. Lim & P.K.L. Ng). DIWPA, Kyoto.

Guzmán-Grajales, S.M. & Walker, L.R. 1991. Differential seedling responses to litter after Hurricane Hugo in the Luquillo Experimental Forest, Puerto Rico. *Biotropica* **23**: 407–13.

Ha, C.O., Sands, V.E., Soepadmo, E. & Jong, K. 1988a. Reproductive patterns of selected understorey trees in the Malaysian rain forest: the sexual species. *Botanical Journal of the Linnean Society* **97**: 295–316.

Ha, C.O., Sands, V.E., Soepadmo, E. & Jong, K. 1988b. Reproductive patterns of selected understorey trees in the Malaysian rain forest: the apomictic species. *Botanical Journal of the Linnean Society* **97**: 317–31.

Hall, J.B. & Swaine, M.D. 1981. *The distribution and ecology of vascular plants in a tropical rain forest: forest vegetation in Ghana*. Junk, The Hague.

Hallé, F. 1986. Modular growth in seed plants. *Philosophical Transactions of the Royal Society of London* B **313**: 77–87.

Hallé, F. 1995. Canopy architecture in tropical trees: a pictorial approach. Pp. 27–44 in *Forest canopies* (ed. M.D. Lowman & N.M. Nadkarni). Academic Press, San Diego.

Hallé, F. & Oldeman, R.A.A. 1970. *Essai sur l'architecture et la dynamique de croissance des arbres tropicaux*. Masson, Paris.

Hallé, F. & Oldeman, R.A.A. 1975. *An essay on the architecture and dynamics of growth of tropical trees*. Penerbit Universiti Malaya, Kuala Lumpur.

Hallé, F., Oldeman, R.A.A. & Tomlinson, P.B. 1978. *Tropical trees and forests: an architectural analysis*. Springer-Verlag, Berlin.

Hammond, D.S. & Brown, V.K. 1995. Seed size of woody plants in relation to disturbance, dispersal, soil type in wet neotropical forests. *Ecology* **76**: 2544–61.

Hammond, D.S. & Brown, V.K. 1998. Disturbance, phenology and life-history characteristics: factors influencing distance/density-dependent attack on tropical seeds and seedlings. Pp. 51–78 in *Dynamics of tropical communities* (ed. D.M. Newbery, H.H.T. Prins & N.D. Brown). Blackwell Science, Oxford.

Hammond, D.S., Brown, V.K. & Zagt, R. 1999. Spatial and temporal patterns of seed attack and germination in a large-seeded neotropical tree species. *Oecologia* **119**: 208–18.

Hamrick, J.L. 1994. Genetic diversity and conservation in tropical forests. Pp. 1–9 in *Proceedings international symposium on genetic conservation and production of tropical forest tree seed* (ed. R.M. Drysdale, S.E.T. John & A.C. Yapa). ASEAN–Canada Forest Tree Seed Centre, Muaklek.

Harborne, J.B. 1993. *Introduction to ecological biochemistry*. Fourth edition. Academic Press, London.

Harms, K.E. & Dalling, J.W. 1997. Damage and herbivory tolerance through resprouting as an advantage of large seed size in tropical trees and lianas. *Journal of Tropical Ecology* **13**: 617–21.

Harms, K.E., Dalling, J.W. & Aizprúa, R. 1997. Regeneration from cotyledons in *Gustavia superba* (Lecythidaceae). *Biotropica* **29**: 234–7.

Hart, T.B. 1995. Seed, seedling and sub-canopy survival in monodominant and mixed forests of the Ituri Forest, Africa. *Journal of Tropical Ecology* **11**: 443–59.

Hart, T.B., Hart, J.A. & Murphy, P.G. 1989. Monodominant and species rich forests of the humid tropics: causes for their co-occurrence. *American Naturalist* **133**: 613–33.

Hartley, T.G., Dunstone, E.A., Fitzgerald, J.S., Johns, S.R. & Lamberton, J.A. 1973. A survey of New Guinea plants for alkaloids. *Lloydia* **36**: 217–319.

Hartshorn, G.S. 1980. Neotropical forest dynamics. *Biotropica* **12** (suppl.): 23–30.

Hawthorne, W.D. 1995. *Ecological profiles of Ghanaian forest trees*. Tropical Forestry Papers 29, Oxford Forestry Institute, Oxford.

Hegde, V., Chandran, M.D.S. & Gadgil, M. 1998. Variation in bark thickness in a tropical forest community of Western Ghats, India. *Functional Ecology* **12**: 313–18.

Hegnauer, R. 1977. Cyanogenic compounds as systematic markers in Tracheophyta. *Plant Systematics and Evolution* (suppl.) **1**: 199–209.

Hegnauer, R. 1988. Biochemistry, distribution and taxonomic relevance of higher plant alkaloids. *Phytochemistry* **27**: 2423–7.

Henderson, A. 1984. Observations on pollination of *Cryosophila albida*. *Principes* **28**: 120–6.

Henderson, A. 1985. Pollination of *Socratea exorrhiza* and *Iriartea ventricosa*. *Principes* **29**: 64–71.

Henderson, A. 1986. A review of pollination studies in the Palmae. *Botanical Review* **52**: 221–59.

Hilty, S.L. 1980. Flowering and fruiting periodicity in a premontane rain forest in Pacific Colombia. *Biotropica* **12**: 292–306.

Hladik, A. & Hladik, C.M. 1977. Signification écologique des teneurs en alcaloïdes des vegetaux de la foret dense: resultats des tests preliminaires effectués au Gabon. *Revue d'Ecologie (La Terre et la Vie)* **31**: 515–55.

Hladik, A. & Miquel, S. 1990. Seedling types and plant establishment in an African rain forest. Pp. 261–82 in *Reproductive ecology of tropical forest plants* (ed. K.S. Bawa & M. Hadley). UNESCO, Paris.

Höft, M., Verpoorte, R. & Beck, E. 1996. Growth and alkaloid contents in leaves of

*Tabernaemontana pachysiphon* Stapf (Apocynaceae) as influenced by light intensity, water and nutrient supply. *Oecologia* **107**: 160–9.

Hogan, K.P., Smith, A.P. & Samaniego, M. 1995. Gas exchange in six tropical semi-deciduous forest canopy tree species during the wet and dry seasons. *Biotropica* **27**: 324–33.

Högberg, P. 1986. Soil nutrient availability, root symbioses and tree species composition in tropical Africa: a review. *Journal of Tropical Ecology* **2**: 359–72.

Holbrook, N.M., Burns, M.J. & Field, C.B. 1995. Negative xylem pressures in plants: a test of the balancing pressure technique. *Science* **270**: 1193–4.

Hopkins, H.C. 1984. Floral biology and pollination ecology of the neotropical species of *Parkia*. *Journal of Ecology* **72**: 1–23.

Hopkins, M.S. & Graham, A.W. 1987. The viability of seeds of rainforest species after experimental soil burials under tropical wet lowland forest in north-eastern Australia. *Australian Journal of Ecology* **12**: 97–108.

Horn, M.H. 1997. Evidence for dispersal of fig seeds by the fruit-eating characid fish *Brycon guatemalensis* Regan in a Costa Rican tropical rain forest. *Oecologia* **109**: 259–64.

House, S.M. 1989. Pollen movement to flowering canopies of pistillate individuals of three rain forest tree species in tropical Australia. *Australian Journal of Ecology* **14**: 77–94.

Howe, H.F. 1985. Gomphothere fruits: a critique. *American Naturalist* **125**: 853–65.

Howe, H.F. 1989. Scatter- and clump-dispersal and seedling demography: hypothesis and implications. *Oecologia* **79**: 417–26.

Howe, H.F. 1990. Survival and growth of juvenile *Virola surinamensis* in Panama: effects of herbivory and canopy closure. *Journal of Tropical Ecology* **6**: 259–80.

Howe, H.F. 1993. Aspects of variation in a neotropical seed dispersal system. *Vegetatio* **107/108**: 149–62.

Howe, H.F. & Primack, R.B. 1975. Differential seed dispersal by birds of the tree *Casearia nitida* (Flacourtiaceae). *Biotropica* **7**: 278–83.

Howe, H.F. & Richter, W.M. 1982. Effects of seed size on seedling size in *Virola surinamensis*: a within and between tree analysis. *Oecologia* **53**: 347–51.

Howe, H.F., Schupp, E.W. & Westley, L.C. 1985. Early consequences of seed dispersal for a neotropical tree (*Virola surinamensis*). *Ecology* **66**: 781–91.

Howe, H.F. & Smallwood, J. 1982. Ecology of seed dispersal. *Annual Review of Ecology and Systematics* **13**: 201–28.

Hubbell, S.P. 1997. A unified theory of biogeography and relative species abundance and its application to tropical rain forests and coral reefs. *Coral Reefs* **16** (suppl.): S9–S21.

Hubbell, S.P. 1998. The maintenance of diversity in a neotropical tree community: conceptual issues, current evidence, and challenges ahead. Pp. 17–44 in *Forest biodiversity research, monitoring and modeling: conceptual background and Old World case studies* (ed. F. Dallmeier & J.A. Comiskey). UNESCO, Paris.

Hubbell, S.P. & Foster, R.A. 1990. Structure, dynamics, and equilibrium status of old-growth forest on Barro Colorado Island. Pp. 522–41 in *Four neotropical rainforests* (ed. A.H. Gentry). Yale University Press, New Haven.

Hubbell, S.P. & Foster, R.B. 1986. Biology, chance and history and the structure of tropical rain forest tree communities. Pp. 314–29 in *Community ecology* (ed. J.M. Diamond & T.J. Case). Harper & Row, New York.

Hubbell, S.P. & Foster, R.B. 1987. La estructura espacial en gran escala de un bosque neotropical. *Revista de Biología Tropical* **35** (suppl. 1): 7–22.

Hubbell, S.P. & Foster, R.B. 1992. Short-term dynamics of a neotropical forest: why ecological research matters to tropical conservation and management. *Oikos* **63**: 48–61.

Hubbell, S.P., Foster, R.B., O'Brien, S.T., Harms, K.E., Condit, R., Wechsler, B., Wright, S.J. & Loo de Lao, S. 1999. Light-gap disturbances, recruitment limitation, and tree diversity in a neotropical forest. *Science* **283**: 554–57.

Hubbell, S.P., Wiemer, D.F. & Adejare, A. 1983. An antifungal terpenoid defends a neotropical tree (*Hymenaea*) against attack by fungus-growing ants (*Atta*). *Oecologia* **60**: 321–7.

Huc, R., Ferhi, A., Guehl, J.M. 1994. Pioneer and late stage tropical rainforest tree species (French Guiana) growing under common conditions differ in leaf gas exchange regulation, carbon isotope discrimination and leaf water potential. *Oecologia* **99**: 297–305.

Huss-Danell, K. 1997. Actinorhizal symbioses and their $N_2$ fixation. *New Phytologist* **136**: 375–405.

Ibarra-Manríquez, G. & Oyama, K. 1992. Ecological correlates of reproductive traits of Mexican rain forest trees. *American Journal of Botany* **79**: 383–94.

Ingleby, K., Munro, R.C., Noor, M., Mason, P.A. & Clearwater, M.J. 1998. Ectomycorrhizal populations and growth of *Shorea parvifolia* (Dipterocarpaceae) seedings regenerating under three different forest canopies following logging. *Forest Ecology and Management* **111**: 171–9.

Irvine, A.K. & Armstrong, J.E. 1990. Beetle pollination in tropical forests of Australia. Pp. 135–49 in *Reproductive ecology of tropical forest plants* (ed. K.S. Bawa & M. Hadley). UNESCO, Paris.

Ishida, A., Nakano, T., Matsumoto, Y., Sakoda, M. & Ang, L.H. 1999c. Diurnal changes in leaf gas exchange and chlorophyll fluorescence in tropical tree species with contrasting light requirements. *Ecological Research* **14**: 77–88.

Ishida, A., Toma, T. & Marjenah 1999a. Leaf gas exchange and chlorophyll fluorescence in relation to leaf angle, azimuth, and canopy position in the tropical pioneer tree, *Macaranga conifera*. *Tree Physiology* **19**: 117–24.

Ishida, A., Toma, T. & Marjenah 1999b. Limitations of leaf carbon gain by stomatal and photochemical processes in the top canopy of *Macaranga conifera*, a tropical pioneer tree. *Tree Physiology* **19**: 467–73.

Ishida, A., Toma, T., Matsumoto, Y., Yap, S.K. & Maruyama, Y. 1996. Diurnal changes in leaf gas exchange characteristics in the uppermost canopy of a rain forest tree, *Dryobalanops aromatica* Gaertn. f. *Tree Physiology* **16**: 779–85.

Ishii, R. & Higashi, M. 1997. Tree coexistence on a slope: an adaptive significance of trunk inclination. *Proceedings of the Royal Society of London* B **264**: 133–40.

Itino, T., Kato, M. & Hotta, M. 1991. Pollination ecology of the two wild bananas, *Musa acuminata* subsp. *halabensis* and *Musa salaccensis*: chiropterophily and ornithophily. *Biotropica* **23**: 151–8.

Itoh, A., Yamakura, T., Ogino, K. & Lee, H.S. 1995. Survivorship and growth of seedlings of four dipterocarp species in a tropical rainforest of Sarawak, East Malaysia. *Ecological Research* **10**: 327–38.

Jackson, P.C., Cavelier, J., Goldstein, G., Meinzer, F.C. & Holbrook, N.M. 1995. Partitioning of water resources among plants of a lowland tropical forest. *Oecologia* **101**: 197–203.

Jacobs, G.H., Neitz, M., Deegan, J.F. & Neitz, J. 1996. Trichromatic colour vision in New World monkeys. *Nature* **382**: 156–8.

Janos, D.P. 1980. Mycorrhizae influence tropical succession. *Biotropica* **12** (suppl.): 56–64.

Janse, J.M. 1896. Les endophytes radicaux de quelques plantes Javanaises. *Annales du Jardin Botanique de Buitenzorg* **14**: 53–212.

Janson, C.H. 1983. Adaptation of fruit morphology to dispersal agents in a neotropical forest. *Science* **219**: 187–9.

Janson, C.H., Terborgh, J. & Emmons, L.H. 1981. Non-flying mammals as pollinating agents in the Amazonian forest. *Biotropica* **13** (suppl.): 1–6.

Janzen, D.H. 1969. Seed-eaters versus seed size, number, toxicity and dispersal. *Evolution* **23**: 1–27.

Janzen, D.H. 1970. Herbivores and the number of tree species in tropical forests. *American Naturalist* **104**: 501–28.

Janzen, D.H. 1971. Seed predation by animals. *Annual Review of Ecology and Systematics* **2**: 465–92.

Janzen, D.H. 1974. Tropical blackwater rivers, animals and mast fruiting by the Dipterocarpaceae. *Biotropica* **6**: 69–103.

Janzen, D.H. 1976a. Why tropical trees have rotten cores. *Biotropica* **8**: 110.

Janzen, D.H. 1976b. Why bamboos wait so long to flower. *Annual Review of Ecology and Systematics* **7**: 347–91.

Janzen, D.H. & Martin, P.S. 1982. Neotropical anachronisms: the fruits the gomphotheres ate. *Science* **215**: 19–27.

Jeník, J. 1978. Roots and root systems in tropical trees: morphologic and ecologic aspects. Pp. 323–49 in *Tropical trees as living systems* (ed. P.B. Tomlinson & M.H. Zimmerman). Cambridge University Press, Cambridge.

Jeronimidis, G. 1980. The fracture behaviour of wood and the relation between toughness and morphology. *Proceedings of the Royal Society of London* B **208**: 447–60.

Johnson, L.A.S. & Wilson, K.L. 1990. General traits of the Cycadales. Pp. 363–8 in *The families and genera of vascular plants 1. Pteridophytes and gymnosperms* (ed. K.U. Kramer & P.S. Green). Springer-Verlag, Berlin.

Johnston, A. 1949. Vesicular-arbuscular mycorrhiza in Sea Island Cotton and other tropical plants. *Tropical Agriculture* **26**: 118–21.

Jones, D.L. 1993. *Cycads of the world.* Reed, Chatswood.

Jordan, C.F. 1982. Productivity of tropical rain forest ecosystems and implications for their use as future wood and energy sources. Pp. 117–36 in *Tropical rain forest ecosystems: structure and function* (ed. F.B. Golley). Elsevier, Amsterdam.

Jordano, P. 1995. Angiosperm fleshy fruits and seed dispersers: a comparative analysis of adaptation and constraints in plant-animal interactions. *American Naturalist* **145**: 163–91.

Julliot, C. 1996. Fruit choice by red howler monkey (*Allouatta seniculus*) in a tropical rain forest. *American Journal of Primatology* **40**: 261–82.

Julliot, C. 1997. Impact of seed dispersal by red howler monkeys *Alouatta seniculus* on the seedling population in the understorey of tropical rain forest. *Journal of Ecology* **85**: 431–40.

Juniper, B.E. 1993. Flamboyant flushes: a re-interpretation of non-green flush colours in leaves. *International Dendrological Society Yearbook* **1993**: 49–57.

Kanzaki, M., Yap, S.K., Kimura, K., Okauchi, Y. & Yamakura, T. 1997. Survival and germination of buried seeds of non-dipterocarp species in a tropical rain forest at Pasoh, West Malaysia. *Tropics* **7**: 9–20.

Kapelle, K. & Leal, M.E. 1996. Changes in leaf morphology and foliar nutrient status along a successional gradient in Costa Rican upper montane *Quercus* forest. *Biotropica* **28**: 331–44.

Kaspari, M. 1993. Removal of seeds from neotropical frugivore droppings. *Oecologia* **95**: 81–8.

Kato, M. 1996. Plant-pollinator interactions in the understory of a lowland mixed dipterocarp forest in Sarawak. *American Journal of Botany* **83**: 732–43.

Kato, M. & Inoue, T. 1994. Origin of insect pollination. *Nature* **368**: 195.

Kato, M., Inoue, T. & Nagamitsu, T. 1995. Pollination biology of *Gnetum* (Gnetaceae) in a lowland mixed dipterocarp forest in Sarawak. *American Journal of Botany* **82**: 862–8.

Kaufman, L. 1988. The role of developmental crises in the formation of buttresses: a unified hypothesis. *Evolutionary Trends in Plants* **21**: 39–51.

Kaufman, S., McKey, D.B., Hossaert-McKey, M. & Horvitz, C.C. 1991. Adaptations for a two-phase seed dispersal system involving vertebrates and ants in a hemiepiphytic fig (*Ficus microcarpa*: Moraceae). *American Journal of Botany* **78**: 971–7.

Keller, R. 1994. Neglected vegetative characters in field identfication at the supraspecific level in woody plants: phyllotaxy, serial buds, syllepsis and architecture. *Botanical Journal of the Linnean Society* **116**: 33–51.

Kelly, C.K. 1995. Seed size in tropical trees: a comparative study of factors affecting seed size in Peruvian angiosperms. *Oecologia* **102**: 377–88.

Kelly, C.K. & Purvis, A. 1993. Seed size and establishment conditions in tropical trees. *Oecologia* **94**: 356–60.

Kelly, D. 1994. The evolutionary ecology of mast seeding. *Trends in Ecology and Evolution* **9**: 465–70.

Kennedy, D.N. & Swaine, M.D. 1992. Germination and growth of colonizing species in artificial gaps of different sizes in dipterocarp rain forest. *Philosophical Transactions of the Royal Society of London* B **335**: 357–66.

Killmann, W. 1983. Some physical properties of the coconut palm stem. *Wood Science and Technology* **7**: 167–85.

King, D.A. 1987. Load bearing capacity of understory treelets of a tropical wet forest. *Bulletin of the Torrey Botanical Club* **114**: 419–28.

King, D.A. 1991a. Tree size. *National Geographic Research and Exploration* **7**: 342–51.

King, D.A. 1991b. Correlations between biomass allocation, relative growth rate and light environment in tropical forest saplings. *Functional Ecology* **5**: 485–92.

King, D.A. 1993. Growth history of a neotropical tree inferred from the spacing of leaf scars. *Journal of Tropical Ecology* **9**: 525–32.

King, D.A. 1994. Influence of light level on the growth and morphology of saplings in a Panamanian forest. *American Journal of Botany* **81**: 948–57.

King, D.A. 1996. Allometry and life history of tropical trees. *Journal of Tropical Ecology* **12**: 25–44.

King, D.A. 1998a. Relationship between crown architecture and branch orientation in rain forest trees. *Annals of Botany* **82**: 1–7.

King, D.A. 1998b. Influence of leaf size on tree architecture: first branch height and crown dimensions in tropical rain forest trees. *Trees* **12**: 438–45.

King, D.A., Leigh, E.G., Condit, R., Foster, R.B. & Hubbell, S.P. 1997. Relationships between branch spacing, growth rate and light in tropical forest saplings. *Functional Ecology* **11**: 627–35.

King, D.A. & Maindonald, J.H. 1999. Tree architecture in relation to leaf dimensions and tree stature in temperate and tropical rain forests. *Functional Ecology* **87**: 1012–24.

Kinsman, S. 1990. Regeneration by fragmentation in tropical montane forest shrubs. *Ameri-*

*can Journal of Botany* **77**: 1626–33.

Kitajima, K. 1992. Relationship between photosynthesis and thickness of cotyledons for tropical tree species. *Functional Ecology* **6**: 582–9.

Kitajima, K. 1994. Relative importance of photosynthetic traits and allocation patterns as correlates of seedling shade tolerance of 13 tropical trees. *Oecologia* **98**: 419–28.

Kitajima, K. 1996. Cotyledon functional morphology, patterns of seed reserve utilization and regeneration niches of tropical tree seedlings. Pp. 193–210 in *The ecology of tropical forest tree seedlings* (ed. M.D. Swaine). UNESCO, Paris.

Kitajima, K. & Augspurger, C.K. 1989. Seed and seedling ecology of a monocarpic tropical tree, *Tachigali versicolor*. *Ecology* **70**: 1102–14.

Kitajima, K., Mulkey, S.S. & Wright, S.J. 1997. Seasonal leaf phenotypes in the canopy of a tropical dry forest: photosynthetic characteristics and associated traits. *Oecologia* **109**: 490–8.

Klinge, H. 1985. Foliar nutrient levels of native tree species from Central Amazonia. II. Campina. *Amazoniana* **9**: 281–95.

Kobe, R.K. 1999. Light gradient partitioning among tropical tree species through differential seedling mortality and growth. *Ecology* **80**: 187–201.

Kobe, R.K., Pacala, S.W., Silander, J.A. & Canham, C.D. 1995. Juvenile tree survivorship as a component of shade tolerance. *Ecological Applications* **5**: 517–32.

Koch, G.W., Amthor, J.S. & Goulden, M.L. 1994. Diurnal patterns of leaf photosynthesis, conductance and water potential at the top of a lowland rain forest canopy in Cameroon: measurements from the *Radeau des Cimes*. *Tree Physiology* **14**: 347–60.

Kochummen, K.M., LaFrankie, J.V. & Manokaran, N. 1990. Floristic composition of Pasoh Forest Reserve, a lowland rain forest in Peninsular Malaysia. *Journal of Tropical Forest Science* **3**: 1–13.

Kohyama, T. 1987. Significance of architecture and allometry in saplings. *Functional Ecology* **1**: 399–404.

Kohyama, T. 1993. Size-structured tree populations in gap-dynamic forest – the forest architecture hypothesis for the stable coexistence of species. *Journal of Ecology* **81**: 131–43.

Kohyama, T. 1994. Size-structure-based models of forest dynamics to interpret population- and community-level mechanisms. *Journal of Plant Research* **107**: 107–16.

Kohyama, T. 1996. The role of architecture in enhancing plant species diversity. Pp. 21–33 in *Biodiversity: an ecological perspective* (ed. T. Abe, S.A. Levin & M. Higashi). Springer-Verlag, New York.

Kohyama, T. & Hotta, M. 1990. Significance of allometry in tropical saplings. *Functional Ecology* **4**: 515–21.

Kohyama, T. & Suzuki, E. 1996. Forest architecture and dynamics in relation to biodiversity. Pp. 103–8 in *Biodiversity and the dynamics of ecosystems* (ed. I.M. Turner, C.H. Diong, S.S.L. Lim & P.K.L. Ng). DIWPA, Kyoto.

Königer, M., Harris, G.C., Virgo, A. & Winter, K. 1995. Xanthophyll-cycle pigments and the photosynthetic capacity in tropical forest species: a comparative field study on canopy, gap and understory plants. *Oecologia* **104**: 280–90.

Korning, J. & Balslev, H. 1994. Growth rates and mortality patterns of tropical lowland tree species and the relation to forest structure in Amazonian Ecuador. *Journal of Tropical Ecology* **10**: 151–66.

Krause, G.H. & Winter, K. 1996. Photoinhibition of photosynthesis in plants growing in natural tropical forest gaps. A chlorophyll fluorescence study. *Acta Botanica* **109**: 456–62.

Krause, G.H., Virgo, A. & Winter, K. 1995. High susceptibility to photoinhibition of young leaves of tropical forest trees. *Planta* **197**: 583–91.

Kress, W.J. & Beach, J.H. 1994. Flowering plant reproductive systems. Pp. 161–82 in *La Selva: ecology and natural history of a neotropical rain forest* (ed. L.A. McDade, K.S. Bawa, H.A. Hespenheide & G.S. Hartshorn). The University of Chicago Press, Chicago.

Kubitzki, K. 1985. The dispersal of forest plants. Pp. 192–206 in *Amazonia* (ed. G.T. Prance & T.E. Lovejoy). Pergamon Press, Oxford.

Kubitzki, K. & Ziburski, A. 1994. Seed dispersal in flood plain forests of Amazonia. *Biotropica* **26**: 30–43.

Kuiters, A.T. 1990. Role of phenolic substances from decomposing forest litter in plant-soil interactions. *Acta Botanica Neerlandica* **39**: 329–48.

Kursar, T.A. & Coley, P.D. 1992a. Delayed greening in tropical leaves: an antiherbivore defense? *Biotropica* **24**: 256–62.

Kursar, T.A. & Coley, P.D. 1992b. The consequences of delayed greening during leaf development for light absorption and light use efficiency. *Plant, Cell and Environment* **15**: 901–9.

Kursar, T.A. & Coley, P.D. 1992c. Delayed development of the photosynthetic apparatus in tropical rain forest species. *Functional Ecology* **6**: 411–22.

Kyereh, B., Swaine, M.D. & Thompson, J. 1999. Effect of light on the germination of forest trees in Ghana. *Journal of Ecology* **87**: 772–83.

Lack, A.J. 1991. Dead logs as a substrate for rain forest trees in Dominica. *Journal of Tropical Ecology* **7**: 401–5.

Laman, T. 1996. *Ficus* seed shadows in a Bornean rain forest. *Oecologia* **107**: 347–55.

Lambert, J.E. & Garber, P.A. 1988. Evolutionary and ecological implications of primate seed dispersal. *American Journal of Primatology* **45**: 9–28.

Langenheim, J.H. 1994. Higher plant terpenoids: a phytocentric overview of their ecological role. *Journal of Chemical Ecology* **20**: 1223–80.

Lawton, R.O. 1984. Ecological constraints on wood density in a tropical montane rain forest. *American Journal of Botany* **71**: 261–7.

Lebreton, P. 1982. Tanins ou alcaloïdes: deux tactiques phytochimiques de dissuasion des herbivores. *Revue d'Ecologie (La Terre et la Vie)* **36**: 539–72.

Lee, D.W., Baskaran, K., Marzalina Mansor, Haris Mohamad & Yap, S.K. 1996. Irradiance and spectral quality affect Asian tropical rain forest tree seedling development. *Ecology* **77**: 568–80.

Lee, D.W., Oberbauer, S.F., Baskaran, K., Marzalina Mansor, Haris Mohamad & Yap, S.K. 1997. Effects of irradiance and spectral quality on seedling development of two Southeast Asian *Hopea* species. *Oecologia* **110**: 1–9.

Lehto, T. & Grace, J. 1994. Carbon balance of tropical tree seedlings: a comparison of two species. *New Phytologist* **127**: 455–63.

Leighton, M. & Leighton, D.R. 1983. Vertebrate responses to fruiting seasonality within a Bornean rain forest. Pp. 181–96 in *Tropical rain forest: ecology and management* (ed. S.L. Sutton, T.C. Whitmore & A.C. Chadwick). Blackwell Scientific Publications, Oxford.

Letourneau, D.K., Arias G., F. & Jebb, M. 1993. Coping with enemy-filled space: herbivores on *Endospermum* in Papua New Guinea. *Biotropica* **25**: 95–9.

Letourneau, D.K. & Barbosa, P. 1999. Ants, stem borers, and pubescense in *Endospermum* in Papua New Guinea. *Biotropica* **31**: 295–302.

Levey, D.J. & Byrne, M.M. 1993. Complex ant-plant interactions: rain-forest ants as secondary dispersers and post-dispersal seed predators. *Ecology* **74**: 1802–12.

Levey, D.J., Moermond, T.C., Denslow, J.S. 1994. Frugivory: an overview. Pp. 282–94 in *La Selva: ecology and natural history of a neotropical rain forest* (ed. L.A. McDade, K.S. Bawa, H.A. Hespenheide & G.S. Hartshorn). The University of Chicago Press, Chicago.

Levin, D.A. 1973. The role of trichomes in plant defense. *Quarterly Review of Biology* **48**: 3–15.

Levin, D.A. 1974. The oil content of seeds: an ecological perspective. *American Naturalist* **108**: 193–206.

Levin, D.A. 1976. Alkaloid-bearing plants: an ecogeographic perspective. *American Naturalist* **110**: 261–84.

Li, M., Lieberman, M. & Lieberman, D. 1996. Seedling demography in undisturbed tropical wet forest in Costa Rica. Pp. 285–314 in *The ecology of tropical forest tree seedlings* (ed. M.D. Swaine). UNESCO, Paris.

Lieberman, D., Hartshorn, G.S., Lieberman, M. & Peralta, R. 1990. Forest dynamics at La Selva Biological Station, 1969–1985. Pp. 509–21 in *Four neotropical rainforests* (ed. A.H. Gentry). Yale University Press, New Haven.

Lieberman, D., Lieberman, M., Hartshorn, G. & Peralta, R. 1985a. Growth rates and age–size relationships of tropical wet forest trees in Costa Rica. *Journal of Tropical Ecology* **1**: 97–109.

Lieberman, D., Lieberman, M., Peralta, R. & Hartshorn, G.S. 1985b. Mortality patterns and stand turnover rates in a wet tropical forest in Costa Rica. *Journal of Ecology* **73**: 915–24.

Lieberman, M., Lieberman, D., Peralta, R. & Hartshorn, G.S. 1995. Canopy closure and the distribution of tropical forest tree species at La Selva, Costa Rica. *Journal of Tropical Ecology* **11**: 161–78.

Lightbody, J.P. 1985. Distribution of leaf shapes of *Piper* sp. in a tropical cloud forest: evidence for the role of drip-tips. *Biotropica* **17**: 339–42.

Lim, W.H.L. & Turner, I.M. 1996. Resource availability and growth responses to defoliation in seedlings of three early-successional, tropical, woody species. *Ecological Research*

**11**: 321–4.

Linhart, Y.B. & Mendenhall, J.A. 1977. Pollen dispersal by hawkmoths in a *Lindenia rivalis* Benth. population in Belize. *Biotropica* **9**: 143.

Linskens, H.F. 1996. No airborne pollen within tropical rain forests. *Proceedings of the Koninklijke Nederlandse Akademie van Wetenschappen, Natural Sciences* **99**: 175–80.

Listabarth, C. 1996. Pollination of *Bactris* by *Phyllotrox* and *Epurea*. Implications of the palm breeding beetles on pollination at the community level. *Biotropica* **28**: 69–81.

Lodge, D.J. 1987. Resurvey of mycorrhizal associations in the El Verde rain forest in Puerto Rico. P. 127 in *Proceedings of the 7th North American conference on mycorrhizae* (ed. D.M. Sylvia, L.L. Hung & J.H. Graham). Institute of Food and Agricultural Science, University of Florida, Gainsville.

Loehle, C. 1986. Phototropism of whole trees: effects of habitat and growth form. *American Midland Naturalist* **116**: 190–6.

Loehle, C. 1997. The adaptive significance of trunk inclination on slopes: a commentary. *Proceedings of the Royal Society of London* B **264**: 1371–4.

Loiselle, B.A., Ribbens, E. & Vargas, O. 1996. Spatial and temporal variation of seed rain in a tropical lowland wet forest. *Biotropica* **28**: 82–95.

Lopes, A.V. & Machado, I.C. 1998. Floral biology and reproductive ecology of *Clusia nemorosa* (Clusiaceae) in northeastern Brazil. *Plant Systematics and Evolution* **213**: 71–90.

Lord, J., Egan, J., Clifford, T., Jurado, E., Leishman, M., Williams, D. & Westoby, M. 1997. Larger seeds in tropical floras: consistent patterns independent of growth form and dispersal mode. *Journal of Biogeography* **24**: 205–11.

Lott, R.H., Harrington, G.N., Irvine, A.K. & McIntyre, S. 1995. Density-dependent seed predation and plant dispersion of the tropical palm *Normanbya normanbyi*. *Biotropica* **27**: 87–95.

Lovelock, C.E., Kursar, T.A., Skillman, J.B. & Winter, K. 1998. Photoinhibition in tropical forest understorey species with short- and long-lived leaves. *Functional Ecology* **12**: 553–60.

Lovelock, C.E., Kyllo, D. & Winter, K. 1996. Growth responses to vesicular-arbuscular mycorrhizae and elevated $CO_2$ in seedlings of a tropical tree, *Beilschmiedia pendula*. *Functional Ecology* **10**: 662–7.

Lucas, P.W., Choong, M.F., Tan, H.T.W., Turner, I.M. & Berrick, A.J. 1991a. The fracture toughness of the leaf of the dicotyledon *Calophyllum inophyllum* L. (Guttiferae). *Philosophical Transactions of the Royal Society of London* B **334**: 95–106.

Lucas, P.W., Darvell, B.W., Lee, P.K.D., Yuen, T.D.B. & Choong, M.F. 1995. The toughness of plant cell walls. *Philosophical Transactions of the Royal Society of London* B **348**: 363–72.

Lucas, P.W., Darvell, B.W., Lee, P.K.D., Yuen, T.D.B. & Choong, M.F. 1998. Colour cues for leaf food selection by long-tailed macaques (*Macaca fascicularis*) with a new suggestion for the evolution of trichromatic colour vision. *Folia Primatologica* **69**: 139–52.

Lucas, P.W., Lowrey, T.K., Pereira, B.P., Sarafis, V. & Kuhn, W. 1991b. The ecology of *Mezzettia leptopoda* (Hk. f. et Thoms.) Oliv. (Annonaceae) seeds as viewed from a mechanical perspective. *Functional Ecology* **5**: 545–53.

Lucas, P.W. & Pereira, B. 1990. Estimation of the fracture toughness of leaves. *Functional Ecology* **4**: 819–22.

Lucas, P.W., Tan, H.T.W. & Chang, P.Y. 1997. The toughness of secondary cell wall and woody tissue. *Philosophical Transactions of the Royal Society of London* B **352**: 341–52.

Lucas, P.W. & Teaford, M.F. 1995. Significance of silica in leaves to long-tailed macaques (*Macaca fascicularis*). *Folia Primatologica* **64**: 30–6.

Lumer, C. 1980. Rodent pollination of *Blakea* (Melastomataceae) in a Costa Rican cloud forest. *Brittonia* **32**: 512–17.

MacArthur, R.H. & Wilson, E.O. 1967. *The theory of island biogeography*. Princeton University Press, Princeton.

Mack, A.L. 1993. The sizes of vertebrate-dispersed fruits: a neotropical-palaeotropical comparison. *American Naturalist* **142**: 840–56.

Mack, A.L. 1995. Distance and non-randomness of seed dispersal by the dwarf cassowary *Casuarius bennetti*. *Ecography* **18**: 286–95.

Mack, A.L. 1998. An advantage of large seed size: tolerating rather than succumbing to seed predators. *Biotropica* **30**: 604–8.

Mack, A.L., Ickes, K., Jessen, J.H., Kennedy, B. & Sinclair, J.R. 1999. Ecology of *Aglaia mackiana* (Meliaceae) seedlings in a New Guinea rain forest. *Biotropica* **31**: 111–20.

Manokaran, N. & Kochummen, K.M. 1987. Recruitment, growth and mortality of tree spe-

cies in a lowland dipterocarp forest in Peninsular Malaysia. *Journal of Tropical Ecology* **3**: 315–30.
Manokaran, N. & Kochummen, K.M. 1994. Tree growth in primary lowland and hill dipterocarp forests. *Journal of Tropical Forest Science* **6**: 332–45.
Manokaran, N., LaFrankie, J.V., Kochummen, K.M., Quah, E.S., Klahn, J.E., Ashton, P.S. & Hubbell, S.P. 1992. Stand table and distribution of species in the fifty hectare research plot at Pasoh Forest Reserve. *FRIM Research Data* **1**: 1–454.
Manokaran, N. & Swaine, M.D. 1994. Population dynamics of trees in dipterocarp forests of Peninsular Malaysia. *Malayan Forest Records* **40**: 1–173.
Marquis, R.J. 1984. Leaf herbivores decrease fitness of a tropical plant. *Science* **226**: 537–9.
Marquis, R.J. 1987. Variación en la herbivoría foliar y su importancia selectiva en *Piper arieianum* (Piperaceae). *Revista de Biología Tropical* **35** (suppl. 1): 133–49.
Marquis, R.J., Young, H.J., Braker, H.E. 1986. The influence of understorey vegetation cover on germination and seedling establishment in a tropical lowland wet forest. *Biotropica* **18**: 273–8.
Marshall, A.G. 1983. Bats, flowers and fruit: evolutionary relationships in the Old World. *Biological Journal of the Linnean Society* **20**: 115–35.
Martínez-Ramos, M. & Alvarez-Buylla, E.R. 1998. How old are tropical rain forest trees? *Trends in Plant Science* **3**: 400–5.
Martínez-Ramos, M., Alvarez-Buylla, E., Sarukhán, J. & Piñero, D. 1988. Treefall age determination and gap dynamics in a tropical forest. *Journal of Ecology* **76**: 700–16.
Martins, E.P. 1996. Conducting phylogenetic comparative studies when the phylogeny is not known. *Evolution* **50**: 12–22.
Masunaga, T., Kubota, D., Hotta, M. & Wakatsuki, T. 1997. Nutritional characteristics of mineral elements in tree species of tropical rain forest, West Sumatra, Indonesia. *Soil Science and Plant Nutrition* **43**: 405–18.
Masunaga, T., Kubota, D., Hotta, M. & Wakatsuki, T. 1998a. Nutritional characteristics of mineral elements in leaves of tree species in tropical rain forest, West Sumatra, Indonesia. *Soil Science and Plant Nutrition* **44**: 315–29.
Masunaga, T., Kubota, D., Hotta, M. & Wakatsuki, T. 1998b. Mineral composition of leaves and bark in aluminum accumulators in a tropical rain forest in Indonesia. *Soil Science and Plant Nutrition* **44**: 347–58.
Mattheck, C. 1993. *Design in der Natur: der Baum als Lehrmeister*. Rombach Verlag, Freiburg. 242 pp.
Mattheck, C. & Kubler, H. 1995. *Wood – the internal optimization of trees*. Springer-Verlag, Berlin. 129 pp.
McKey, D. 1994. Legumes and nitrogen: the evolutionary ecology of a nitrogen demanding lifestyle. Pp. 211–28 in *Advances in legume systematics 5: the nitrogen factor* (ed. J.T. Sprent & D. McKey). Royal Botanic Gardens, Kew.
McKey, D.A. & Davidson, D.W. 1993. Ant-plant symbiosis in Africa and the neotropics: history, biogeography and biodiversity. Pp. 568–606 in *Biological relationships between Africa and South America* (ed. P. Goldblatt). Yale University Press, New Haven.
McKey, D.B., Gartlan, S.J., Waterman, P.G. & Choo, G.M. 1981. Food selection by black colobus monkeys (*Colobus satanus*) in relation to plant chemistry. *Biological Journal of the Linnean Society* **16**: 115–46.
Medina, E. 1984. Nutrient balance and physiological processes at the level of the leaf. Pp. 134–54 in *Physiological ecology of plants in the wet tropics* (ed. E. Medina, H.A. Mooney & C. Vazquez-Yanes). Junk, The Hague.
Medina, E. 1986. Forests, savannas and montane tropical environments. Pp. 139–71 in *Photosynthesis in contrasting environments* (ed. N.R. Baker & S.P. Long). Elsevier Science Publishers B.V. (Biomedical Division), Amsterdam.
Medina, E. 1996. CAM and $C_4$ plants in the humid tropics. Pp. 56–88 in *Tropical forest plant ecophysiology* (ed. S.S. Mulkey, R.L. Chazdon & A.P. Smith). Chapman & Hall, New York.
Medina, E., Garcia, V. & Cuevas, E. 1990. Sclerophylly and oligotrophic environments: relationships between leaf structure, mineral nutrient content, and drought resistance in tropical rain forests of the Upper Rio Negro Region. *Biotropica* **22**: 51–64.
Meinzer, F.C., Andrade, J.L., Goldstein, G., Holbrook, N.M., Cavelier, J. & Jackson, P. 1997. Control of transpiration from the upper canopy of a tropical forest: the role of stomatal, boundary layer and hydraulic architecture components. *Plant, Cell and Environment* **20**: 1242–52.
Meinzer, F.C., Goldstein, G., Jackson, P., Holbrook, N.M., Gutiérrez, M.V. & Cavelier, J.

1995. Environmental and physiological regulation of transpiration in tropical forest gap species: the influence of boundary layer and hydraulic properties. *Oecologia* **101**: 514–22.

Mendoza, A., Piñero, D. & Sarukhán, J. 1987. Effects of experimental defoliation on growth, reproduction and survival of *Astrocaryum mexicanum*. *Journal of Ecology* **75**: 545–54.

Metcalfe, D.J. 1996. Germination of small-seeded tropical rain forest plants exposed to different spectral compositions. *Canadian Journal of Botany* **74**: 516–20.

Metcalfe, D.J. & Grubb, P.J. 1995. Seed mass and light requirements for regeneration in Southeast Asian rain forest. *Canadian Journal of Botany* **73**: 817–26.

Metcalfe, D.J. & Grubb, P.J. 1997. The responses to shade of seedlings of very small-seeded tree and shrub species from tropical rain forest in Singapore. *Functional Ecology* **11**: 215–21.

Metcalfe, D.J., Grubb, P.J. & Turner, I.M. 1998. The ecology of very small-seeded shade-tolerant trees and shrubs in lowland rain forest in Singapore. *Plant Ecology* **134**: 131–49.

Metcalfe, D.J. & Turner, I.M. 1998. Soil seed bank from lowland rain forest in Singapore: canopy-gap and litter-gap demanders. *Journal of Tropical Ecology* **14**: 103–8.

Milburn, J.A. 1996. Sap ascent in vascular plants: challengers to the cohesion tension theory ignore the significance of immature xylem and the recycling of Münch water. *Annals of Botany* **78**: 399–407.

Miller, I.M. 1990. Bacterial leaf nodule symbiosis. *Advances in Botanical Research* **17**: 163–234.

Milton, K. 1979. Factors influencing leaf choice by howler monkeys: a test of some hypotheses of food selection by generalist herbivores. *American Naturalist* **114**: 362–78.

Miquel, S. 1987. Morphologie fonctionelle de plantules d'espèces forestières du Gabon. *Bulletin du Muséum d'Histoire Naturelle, Adansonia* **9**: 101–21.

Mirmanto, E., Proctor, J., Green, J., Nagy, L. & Suriantata 1999. Effects of nitrogen and phosphorus fertilization in a lowland evergreen rainforest. *Philosophical Transactions of the Royal Society of London* B **354**: 1825–9.

Moiroud, A. 1996. Diversité et écologie des plantes actinorhiziennes. *Acta Botanica Gallica* **143**: 651–61.

Mole, S. 1993. The systematic distribution of tannins in the leaves of angiosperms: a tool for ecological studies. *Biochemical Systematics and Ecology* **21**: 833–46.

Mole, S., Ross, J.A.M. & Waterman, P.G. 1988. Light-induced variation in phenolic levels in foliage of rain-forest plants. I. Chemical changes. *Journal of Chemical Ecology* **14**: 1–21.

Moll, D. & Jansen, K.P. 1995. Evidence for a role in seed dispersal by two tropical herbivorous turtles. *Biotropica* **27**: 121–7.

Molofsky, J. & Augspurger, C.K. 1992. The effect of leaf litter on early seedling establishment in a tropical forest. *Ecology* **73**: 68–77.

Molofsky, J. & Fisher, B.L. 1993. Habitat and predation effects on seedling survival and growth in shade-tolerant tropical trees. *Ecology* **74**: 261–5.

Momose, K., Hatada, A., Yamaoka, R., Inoue, T. 1998a. Pollination biology of the genus *Artocarpus*, Moraceae. *Tropics* **7**: 165–72.

Momose, K., Ishii, R., Sakai, S. & Inoue, T. 1998b. Plant reproductive intervals and pollinators in the aseasonal tropics: a new model. *Proceedings of the Royal Society of London* B **265**: 2333–9.

Momose, K., Nagamitsu, T. & Inoue, T. 1996. The reproductive ecology of an emergent dipterocarp in a lowland rain forest in Sarawak. *Plant Species Biology* **11**: 189–98.

Momose, K., Nagamitsu, T. & Inoue, T. 1998c. Thrips cross-pollination of *Popowia pisocarpa* (Annonaceae) in a lowland dipterocarp forest in Sarawak. *Biotropica* **30**: 444–8.

Momose, K., Yumoto, T., Nagamitsu, T., Kato, M., Nagamasu, H., Sakai, S., Harrison, R.D., Itioka, T., Hamid, A.A. & Inoue, T. 1998d. Pollination biology in a lowland dipterocarp forest in Sarawak, Malaysia. I. Characteristics of the plant-pollinator community in a lowland dipterocarp forest. *American Journal of Botany* **85**: 1477–501.

Montagnini, F. & Sancho, F. 1990. Impacts of native trees on tropical soils: a study in the Atlantic lowlands of Costa Rica. *Ambio* **19**: 386–90.

Morawetz, W., Henzl, M., Wallnöfer, B. 1992. Tree killing by herbicide-producing ants for the establishment of pure *Tococa occidentalis* populations in the Peruvian Amazon. *Biodiversity and Conservation* **1**: 19–33.

Mori, S.A., Boom, B.M., de Carvalho, A.M. & dos Santos, T.S. 1983. Southern Bahian moist forests. *Botanical Review* **49**: 155–232.

Mori, S.A. & Brown, J.L. 1994. Report on wind

dispersal in a lowland moist forest in central French Guiana. *Brittonia* **46**: 105–25.
Mori, S.A. & Brown, J.L. 1998. Epizoochorous dispersal by barbs, hooks and spines in a lowland moist forest in central French Guiana. *Brittonia* **50**: 165–73.
Mosbrugger, V. 1990. *The tree habit in land plants*. Springer-Verlag, Berlin. Lecture Notes in Earth Sciences No. 28. 161 pp.
Moyersoen, B. 1993. *Ectomicorrizas y micorrizas vesículo-arbusculares en Caatinga Amazónica del Sur de Venezuela*. CONICIT, Caracas. *Scientia Guaianae* **3**: 1–82.
Mulkey, S.S., Wright, S.J. & Smith, A.P. 1993. Comparative physiology and demography of three Neotropical forest shrubs: alternative shade-adaptive character syndromes. *Oecologia* **96**: 526–36.
Murawski, D. 1998. Genetic variation within tropical tree crowns. Pp. 104–13 in *Biologie d'une canopée de forêt équatoriale – III* (ed. F. Hallé). Pro-Natura International, Paris, and Opération Canopée, Lyon.
Murawski, D.A. 1995. Reproductive biology and genetics of tropical trees from a canopy perspective. Pp. 457–93 in *Forest Canopies* (ed. M.D. Lowman & N.M. Nadkarni). Academic Press, San Diego.
Murawski, D.A. & Hamrick, J.L. 1992. The mating system of *Cavanillesia platanifolia* under extremes of flowering-tree density. *Biotropica* **24**: 99–101.
Murray, K.G. 1988. Avian seed dispersal of three neotropical gap-dependent plants. *Ecological Monographs* **58**: 271–98.
Murray, K.G., Russell, S., Picone, C.M., Winnett-Murray, K., Sherwood, W. & Kuhlmann, M.L. 1994. Fruit laxatives and seed passage rates in frugivores: consequences for plant reproductive success. *Ecology* **75**: 989–94.
Myers, B.J., Robichaux, R.H., Unwin, G.L. & Craig, L.E. 1987. Leaf water relations and anatomy of a tropical rainforest tree species vary with crown position. *Oecologia* **74**: 81–5.
Nadkarni, N.M. 1981. Canopy roots: convergent evolution in rainforest nutrient cycles. *Science* **214**: 1023–4.
Nadkarni, N.M. 1994. Factors affecting the initiation and growth of aboveground adventitious roots in a tropical cloud forest tree: an experimental approach. *Oecologia* **100**: 94–7.
Nagamitsu, T. & Inoue, T. 1997. Cockroach pollination and breeding system of *Uvaria elmeri* (Annonaceae) in a lowland mixed-dipterocarp forest in Sarawak. *American Journal of Botany* **84**: 208–13.
Nascimento, J.C. & Langenheim, J.H. 1986. Leaf sesquiterpenes and phenolics in *Copaifera multijuga* on contrasting soil types in a Central Amazonian rain forest. *Journal of Chemical Ecology* **14**: 615–24.
Nason, J.D. & Hamrick, J.L. 1997. Reproductive and genetic consequences of forest fragmentation: two case studies of neotropical canopy trees. *Journal of Heredity* **88**: 264–76.
Nason, J.D., Herre, E.A. & Hamrick, J.L. 1998. The breeding structure of a tropical keystone plant resource. *Nature* **391**: 685–7.
Negrelle, R.R.B. 1995. Sprouting after uprooting of canopy trees in the Atlantic rain forest of Brazil. *Biotropica* **27**: 448–54.
Nepstad, D.C., de Carvalho, C.R., Davidson, E.A., Jipp, P.H., Lefebvre, P.A., Negreiros, G.H., da Silva, E.D., Stone, T.A., Trumbore, S.E. & Vieira, S. 1994. The role of deep roots in the hydrological and carbon cycles of Amazonian forests and pastures. *Nature* **372**: 666–9.
Neuwinger, H.D. 1996. *African ethnobotany: poisons and drugs*. Chapman & Hall, London.
Newbery, D.M., Alexander, I.J. & Rother, J.A. 1997. Phosphorus dynamics in a lowland African rain forest: the influence of ectomycorrhizal trees. *Ecological Monographs* **67**: 367–409.
Newbery, D.M. & Gartlan, J.S. 1996. A structural analysis of rain forest at Korup and Douala-Edea, Cameroon. *Proceedings of the Royal Society of Edinburgh* B **104**: 177–224.
Newell, E.A., McDonald, E.P., Strain, B.R. & Denslow, J.S. 1993. Photosynthetic responses of *Miconia* species to canopy openings in a lowland tropical rainforest. *Oecologia* **94**: 49–56.
Newstrom, L.E., Frankie, G.W., Baker, H.G. & Colwell, R.K. 1993. Diversity of long-term flowering patterns. Pp. 142–60 in *La Selva: ecology and natural history of a neotropical rain forest* (ed. L.A. McDade, K.S. Bawa, H.A. Hespenheide & G.S. Hartshorn). The University of Chicago Press, Chicago.
Newton, A.C. & Healey, J.R. 1989. Establishment of *Clethra occidentalis* on stems of the tree-fern *Cyathea pubescens* in a Jamaican montane rain forest. *Journal of Tropical Ecology* **5**: 441–5.
Ng, F.S.P. 1978. Strategies of establishment in Malayan forest trees. Pp. 129–62 in *Tropical trees as living systems* (ed. P.B. Tomlinson & M.H. Zimmermann). Cambridge University

Press, Cambridge.
Ng, F.S.P. 1980. Germination ecology of Malaysian woody plants. *Malaysian Forester* **43**: 406–37.
Ng, F.S.P. & Tang, H.T. 1974. Comparative growth rates of Malaysian trees. *Malaysian Forester* **37**: 2–23.
Nichols-Orians, C.M. 1991. Environmentally induced differences in plant traits: consequences for susceptibility to a leaf-cutter ant. *Ecology* **72**: 1609–23.
Nickol, M. 1993. Melastomataceen als Myrmekophyten: eine Möglichkeit der Anpassung an der Lebensraum. Pp. 125–37 in *Animal–plant interactions in tropical environments* (ed. W. Barthlott, C.M. Naumann, K. Schmidt-Loske & K.-L. Schuchmann). Zoologisches Forschungsinstitut und Museum Alexander König, Bonn.
Nilsson, L.A., Rabakonandrianina, E., Pettersson, B. & Ranaivo, J. 1990. 'Ixoroid' secondary pollen presentation and pollination by small moths in the Malagasy treelet *Ixora platythyrsa* (Rubiaceae). *Plant Systematics and Evolution* **170**: 161–75.
Northup, R.R., Dahlgren, R.A. & Yu, Z. 1995. Intraspecific variation of conifer phenolic concentration on a maritime terrace soil acidity gradient; a new interpretation. *Plant and Soil* **171**: 255–62.
Notman, E., Gorchov, D.L. & Cornejo, F. 1996. Effect of distance, aggregation, and habitat on levels of seed predation for two mammal-dispersed neotropical rain forest tree species. *Oecologia* **106**: 221–7.
Núñez-Farfán, J. & Dirzo, R. 1991. Effects of defoliation on the saplings of a gap-colonizing neotropical tree. *Journal of Vegetation Science* **2**: 459–64.
Nyffeler, R. 1999. A new ordinal classification of the flowering plants. *Trends in Ecology and Evolution* **14**: 168–9.
O'Brien, S.T., Hubbell, S.P., Spiro, P., Condit, R. & Foster, R.B. 1995. Diameter, height, crown, and age relationships in eight neotropical tree species. *Ecology* **76**: 1926–39.
O'Brien, T.G., Kinnaird, M.F., Dierenfeld, E.S., Conklin-Brittain, N.L., Wrangham, R.W. & Silver, S.C. 1998. What's so special about figs? *Nature* **392**: 668.
O'Dowd, D.J. & Lake, P.S. 1991. Red crabs in rain forest, Christmas Island, Indian Ocean: removal and fate of fruits and seeds. *Journal of Tropical Ecology* **7**: 113–22.
Oates, J.F., Waterman, P.G. & Choo, G.M. 1980. Food selection by the South Indian leaf-monkey, *Presbytis johnii*, in relation to leaf chemistry. *Oecologia* **45**: 45–56.
Oates, J.F., Whitesides, G.H., Davies, A.G., Waterman, P.G., Green, S.M., Dasilva, G.L. & Mole, S. 1990. Determinants of variation in tropical forest primate biomass: new evidence from West Africa. *Ecology* **71**: 328–43.
Oberbauer, S.F. & Strain, B.R. 1986. Effects of canopy position and irradiance on the leaf physiology and morphology of *Pentaclethra macroloba* (Mimosaceae). *American Journal of Botany* **73**: 409–16.
Okuda, T., Kachi, N., Yap, S.K. & Manokaran, N. 1997. Tree distribution pattern and fate of juveniles in a lowland tropical rain forest – implications for regeneration and maintenance of species diversity. *Plant Ecology* **131**: 155–71.
Oldeman, R.A.A. & van Dijk, J. 1991. Diagnosis of the temperament of tropical rain forest trees. Pp. 21–65 in *Rain forest regeneration and management* (ed. A. Gómez-Pompa, T.C. Whitmore & M. Hadley). UNESCO, Paris.
Olesen, J.M. & Balslev, H. 1990. Flower biology and pollinators of the Amazonian monoecious palm, *Geonoma macrostachys*: a case of Bakerian mimicry. *Principes* **34**: 181–90.
Oliveira, P.S., Galetti, M., Pedroni, F. & Morellato, L.P.C. 1995. Seed cleaning by *Mycocepurus goeldii* ants (Attini) facilitates germination in *Hymenaea courbaril* (Caesalpiniaceae). *Biotropica* **27**: 518–22.
Opler, P.A., Baker, H.G & Frankie, G.W. 1980. Plant reproductive characteristics during secondary succession in neotropical lowland forest ecosystems. *Biotropica* **12** (suppl.): 40–6.
Orozco-Segovia, A., Vázquez-Yanes, C., Coates-Estrada, R. & Pérez-Nasser, N. 1987. Ecophysiological characteristics of the seed of the tropical forest pioneer *Urera caracasana* (Urticaceae). *Tree Physiology* **3**: 375–86.
Osaki, M., Watanabe, T. & Tadano, T. 1997. Beneficial effect of aluminum on growth of plants adapted to low pH soils. *Soil Science and Plant Nutrition* **43**: 551–63.
Osorio, D. & Vorobyev, M. 1996. Colour vision as an adaptation to frugivory in primates. *Proceedings of the Royal Society of London* B **263**: 593–99.
Osunkoya, O.O. 1996. Light requirements for regeneration in tropical forest plants: taxon-level and ecological attribute effects. *Australian Journal of Ecology* **21**: 429–41.
Osunkoya, O.O. 1999. Population structure and

breeding biology in relation to conservation in the dioecious *Gardenia actinocarpa* (Rubiaceae) – a rare shrub of North Queensland rainforest. *Biological Conservation* **88**: 347–59.

Osunkoya, O.O. & Ash, J.E. 1991. Acclimation to a change in light regime in seedlings of six Australian rainforest tree species. *Australian Journal of Botany* **39**: 591–605.

Osunkoya, O.O., Ash, J.E., Hopkins, M.S. & Graham, A.W. 1992. Factors affecting survival of tree seedlings in North Queensland rainforests. *Oecologia* **91**: 569–78.

Osunkoya, O.O., Ash, J.E., Hopkins, M.S. & Graham, A.W. 1994. Influence of seed size and seedling ecological attributes on shade-tolerance of rain-forest tree species in northern Queensland. *Journal of Ecology* **82**: 149–63.

Ovington, J.B. & Olson, J.S. 1970. Biomass and chemical content of El Verde lower montane rain forest plants. Pp. H53–7 in *A tropical rain forest* (ed. H.T. Odum & R.F. Pigeon). United States Atomic Energy Corporation, Oak Ridge.

Oyama, K. & Mendoza, A. 1990. Effects of defoliation on growth, reproduction, and survival of a Neotropical dioecious palm, *Chamaedora tepejilote*. *Biotropica* **22**: 119–23.

Pannell, C.M. & Kozioł, M.J. 1987. Ecological and phytochemical diversity of arillate seeds in *Aglaia* (Meliaceae): a study of vertebrate dispersal in tropical trees. *Philosophical Transactions of the Royal Society of London* B **316**: 303–33.

Peace, W.J.H. & MacDonald, F.D. 1981. An investigation of the leaf anatomy, foliar mineral levels, and water relations of trees of a Sarawak forest. *Biotropica* **13**: 100–9.

Peres, C.A. & van Roosmalen, M.G.M. 1996. Avian dispersal of 'mimetic seeds' of *Ormosia lignivalvis* by terrestrial granivores: deception or mutualism? *Oikos* **75**: 249–58.

Peres, C.A., Schiesari, L.C. & Dias-Leme, C.L. 1997. Vertebrate predation of brazil-nuts (*Bertholletia excelsa*, Lecythidaceae), an agouti-dispersed Amazonian seed crop: a test of the escape hypothesis. *Journal of Tropical Ecology* **13**: 69–79.

Perry, D.R. & Starrett, A. 1980. The pollination ecology and blooming strategy of a neotropical emergent tree, *Dipteryx panamensis*. *Biotropica* **12**: 307–13.

Philipson, W.R. 1979. Araliaceae – I. *Flora Malesiana, series I* **9**(1): 1–105.

Phillips, O.L., Hall, P., Gentry, A.H., Sawyer, S.A. & Vázquez, R. 1994. Dynamics and species richness of tropical rain forests. *Proceedings of the National Academy of Sciences of the United States of America* **91**: 2805–9.

Phua, P.B. & Corlett, R.T. 1989. Seed dispersal by the lesser short-nosed fruit bat (*Cynopteris brachyotis*, Pteropodidae, Megachiroptera). *Malayan Nature Journal* **42**: 251–6.

Pianka, E.R. 1970. On *r*- and *K*-selection. *American Naturalist* **104**: 592–7.

Pizo, M.A. & Oliveira, P.S. 1998. Interaction between ants and seeds of a nonmyrmecochorous neotropical tree, *Cabralea canjerana* (Meliaceae), in the Atlantic forest of southeast Brazil. *American Journal of Botany* **85**: 669–74.

Pockman, W.T., Sperry, J.S. & O'Leary, J.W. 1995. Sustained and significant negative water pressure in xylem. *Nature* **378**: 715–16.

Poorter, L. 1999. Growth responses of 15 rainforest tree species to a light gradient: the relative importance of morphological and physiological traits. *Functional Ecology* **13**: 396–410.

Poorter, L. & Oberbauer, S.F. 1993. Photosynthetic induction responses of two rainforest tree species in relation to light environment. *Oecologia* **96**: 193–9.

Poorter, L., Oberbauer, S.F. & Clark, D.A. 1995. Leaf optical properties along a vertical gradient in a tropical rain forest canopy in Costa Rica. *American Journal of Botany* **82**: 1257–63.

Popma, J. & Bongers, F. 1988. The effect of canopy gaps on growth and morphology of seedlings of rain forest species. *Oecologia* **75**: 625–32.

Popma, J. & Bongers, F. 1991. Acclimation of seedlings of three Mexican tropical rain forest tree seedlings to a change in light availability. *Journal of Tropical Ecology* **7**: 85–97.

Popma, J., Bongers, F. & Meave del Castillo, J. 1988. Patterns in the vertical structure of the tropical lowland rain forest of Los Tuxtlas, Mexico. *Vegetatio* **74**: 81–91.

Popma, J., Bongers, F. & Werger, M.J.A. 1992. Gap-dependence and leaf characteristics of trees in a tropical lowland rain forest in Mexico. *Oikos* **63**: 207–14.

Portnoy, S. & Willson, M.F. 1993. Seed dispersal curves: behavior of the tail of the distribution. *Evolutionary Ecology* **7**: 25–44.

Prance, G.T. 1985. The pollination of Amazonian plants. Pp. 166–91 in *Amazonia* (ed. G.T. Prance & T.E. Lovejoy). Pergamon

Press, Oxford.

Pratt, T.K. & Stiles, E.W. 1985. The influence of fruit size and structure on composition of frugivore assemblages in New Guinea. *Biotropica* **17**: 314–21.

Press, M.C., Brown, N.D., Barker, M.G. & Zipperlen, S.W. 1996. Photosynthetic responses to light in tropical rain forest tree seedlings. Pp. 41–58 in *The ecology of tropical forest tree seedlings* (ed. M.D. Swaine). UNESCO, Paris.

Prevost, M.-F. 1983. Les fruits et les graines des espèces végétales pionnières de Guyane Française. *Revue d'Ecologie (La Terre et la Vie)* **38**: 121–45.

Primack, R.B. & Lee, H.S. 1991. Population dynamics of pioneer (*Macaranga*) trees and understorey (*Mallotus*) trees (Euphorbiaceae) in primary and selectively logged Bornean rain forests. *Journal of Tropical Ecology* **7**: 439–58.

Putz, F.E. 1979. Aseasonality in Malaysian tree phenology. *Malaysian Forester* **42**: 1–24.

Putz, F.E. & Appanah, S. 1987. Buried seeds, newly dispersed seeds, and the dynamics of a lowland forest in Malaysia. *Biotropica* **19**: 326–33.

Putz, F.E. & Brokaw, N.V.L. 1989. Sprouting of broken trees on Barro Colorado Island, Panama. *Ecology* **70**: 508–12.

Putz, F.E., Coley, P.D., Lu, K., Montalvo, A. & Aiello, A. 1983. Uprooting and snapping of trees: structural determinants and ecological consequences. *Canadian Journal of Forest Research* **13**: 1011–20.

Putz, F.E. & Holbrook, N.M. 1986. Notes on the natural history of hemiepiphytes. *Selbyana* **9**: 61–9.

Pyykkö, M. 1979. Morphology and anatomy of leaves from some woody plants in a humid tropical forest of Venezuelan Guayana. *Acta Botanica Fennica* **112**: 1–41.

Quénéhervé, P., Bereau, M. & van den Berg, E. 1996. Plant-feeding nematodes associated with *Dicorynia guianensis* Amshoff (sub-family Caesalpinioidae) seedlings in a primary rain forest near Paracou, French Guiana. *European Journal of Soil Biology* **32**: 187–93.

Raaimakers, D., Boot, R.G.A., Dijkstra, P., Pot, S. & Pons, T. 1995. Photosynthetic rates in relation to leaf phosphorus content in pioneer versus climax tropical rainforest trees. *Oecologia* **102**: 120–5.

Raaimakers, D. & Lambers, H. 1996. Response to phosphorus supply of tropical tree seedlings: a comparison between a pioneer species *Tapirira obtusa* and a climax species *Lecythis corrugata*. *New Phytologist* **132**: 97–102.

Raich, J.W. 1983. Understory palms as nutrient traps: a hypothesis. *Brenesia* **21**: 119–29.

Raich, J.W. & Gong, W.K. 1990. Effects of canopy openings on tree seed germination in a Malaysian dipterocarp forest. *Journal of Tropical Ecology* **6**: 203–17.

Raich, J.W., Russel, A.E., Crews, T.E., Farrington, H. & Vitousek, P.M. 1996. Both nitrogen and phosphorus limit plant production on young Hawaiian lava flows. *Biogeochemistry* **32**: 1–14.

Ramirez, N. & Arroyo, M.K. 1987. Variación espacial y temporal en la depredación de semillas de *Copaifera pubiflora* Benth. en Venezuela. *Biotropica* **19**: 32–9.

Ramos, J. & Grace, J. 1990. The effects of shade on the gas exchange of seedlings of four tropical trees from Mexico. *Functional Ecology* **4**: 667–77.

Rankin-de-Merona, J.M., Hutchings, H.R.W. & Lovejoy, T.E. 1990. Tree mortality and recruitment over a five-year period in undisturbed upland rainforest of the Central Amazon. Pp. 573–84 in *Four neotropical rainforests* (ed. A.H. Gentry). Yale University Press, New Haven.

Raunkiaer, C. 1916. Om Bladstorrelsens Anvendelse i den biologiske Plantegeografi. *Botanisk Tidsskrift* **33**: 225–40.

Raunkiaer, C. 1934. *The life forms of plants and statistical plant geography*. Clarendon Press, Oxford.

Redhead, J.F. 1968. Mycorrhizal associations in some Nigerian forest trees. *Transactions of the British Mycological Society* **51**: 377–87.

Reekie, E.G. & Wayne, P. 1992. Leaf canopy display, stomatal conductance, and photosynthesis in seedlings of three pioneer tree species subjected to drought. *Canadian Journal of Botany* **70**: 2334–8.

Rehr, S.S., Feeny, P.P. & Janzen, D.H. 1973. Chemical defence in Central American non-ant-acacias. *Journal of Animal Ecology* **42**: 405–16.

Reich, P.B., Ellsworth, D.S. & Uhl, C. 1995. Leaf carbon and nutrient assimilation and conservation in species of differing successional status in an oligotrophic Amazonian forest. *Functional Ecology* **9**: 65–76.

Reich, P.B., Uhl, C., Walters, M.B. & Ellsworth, D.S. 1991. Leaf lifespan as a determinant of leaf structure and function among 23 amazonian tree species. *Oecologia* **86**: 16–24.

Reitsma, J.M. 1988. Forest vegetation of Gabon. *Tropenbos Technical Series* **1**: 1–103.
Renner, S.S. & Feil, J.P. 1993. Pollinators of tropical dioecious angiosperms. *American Journal of Botany* **80**: 1100–7.
Renner, S.S. & Ricklefs, R.E. 1998. Herbicidal activity of domatia-inhabiting ants in patches of *Tococa guianensis* and *Clidemia heterophylla*. *Biotropica* **30**: 324–7.
Reyes, G., Brown, S., Chapman, J., Lugo, A.E. 1992. Wood densities of tropical tree species. *U.S.D.A. Forest Service, Southern Forest Experiment Station General Technical Report* SO–88.
Rich, P.M. 1987. Mechanical structure of the stem of arborescent palms. *Botanical Gazette* **148**: 42–50.
Rich, P.M., Helenurm, K., Kearns, D., Morse, S.R., Palmer, M.W. & Short, L. 1986. Height and stem diameter relationships for dicotyledonous trees and arborescent palms of Costa Rican tropical wet forest. *Bulletin of the Torrey Botanical Club* **113**: 241–6.
Richards, A.J. 1990a. Studies in *Garcinia*, dioecious tropical forest trees: agamospermy. *Botanical Journal of the Linnean Society* **103**: 233–50.
Richards, A.J. 1990b. Studies in *Garcinia*, dioecious tropical forest trees: the phenology, pollination biology and fertilization of *G. hombroniana* Pierre. *Botanical Journal of the Linnean Society* **103**: 251–61.
Richards, P.W. 1952. *The tropical rain forest*. First edition. Cambridge University Press, Cambridge.
Richards, P.W. 1996. *The tropical rain forest*. Second edition. Cambridge University Press, Cambridge.
Riera, B. 1985. Importance des buttes de deracinement dans la regeneration forestiere en Guyane Française. *Revue d'Ecologie (La Terre et la Vie)* **40**: 321–9.
Riswan, S. 1988. Leaf nutrient status in the lowland dipterocarp forest. *Reinwardtia* **10**: 425–37.
Roberts, E.H. 1973. Predicting the storage life of seeds. *Seed Science and Technology* **1**: 499–574.
Roberts, R.T. & Heithaus, E.R. 1986. Ants rearrange the vertebrate-generated seed shadow of a neotropical fig tree. *Ecology* **67**: 1046–51.
Robinson, D.F. 1996. A symbolic framework for the description of tree architecture models. *Botanical Journal of the Linnean Society* **121**: 243–61.
Rogers, M.E., Maisels, F., Williamson, E.A., Fernandez, M., Tutin, C.E.G. 1990. Gorilla diet in the Lopé Reserve, Gabon: a nutritional analysis. *Oecologia* **84**: 326–39.
Rogers, M.E., Voysey, B.C., McDonald, K.E., Parnell, R.J. & Tutin, C.E.G. 1998. Lowland gorillas and seed dispersal: the importance of nest sites. *American Journal of Primatology* **45**: 45–68.
Rogstad, S.H. 1989. The biosystematics and evolution of the *Polyalthia hypoleuca* complex (Annonaceae) of Malesia, I. Systematic treatment. *Journal of the Arnold Arboretum* **70**: 153–246.
Rollet, B. 1969. Etudes quantitative d'une foret dense humide sempervirente de plaine en Guyane Vénézuélienne. *Travaux du Laboratoire Forestier de Toulouse* **8** (art. 1): 1–36.
Rollet, B. 1990. Leaf morphology. Pp. 1–75 in *Stratification of tropical forests as seen in leaf structure*, part 2 (ed. B. Rollet, C. Högermann & I. Roth). Kluwer Academic Publishers, Dordrecht.
Roth, A., Mosbrugger, V., Belz, G. & Neugebauer, H.J. 1995. Hydrodynamic modelling study of angiosperm leaf venation types. *Botanica Acta* **108**: 121–6.
Roth, I. 1981. *Structural patterns of tropical barks*. Encyclopedia of Plant Anatomy 9(3). Gebrüder Borntraeger, Berlin.
Roth, I. 1984. *Stratification of tropical forests as seen in leaf structure*. Dr W. Junk, The Hague.
Roth, I. 1987. *Stratification of a tropical forest as seen in dispersal types*. Dr W. Junk Publishers, Dordrecht.
Roth, I. 1996. *Microscopic venation patterns of leaves and their importance in the distinction of (tropical) species*. Encyclopedia of Plant Anatomy 14(4). Gebrüder Borntraeger, Berlin.
Roy, J. & Salager, J. 1992. Midday depression of net $CO_2$ exchange of leaves of an emergent tree in French Guiana. *Journal of Tropical Ecology* **8**: 499–504.
Rueda, R. & Williamson, G.B. 1992. Radial and vertical wood specific gravity in *Ochroma pyramidale* (Cav. ex Lam.) Urb. (Bombacaceae). *Biotropica* **24**: 512–18.
Ryan, M.G., Hubbard, R.M., Clark, D.A. & Sanford, R.L. 1994. Woody-tissue respiration for *Simarouba amara* and *Minquartia guianensis*, two tropical wet forest trees with different growth habits. *Oecologia* **100**: 213–20.
Sagers, C.L. 1993. Reproduction in neotropical

shrubs: the occurrence and some mechanisms of asexuality. *Ecology* **74**: 615–18.

Sagers, C.L. & Coley, P.D. 1995. Benefits and costs of defense in a neotropical shrub. *Ecology* **76**: 1835–43.

Sakai, S., Momose, K., Yumoto, T., Kato, M. & Inoue, T. 1999. Beetle pollination of *Shorea parvifolia* (section *Mutica*, Dipterocarpaceae) in a general flowering period in Sarawak, Malaysia. *American Journal of Botany* **86**: 62–9.

Sánchez-Cordera, V. & Martínez-Gallardo, R. 1998. Postdispersal fruit and seed removal by forest-dwelling rodents in a lowland rainforest in Mexico. *Journal of Tropical Ecology* **14**: 139–51.

Sazima, M., Buzato, S. & Sazima, I. 1999. Bat-pollinated flower assemblages and bat visitors at two Atlantic forest sites in Brazil. *Annals of Botany* **83**: 705–12.

Schatz, G.E. 1990. Some aspects of pollination in Central America. Pp. 69–84 in *Reproductive ecology of tropical forest plants* (ed. K.S. Bawa & M. Hadley). UNESCO, Paris.

Schatz, G.E., Williamson, G.B., Cogswell, C.M. & Stam, A.C. 1985. Stilt roots and growth of arboreal palms. *Biotropica* **17**: 206–9.

Schmid, R. 1970. Notes on the reproductive biology of *Asterogyne martiana* (Palmae). II. Pollination by syrphid flies. *Principes* **14**: 39–49.

Schupp, E.W. 1992. The Janzen–Connell model for tropical tree diversity: population implications and the importance of spatial scale. *American Naturalist* **140**: 526–30.

Schupp, E.W. & Feener, D.H. 1991. Phylogeny, lifeform, and habitat dependence of ant-defended plants in a Panamanian forest. Pp. 175–97 in *Ant–plant interactions* (ed. C.A. Huxley & D.F. Cutler). Oxford University Press, Oxford.

Setten, G.G.K. 1953. The incidence of buttressing among Malayan tree species when of commercial size. *Malayan Forester* **16**: 219–21.

Shepherd, V.E. & Chapman, C.A. 1998. Dung beetles as secondary seed dispersers: impact on seed predation and germination. *Journal of Tropical Ecology* **14**: 199–215.

Shugart, H.H. 1987. Dynamic ecosystem consequences of tree birth and death patterns. *BioScience* **37**: 596–602.

Silva Matos, D.M., Freckleton, R.P. & Watkinson, A.R. 1999. The role of density dependence in the population dynamics of a tropical palm. *Ecology* **80**: 2635–50.

Silver, W.L. 1994. Is nutrient availability related to plant nutrient use in humid tropical forests? *Oecologia* **98**: 336–43.

Sim, J.W.S., Tan, H.T.W. & Turner, I.M. 1992. Adinandra belukar: an anthropogenic heath forest in Singapore. *Vegetatio* **102**: 125–37.

Simard, S.W., Perry, D.A., Jones, M.D., Myrold, D.D., Durall, M.D. & Molina, R. 1997. Net transfers of carbon between ectomycorrhizal tree species in the field. *Nature* **388**: 579–82.

Sinha, A. & Davidar, P. 1992. Seed dispersal ecology of a wind dispersed rain forest tree in the Western Ghats, India. *Biotropica* **24**: 519–26.

Smith, A.M. 1994. Xylem transport and the negative pressures sustainable by water. *Annals of Botany* **74**: 647–51.

Smith, S.E. & Read, D.J. 1997. *Mycorrhizal symbiosis*. Academic Press, San Diego.

Smythe, N. 1989. Seed survival in the palm *Astrocaryum standleyanum*: evidence for dependence upon its seed dispersers. *Biotropica* **21**: 50–6.

Snow, D.W. 1981. Tropical frugivorous birds and their food plants: a world survey. *Biotropica* **13**: 1–14.

Sobrado, M.A. & Medina, E. 1980. General morphology, anatomical structure, and nutrient content on sclerophyllous leaves of the 'Bana' vegetation of Amazonas. *Oecologia* **45**: 341–5.

Sobrevila, C. & Arroyo, M.T.K. 1982. Breeding systems in a montane tropical cloud forest in Venezuela. *Plant Systematics and Evolution* **140**: 19–38.

Sollins, P. 1998. Factors influencing species composition in tropical lowland rain forest: does soil matter? *Ecology* **79**: 23–30.

Soltis, D.E., Soltis, P.S., Morgan, D.R., Swensen, S.M., Mullin, B.C., Dowd, J.M. & Martin, P.G. 1995. Chloroplast gene sequence data suggest a single origin of the predisposition for symbiotic nitrogen fixation in angiosperms. *Proceedings of the National Academy of Sciences of the United States of America* **92**: 2647–51.

Sork, V.L. 1987. Effects of predation and light on seedling establishment in *Gustavia superba*. *Ecology* **68**: 1341–50.

Sprent, J.I. 1994. Legume trees and shrubs in the tropics: $N_2$ fixation in perspective. *Soil Biology and Biochemistry* **27**: 401–7.

St. John, T.V. 1980. A survey of micorrhizal [sic] infection in an amazonian rain forest. *Acta Amazonica* **10**: 527–33.

St. John, T.V. & Uhl, C. 1983. Mycorrhizae in

the rain forest at San Carlos de Río Negro, Venezuela. *Acta Científica Venezolana* **34**: 233–7.

Stacy, E.A., Hamrick, J.L., Nason, J.D., Hubbell, S.P., Foster, R.B. & Condit, R. 1996. Pollen dispersal in low-density populations of three neotropical tree species. *American Naturalist* **148**: 275–98.

Stanley, W.G. & Montagnini, F. 1999. Biomass and nutrient accumulation in pure and mixed plantations of indigenous tree species grown on poor soils in the humid tropics of Costa Rica. *Forest Ecology and Management* **113**: 91–103.

Stiles, F.G. 1975. Ecology, flowering phenology and hummingbird pollination of some Costa Rican *Heliconia* species. *Ecology* **56**: 285–310.

Stiles, F.G. 1978. Temporal organization of flowering among the hummingbird food plants of a tropical wet forest. *Biotropica* **10**: 194–210.

Stiles, F.G. 1981. Geographical aspects of bird-flower coevolution, with particular reference to Central America. *Annals of the Missouri Botanical Garden* **68**: 323–51.

Stiller, V. & Sperry, J.S. 1999. Canny's compensating pressure theory fails a test. *American Journal of Botany* **86**: 1082–6.

Stocker, G.C. & Irvine, A.K. 1983. Seed dispersal by cassowaries (*Casuarius casuarius*) in North Queensland's rainforests. *Biotropica* **15**: 170–6.

Strauss-Debenedetti, S. & Bazzaz, F.A. 1991. Plasticity and acclimation to light in tropical Moraceae of different sucessional [sic] positions. *Oecologia* **87**: 377–87.

Strauss-Debenedetti, S. & Berlyn, G.P. 1994. Leaf anatomical responses to light in five tropical Moraceae of different successional status. *American Journal of Botany* **81**: 1582–91.

Sugden, A.M. 1985. Leaf anatomy in a Venezuelan montane forest. *Botanical Journal of the Linnean Society* **90**: 231–41.

Sun, C., Ives, A.R., Kraeuter, H.J. & Moermond, T.C. 1997. Effectiveness of three turacos as seed dispersers in a tropical montane forest. *Oecologia* **112**: 94–103.

Suzuki, E. 1999. Diversity in specific gravity and water content of wood among Bornean tropical rainforest trees. *Ecological Research* **14**: 211–24.

Suzuki, E. & Ashton, P.S. 1996. Sepal and nut size ratio of fruits of Asian Dipterocarpaceae and its implications for dispersal. *Journal of Tropical Ecology* **12**: 853–70.

Suzuki, E. & Kohyama, T. 1991. Spatial distribution of wind-dispersed fruits and trees of *Swintonia schwenkii* (Anacardiaceae) in a tropical forest of West Sumatra. *Tropics* **1**: 131–42.

Swaine, M.D. 1983. Stilt roots and ephemeral germination sites. *Biotropica* **15**: 240.

Swaine, M.D. & Beer, T. 1977. Explosive seed dispersal in *Hura crepitans* L. (Euphorbiaceae). *New Phytologist* **78**: 695–708.

Swaine, M.D., Lieberman, D. & Putz, F.E. 1987. The dynamics of tree populations in tropical forest: a review. *Journal of Tropical Ecology* **3**: 359–66.

Swaine, M.D. & Whitmore, T.C. 1988. On the definition of ecological species groups in tropical rain forests. *Vegetatio* **75**: 81–6.

Tan, G.C.-H., Ong, B.-L. & Turner, I.M. 1994. The photosynthetic performance of six early successional tropical tree species. *Photosynthetica* **30**: 201–6.

Tanner, E.V.J. 1977. Four montane rain forests of Jamaica: a quantitative characterization of the floristics, the soils and foliar mineral levels, and a discussion of the interrelations. *Journal of Ecology* **65**: 883–918.

Tanner, E.V.J. 1982. Species diversity and reproductive mechanisms in Jamaican trees. *Biological Journal of the Linnean Society* **18**: 263–78.

Tanner, E.V.J. & Kapos, V. 1982. Leaf structure of Jamaican upper montane rain-forest trees. *Biotropica* **14**: 16–24.

Tanner, E.V.J., Kapos, V. & Franco, W. 1992. Nitrogen and phosphorus fertilization effects on Venezuelan montane forest trunk growth and litterfall. *Ecology* **73**: 78–86.

Tanner, E.V.J., Kapos, V., Freskos, S., Healey, J. & Theobold, A.M. 1990. Nitrogen and phosphorus fertilization of Jamaican montane forest trees. *Journal of Tropical Ecology* **6**: 231–8.

Temple, S.A. 1977. Plant–animal mutualism: coevolution with dodo leads to near extinction of plant. *Science* **197**: 885–6.

ter Steege, H. 1994a. Seedling growth of *Mora gongrijpii*, a large seeded climax species under different soil and light conditions. *Vegetatio* **112**: 161–70.

ter Steege, H. 1994b. Flooding and drought tolerance in seeds and seedlings of two *Mora* species segregated along a soil hydrological gradient in the tropical rain forest of Guyana. *Oecologia* **100**: 356–67.

ter Welle, B.J.H. 1976. Silica grains in woody

plants of the neotropics, especially Surinam. *Leiden Botanical Series* **3**: 107–42.

Terborgh, J. 1986. Keystone plant resources in the tropical forest. Pp. 330–44 in *Conservation biology: the science of scarcity and diversity* (ed. M.E. Soulé). Sinauer Associates, Sunderland.

Terborgh, J., Foster, R.B. & Nuñez, V.P. 1996. Tropical tree communities: a test of the nonequilibrium hypothesis. *Ecology* **77**: 561–7.

Terborgh, J. & Mathews, J. 1999. Partitioning of the understorey light environment by two Amazonian treelets. *Journal of Tropical Ecology* **15**: 751–63.

Terborgh, J., Losos, E., Riley, M.P. & Bolaños Riley, M. 1993. Predation by vertebrates and invertebrates on the seeds of five canopy tree species of an Amazonian forest. *Vegetatio* **107/108**: 373–86.

Thiele, A., Krause, G.H. & Winter, K. 1998. *In situ* study of photoinhibition of photosynthesis and xanthophyll cycle activity in plants growing in natural gaps of the tropical forest. *Australian Journal of Plant Physiology* **25**: 189–95.

Thien, L.B. 1980. Patterns of pollination in the primitive angiosperms. *Biotropica* **12**: 1–13.

Thomas, D.W. 1988. The influence of aggressive ants on fruit removal in the tropical tree, *Ficus capensis* (Moraceae). *Biotropica* **20**: 49–53.

Thomas, S.C. 1996a. Asymptotic height as a predictor of growth and allometric characteristics in Malaysian rain forest trees. *American Journal of Botany* **83**: 556–66.

Thomas, S.C. 1996b. Relative size at onset of maturity in rain forest trees: a comparative analysis of 37 Malaysian species. *Oikos* **76**: 145–54.

Thomas, S.C. 1996c. Reproductive allometry in Malaysian rain forest trees: biomechanics versus optimal allocation. *Evolutionary Ecology* **10**: 517–30.

Thomas, S.C. 1997. Geographic parthenogenesis in a tropical forest tree. *American Journal of Botany* **84**: 1012–15.

Thomas, S.C. & Bazzaz, F.A. 1999. Asymptotic height as a predictor of photosynthetic characteristics in Malaysian rain forest trees. *Ecology* **80**: 1607–22.

Thomas, S.C. & Ickes, K. 1995. Ontogenetic changes in leaf size in Malaysian rain forest trees. *Biotropica* **27**: 427–34.

Thomas, S.C. & LaFrankie, J.V. 1993. Sex, size, and interyear variation in flowering among dioecious trees of the Malayan rain forest. *Ecology* **74**: 1529–37.

Thompson, W.A., Huang, L.-K. & Kriedemann, P.E. 1992a. Photosynthetic response to light and nutrients in sun-tolerant and shade-tolerant rainforest trees. II. Leaf gas exchange and component processes of photosynthesis. *Australian Journal of Plant Physiology* **19**: 19–42.

Thompson, W.A., Kriedemann, P.E. & Craig, I.E. 1992b. Photosynthetic responses to light and nutrients in sun-tolerant and shade-tolerant rainforest trees. I. Growth, leaf anatomy and nutrient content. *Australian Journal of Plant Physiology* **19**: 1–18.

Thompson, W.A., Stocker, G.C. & Kriedemann, P.E. 1988. Growth and photosynthetic response to light and nutrients of *Flindersia brayleyana* F. Muell., a rainforest tree with broad tolerance to sun and shade. *Australian Journal of Plant Physiology* **15**: 299–315.

Thomsen, K. & Brimer, L. 1997. Cyanogenic constituents in woody plants in natural lowland rain forest in Costa Rica. *Botanical Journal of the Linnean Society* **124**: 273–94.

Thomson, J.D., Herre, E.A., Hamrick, J.L. & Stone, J.L. 1991. Genetic mosaics in strangler fig trees: implications for tropical conservation. *Science* **254**: 1214–16.

Thurston, E.L. & Lersten, N.R. 1969. The morphology and toxicology of plant stinging hairs. *Botanical Review* **35**: 393–412.

Tinoco-Ojanguren, C. & Pearcy, R.W. 1992. Dynamic stomatal behavior and its role in carbon gain during lightflecks of a gap phase and an understory *Piper* species acclimated to high and low light. *Oecologia* **92**: 222–8.

Tinoco-Ojanguren, C. & Pearcy, R.W. 1995. A comparison of light quality and quantity effects on the growth and steady-state and dynamic photosynthetic characteristics of three tropical tree species. *Functional Ecology* **9**: 222–30.

Toledo, V.M. 1977. Pollination of some rain forest plants by non-hoveringbirds in Veracruz, Mexico. *Biotropica* **9**: 262–7.

Torti, S.D. & Coley, P.D. 1999. Tropical monodominance: a preliminary test of the ectomycorrhizal hypothesis. *Biotropica* **31**: 220–8.

Torti, S.D., Coley, P.D. & Janos, D.P. 1997. Vesicular-arbuscular mycorrhizae in two tropical monodominant trees. *Journal of Tropical Ecology* **13**: 623–9.

Toy, R.J. 1991. Interspecific flowering patterns in the Dipterocarpaceae in West Malaysia:

implications for predator satiation. *Journal of Tropical Ecology* **7**: 49–57.

Turner, I.M. 1990a. The seedling survivorship and growth of three *Shorea* species in a Malaysian tropical rain forest. *Journal of Tropical Ecology* **6**: 469–78.

Turner, I.M. 1990b. Tree seedling growth and survival in a Malaysian rain forest. *Biotropica* **22**: 146–54.

Turner, I.M. 1994. Sclerophylly: primarily protective? *Functional Ecology* **8**: 669–75.

Turner, I.M. 1995a. Foliar defences and habitat adversity of three woody plant communities in Singapore. *Functional Ecology* **9**: 279–84.

Turner, I.M. 1995b. The foliar nitrate reductase activity of some tropical secondary forest trees in Singapore. *Asian Journal of Tropical Biology* **1**: 31–5.

Turner, I.M. 2001. An overview of the plant diversity of South-East Asia. *Asian Journal of Tropical Biology* **4(2)**: 1–20.

Turner, I.M., Brown, N.D. & Newton, A.C. 1993a. The effect of fertilizer application on dipterocarp seedling growth and mycorrhizal infection. *Forest Ecology and Management* **57**: 329–37.

Turner, I.M., Lucas, P.W., Becker, P., Wong, S.C., Yong, J.W.H., Choong, M.F. & Tyree, M.T. 2000. Tree leaf form in Brunei: a heath forest and a mixed dipterocarp forest compared. *Biotropica* **32**: 53–61.

Turner, I.M. & Newton, A.C. 1990. The initial responses of some tropical rain forest tree seedlings to a large gap environment. *Journal of Applied Ecology* **27**: 605–8.

Turner, I.M. & Tan, H.T.W. 1991. Habitat-related variation in tree leaf form in four tropical forest types on Pulau Ubin, Singapore. *Journal of Vegetation Science* **2**: 691–8.

Turner, I.M., Choong, M.F., Tan, H.T.W. & Lucas, P.W. 1993b. How tough are sclerophylls? *Annals of Botany* **71**: 343–6.

Tutin, C.E.G., Williamson, E.A., Rogers, M.E. & Fernandez, M. 1991. A case study of a plant–animal relationship: *Cola lizae* and lowland gorillas in the Lopé Reserve, Gabon. *Journal of Tropical Ecology* **7**: 181–99.

Tyree, M.T., Davis, S.D. & Cochard, H. 1994. Biophysical perspectives of xylem evolution: Is there a tradeoff of hydraulic efficiency for vulnerability to dysfunction? *IAWA Journal* **15**: 335–60.

Tyree, M.T. & Ewers, F.W. 1996. Hydraulic architecture of woody tropical plants. Pp. 217–43 in *Tropical forest plant ecophysiology* (ed. S.S. Mulkey, R.L. Chazdon & A.P. Smith). Chapman & Hall, New York.

Tyree, M.T., Patiño, S. & Becker, P. 1998a. Vulnerability to drought-induced embolism of Bornean heath and dipterocarp forest trees. *Tree Physiology* **18**: 583–8.

Tyree, M.T., Sobrado, M.A., Stratton, L.J. & Becker, P. 1999. Diversity of hydraulic conductance in leaves of temperate and tropical species: possible causes and consequences. *Journal of Tropical Forest Science* **11**: 47–60.

Tyree, M.T., Velez, V. & Dalling, J.W. 1998b. Growth dynamics of root and shoot hydraulic conductance in seedling of five neotropical tree species: scaling to show possible adaptation to different light regimes. *Oecologia* **114**: 293–8.

Ungar, P.S. 1995. Fruit preferences of four sympatric primate species at Ketambe, Northern Sumatra, Indonesia. *International Journal of Primatology* **16**: 221–45.

Valladares, F., Allen, M.T. & Pearcy, R.W. 1997. Photosynthetic responses to dynamic light under field conditions in six tropical rainforest shrubs occuring along a light gradient. *Oecologia* **111**: 505–14.

van der Graaff, N.A. & Baas, P. 1974. Wood anatomical variation in relation to latitude and altitude. *Blumea* **22**: 101–21.

van der Meer, P.J., Sterck, F.J. & Bongers, F. 1998. Tree seedling performance in canopy gaps in a tropical rain forest at Nouragues, French Guiana. *Journal of Tropical Ecology* **14**: 119–37.

van der Pijl, L. 1982. *Principles of dispersal in higher plants*. Third edition. Springer-Verlag, Berlin.

van Steenis, C.G.G.J. 1956. De biologische nomaden-theorie. *Vakblad voor Biologen* **36**: 165–72.

Vander Wall, S.B. 1990. *Food hoarding in animals*. The University of Chicago Press, Chicago.

Vasconcelos, H.L. 1991. Mutualism between *Maieta guianensis* Aubl., a myrmecophytic melastome, and one of its ant inhabitants: ant protection against insect herbivores. *Oecologia* **87**: 295–8.

Vázquez-Yanes, C. & Orozco-Segovia, A. 1992. Effects of litter from tropical rainforest on tree seed germination and establishment under controlled conditions. *Tree Physiology* **11**: 391–400.

Vázquez-Yanes, C. & Orozco-Segovia, A. 1993. Patterns of seed longevity and germination in the tropical rainforest. *Annual Review of*

*Ecology and Systematics* **24**: 69–87.
Vázquez-Yanes, C. & Orozco-Segovia, A. 1996a. Physiological ecology of seed dormancy and longevity. Pp. 535–58 in *Tropical forest plant ecophysiology* (ed. S.S. Mulkey, R.L. Chazdon & A.P. Smith). Chapman & Hall, New York..
Vázquez-Yanes, C. & Orozco-Segovia, A. 1996b. Comparative longevity of seeds of five tropical rain forest woody species stored under different moisture conditions. *Canadian Journal of Botany* **74**: 1635–9.
Vázquez-Yanes, C., Orozco-Segovia, A., Rincón, E., Sánchez-Coronado, M.E., Huante, P., Toledo, J.R. & Barradas, V.L. 1990. Light beneath the litter in a tropical forest: effect on seed germination. *Ecology* **71**: 1952–8.
Veenendaal, E.M., Swaine, M.D., Lecha, R.T., Walsh, M.F., Abebrese, I.K. & Owusu-Afriyie, K. 1996. Responses of West African forest tree seedlings to irradiance and soil fertility. *Functional Ecology* **10**: 501–11.
Veneklaas, E.J., Poorter, L. 1998. Growth and carbon partitioning of trpoical tree seedlings in contrasting light environments. Pp. 337–61 in *Inherent variation in plant growth. Physiological mechanisms and ecological consequences* (ed. H. Lambers, H. Poorter & M.M.I. van Vuuren). Backhuys Publishers, Leiden.
Vester, H.F.M. & Cleef, A.M. 1998. Tree architecture and secondary tropical rain forest development. *Flora* **193**: 75–97.
Vicentini, A. & Fischer, E.A. 1999. Pollination of *Moronobea coccinea* (Clusiaceae) by the golden-winged parakeet in the Central Amazon. *Biotropica* **31**: 692–6.
Vitousek, P.M. 1984. Litterfall, nutrient cycling, and nutrient limitation in tropical forests. *Ecology* **65**: 285–98.
Vitousek, P.M. & Sanford, R.L. 1986. Nutrient cycling in moist tropical forest. *Annual Review of Ecology and Systematics* **17**: 137–67.
Vitousek, P.M., Walker, L.R., Whiteaker, L.D. & Matson, P.A. 1993. Nutrient limitations to plant growth during primary succession in Hawaii Volcanoes National Park. *Biogeochemistry* **23**: 197–215.
Voysey, B.C., McDonald, K.E., Rogers, M.E., Tutin, C.E.G. & Parnell, R.J. 1999. Gorillas and seed dispersal in the Lopé Reserve, Gabon. II: Survival and growth of seedlings. *Journal of Tropical Ecology* **15**: 39–60.
Walter, D.E. 1996. Living on leaves: mites, tomenta, and leaf domatia. *Annual Review of Entomology* **41**: 101–14.
Wang, D., Bormann, F.H., Lugo, A.E. & Bowden, R.D. 1991. Comparison of nutrient-use efficiency and biomass production in five tropical tree taxa. *Forest Ecology and Management* **46**: 1–21.
Watanabe, T., Osaki, M. & Tadano, T. 1997. Aluminum-induced growth stimulation in relation to calcium, magnesium, and silicate nutrition in *Melastoma malabathricum* L. *Soil Science and Plant Nutrition* **43**: 827–37.
Watanabe, T., Osaki, M. & Tadano, T. 1998. Effects of nitrogen source and aluminum on growth of tropical tree seedlings adapted to low pH soils. *Soil Science and Plant Nutrition* **44**: 655–66.
Waterman, P.G. 1996. Secondary metabolites. *Proceedings of the Royal Society of Edinburgh* **104B**: 225–42.
Waterman, P.G. & McKey, D. 1989. Herbivory and secondary compounds in rain-forest plants. Pp. 513–36 in *Ecosystems of the world 14B, tropical rain forest ecosystems: biogeographical and ecological studies* (ed. H. Lieth & M.J.A. Werger). Elsevier, Amsterdam.
Waterman, P.G. & Mole, S. 1989. Soil nutrients and plant secondary compounds. Pp. 241–54 in *Mineral nutrients in tropical forest and savanna ecosystems* (ed. J. Proctor). Blackwell Scientific Publications, Oxford.
Waterman, P.G., Ross, J.A.M., Bennett, E.L. & Davies, A.G. 1988. A comparison of the floristics and leaf chemistry of the tree flora in two Malaysian rain forests and the influence of leaf chemistry on populations of colubine monkeys in the Old World. *Biological Journal of the Linnean Society* **34**: 1–32.
Webb, C.O. & Peart, D.R. 1999. Seedling density dependence promotes coexistence of Bornean rain forest trees. *Ecology* **80**: 2006–17.
Webb, L.J. 1959. A physiognomic classification of Australian rain forests. *Journal of Ecology* **47**: 551–70.
Webber, A.C. & Gottsberger, G. 1995. Floral biology and pollination of *Bocageopsis multiflora* and *Oxandra euneura* in Central Amazonia, with remarks on the evolution of stamens in the Annonaceae. *Feddes Repertorium* **106**: 515–24.
Wei, C., Steudle, E. & Tyree, M.T. 1999. Water ascent in plants: do ongoing controversies have a sound basis? *Trends in Plant Science* **4**: 372–5.
Welden, C.W., Hewett, S.W., Hubbell, S.P. & Foster, R.B. 1991. Sapling survival, growth, and recruitment: relationship to canopy

height in a neotropical forest. *Ecology* **72**: 35–50.

Wenny, D.G. 1999. Two-stage dispersal of *Guarea glabra* and *G. kunthiana* (Meliaceae) in the Monteverde, Costa Rica. *Journal of Tropical Ecology* **15**: 481–96.

Wenny, D.G. & Levey, D.J. 1998. Directed seed dispersal by bellbirds in a tropical cloud forest. *Proceedings of the National Academy of Sciences of the United States of America* **95**: 6204–7.

Westerkamp, C. 1990. Bird-flowers: hovering versus perching exploitation. *Botanica Acta* **103**: 366–71.

Wheelwright, N.T. & Janson, C.H. 1985. Colors of fruit displays of bird-dispersed plants in two tropical forests. *American Naturalist* **126**: 777–99.

White, L.J.T., Tutin, C.E.G. & Fernandez, M. 1993. Group composition and diet of forest elephants, *Loxodonta africana cyclotis* Matschie 1900, in the Lopé Forest, Gabon. *African Journal of Ecology* **31**: 181–99.

Whitmore, T.C. 1975. *Tropical rain forests of the Far East*. First edition. Clarendon Press, Oxford.

Whitmore, T.C. 1984. *Tropical rain forests of the Far East*. Second edition. Clarendon Press, Oxford.

Whitmore, T.C. 1989. Changes of twenty-one years in the Kolombangara rain forests. *Journal of Ecology* **77**: 469–83.

Whitmore, T.C. 1996. A review of some aspects of tropical rain forest seedling ecology with suggestions for further enquiry. Pp. 3–39 in *The ecology of tropical forest tree seedlings* (ed. M.D. Swaine). UNESCO, Paris.

Whitmore, T.C. 1998. *An introduction to tropical rain forests*. Second edition. Oxford University Press, Oxford.

Whitmore, T.C. & Brown, N.D. 1996. Dipterocarp seedling growth in rain forest canopy gaps during six and a half years. *Philosophical Transactions of the Royal Society of London* B **351**: 1195–203.

Whitmore, T.C. & Gong, W.-K. 1983. Growth analysis of the seedlings of balsa, *Ochroma lagopus*. *New Phytologist* **95**: 305–11.

Whitmore, T.C. & Silva, J.N.M. 1990. Brazil rain forest timbers are mostly very dense. *Commonwealth Forestry Review* **69**: 87–90.

Whittaker, R.J., Bush, M.B. & Richards, K. 1989. Plant recolonization and vegetation succession on the Krakatau Islands, Indonesia. *Ecological Monographs* **59**: 59–123.

Wiemann, M.C., Manchester, S.R., Dilcher, D.L., Hinojosa, L.F. & Wheeler, E.A. 1998. Estimation of temperature and precipitation from morphological characters of dicotyledonous leaves. *American Journal of Botany* **85**: 1796–802.

Wiemann, M.C. & Williamson, G.B. 1989. Radial gradients in the specific gravity of wood in some tropical and temperate trees. *Forest Science* **35**: 197–210.

Williams, G. & Adam, P. 1993. Ballistic pollen release in Australian members of the Moraceae. *Biotropica* **25**: 478–80.

Williams, K., Field, C.B. & Mooney, H.A. 1989. Relationships among leaf construction cost, leaf longevity, and light environment in rainforest plants of the genus *Piper*. *American Naturalist* **133**: 198–211.

Williamson, G.B. 1984. Gradients in wood specific gravity of trees. *Bulletin of the Torrey Botanical Club* **111**: 51–5.

Williamson, G.B., Costa, F. & Minte Vera, C.V. 1999. Dispersal of Amazonian trees: hydrochory in *Swartzia polyphylla*. *Biotropica* **31**: 460–5.

Wills, C., Condit, R., Foster, R.B. & Hubbell, S.P. 1997. Strong density- and diversity-related effects help to maintain species diversity in a neotropical forest. *Proceedings of the National Academy of Sciences of the United States of America* **94**: 1252–7.

Willson, M.F., Irvine, A.K. & Walsh, N.G. 1989. Vertebrate dispersal syndromes in some Australian and New Zealand plant communities, with geographic comparisons. *Biotropica* **21**: 133–47.

Wilson, D.E. & Janzen, D.H. 1972. Predation on *Scheelea* palm seeds by bruchid beetles: seed density and distance from the parent palm. *Ecology* **53**: 954–9.

Witmer, M.C. 1996. Do some bird-dispersed fruits contain natural laxatives? A comment. *Ecology* **77**: 1947–8.

Witmer, M.C. & Cheke, A.S. 1991. The dodo and the tambalocoque tree: an obligate mutualism reconsidered. *Oikos* **61**: 133–7.

Wong, K.M. 1996. 'Species capture' through taxonomic inventories: what wisdoms do the tallies suggest? Some case studies relevant to Borneo. Pp. 317–22 in *Biodiversity and the dynamics of ecosystems* (ed. I.M. Turner, C.H. Diong, S.S.L. Lim & P.K.L. Ng). DIWPA, Kyoto.

Wrangham, R.W., Chapman, C.A. & Chapman, L.J. 1994. Seed dispersal by forest chimpanzees in Uganda. *Journal of Tropical Ecology* **10**: 355–68.

Wright, S.J. 1983. The dispersion of eggs by a bruchid beetle among *Scheelea* palm seeds and the effect of distance to the parent palm. *Ecology* **64**: 1016–21.

Yamada, T. & Suzuki, E. 1996. Ontogenetic changes in leaf shape and crown form of a tropical tree, *Scaphium macropodum* (Sterculiaceae) in Borneo. *Journal of Plant Research* **109**: 211–17.

Yamada, T. & Suzuki, E. 1997. Changes in spatial distribution during the life history of a tropical tree, *Scaphium macropodum* (Sterculiaceae) in Borneo. *Journal of Plant Research* **110**: 179–86.

Yoda, K. 1978. Three-dimensional distribution of light intensity in a tropical rain forest of West Malaysia. *Malayan Nature Journal* **30**: 161–77.

Yoneyama, T., Muraoka, T., Murakami, T. & Boonkerd, N. 1993. Natural abundance of $^{15}N$ in tropical plants with emphasis on tree legumes. *Plant and Soil* **153**: 295–304.

Young, A.M. 1982. Effects of shade cover and availability of midge breeding sites on pollinating midge populations and fruit set in two cocoa farms. *Journal of Applied Ecology* **19**: 47–63.

Young, T.P. & Augspurger, C.K. 1991. Ecology and evolution of long-lived semelparous plants. *Trends in Ecology and Evolution* **6**: 285–9.

Young, T.P. & Hubbell, S.B. 1991. Crown asymmetry, treefalls, and repeat distrubance of broad-leaved forest gaps. *Ecology* **72**: 1464–71.

Young, T.P. & Perkocha, V. 1994. Treefalls, crown asymmetry, and buttresses. *Journal of Ecology* **82**: 319–24.

Yu, D.W. & Pierce, N.E. 1998. A castration parasite of an ant–plant mutualism. *Proceedings of the Royal Society of London* B **265**: 375–82.

Yumoto, T. 1999. Seed dispersal by Salvin's curassow, *Mitu salvini* (Cracidae), in a tropical forest of Colombia: direct measurements of dispersal distance. *Biotropica* **31**: 654–60.

Yumoto, T., Kimura, K. & Nishimura, A. 1999. Estimation of the retention times and distances of seed dispersed by two monkey species, *Alouatta seniculus* and *Lagothrix lagotricha*, in a Colombian forest *Ecological Research* **14**: 179–92.

Yumoto, T., Maruhashi, T., Yamagiwa, J. & Mwanza, N. 1995. Seed-dispersal by elephants in a tropical rain forest in Kahuzi-Biega National Park, Zaire. *Biotropica* **27**: 526–30.

Zagt, R.J. & Werger, M.J.A. 1997. Spatial components of dispersal and survival for seeds and seedlings of two codominant tree species in the tropical rain forest of Guyana. *Tropical Ecology* **38**: 343–55.

Zagt, R.J. & Werger, M.J.A. 1998. Community structure and the demography of primary forest species in tropical rainforest. Pp. 193–219 in *Dynamics of tropical communities* (ed. D.M. Newbery, H.H.T. Prins & N.D. Brown). Blackwell Science, Oxford.

Zimmerman, J.K., Everham, E.M., Waide, R.B., Lodge, D.J., Taylor, C.M. & Brokaw, N.V.L. 1994. Responses of tree species to hurricane winds in subtropical wet forests in Puerto Rico: implications for tropical tree life histories. *Journal of Ecology* **82**: 911–22.

Zimmermann, U., Meinzer, F.C., Benkert, R., Zhu, J.J., Schneider, H., Goldstein, G., Kuchenbrod, E. & Haase, A. 1994. Xylem water transport: is the available evidence consistent with the cohesion theory? *Plant, Cell and Environment* **17**: 1169–81.

Zipperlen, S.W. & Press, M.C. 1996. Photosynthesis in relation to growth and seedling ecology of two dipterocarp rain forest tree species. *Journal of Ecology* **84**: 863–76.

Zipperlen, S.W. & Press, M.C. 1997. Photosynthetic induction and stomatal oscillations in relation to the light environment of two dipterocarp rain forest tree species. *Journal of Ecology* **85**: 491–503.

Zotz, G., Harris, G., Königer, M. & Winter, K. 1995. High rates of photosynthesis in the tropical pioneer tree, *Ficus insipida* Willd. *Flora* **190**: 265–72.

Zotz, G. & Winter, K. 1994. Photosynthesis of a tropical canopy tree, *Ceiba pentandra*, in a lowland forest in Panama. *Tree Physiology* **14**: 1291–301.

# INDEX OF SCIENTIFIC NAMES OF PLANTS

Page numbers in italics indicate relevant entries in figures or tables.



[1] Currently considered a synonym of *Aglaia odoratissima*.

[2] Recently transferred to the genus *Hydrochorea*.

[3] A synonym of *Piper trigonum*.
[4] Recently transferred to the genus *Balizia*.
[5] A synonym of *Pourouma bicolor*.

[6] *Rheedia* is not considered distinct from *Garcinia* by most authorities, making the correct name for this species *Garcinia madruno*.

[7] Recently transferred to the genus *Jessea*.

[8] A synonym of *Xerospermum noronhianum*.

# INDEX OF SCIENTIFIC NAMES OF ANIMALS

Page numbers in italics indicate relevant entries in figures or tables.

# GENERAL INDEX

Page numbers in italics indicate relevant entries in figures or tables.